Activated Sludge and Aerobic Biofilm Reactors

Biological Wastewater Treatment Series

The *Biological Wastewater Treatment* series is based on the book *Biological Wastewater Treatment in Warm Climate Regions* and on a highly acclaimed set of best selling textbooks. This international version is comprised by six textbooks giving a state-of-the-art presentation of the science and technology of biological wastewater treatment.

Titles in the *Biological Wastewater Treatment* series are:

Volume 1: *Wastewater Characteristics, Treatment and Disposal*
Volume 2: *Basic Principles of Wastewater Treatment*
Volume 3: *Waste Stabilisation Ponds*
Volume 4: *Anaerobic Reactors*
Volume 5: *Activated Sludge and Aerobic Biofilm Reactors*
Volume 6: *Sludge Treatment and Disposal*

Biological Wastewater Treatment Series

VOLUME FIVE

Activated Sludge and Aerobic Biofilm Reactors

Marcos von Sperling

Department of Sanitary and Environmental Engineering
Federal University of Minas Gerais, Brazil

Publishing
LONDON • NEW YORK

Published by IWA Publishing, Alliance House, 12 Caxton Street, London SW1H 0QS, UK

Telephone: +44 (0) 20 7654 5500; Fax: +44 (0) 20 7654 5555; Email: publications@iwap.co.uk
Website: **www.iwapublishing.com**

First published 2007
© 2007 IWA Publishing

Copy-edited and typeset by Aptara Inc., New Delhi, India

British Library Cataloguing in Publication Data
A CIP catalogue record for this book is available from the British Library

Library of Congress Cataloguing-in-Publication Data
A catalogue record for this book is available from the Library of Congress

ISBN: 1 84339 165 1
ISBN 13: 9781843391654

Contents

Preface

The present series of books has been produced based on the book *"Biological wastewater treatment in warm climate regions"*, written by the same authors and also published by IWA Publishing. The main idea behind this series is the subdivision of the original book into smaller books, which could be more easily purchased and used.

The implementation of wastewater treatment plants has been so far a challenge for most countries. Economical resources, political will, institutional strength and cultural background are important elements defining the trajectory of pollution control in many countries. Technological aspects are sometimes mentioned as being one of the reasons hindering further developments. However, as shown in this series of books, the vast array of available processes for the treatment of wastewater should be seen as an incentive, allowing the selection of the most appropriate solution in technical and economical terms for each community or catchment area. For almost all combinations of requirements in terms of effluent quality, land availability, construction and running costs, mechanisation level and operational simplicity there will be one or more suitable treatment processes.

Biological wastewater treatment is very much influenced by climate. Temperature plays a decisive role in some treatment processes, especially the natural-based and non-mechanised ones. Warm temperatures decrease land requirements, enhance conversion processes, increase removal efficiencies and make the utilisation of some treatment processes feasible. Some treatment processes, such as anaerobic reactors, may be utilised for diluted wastewater, such as domestic sewage, only in warm climate areas. Other processes, such as stabilisation ponds, may be applied in lower temperature regions, but occupying much larger areas and being subjected to a decrease in performance during winter. Other processes, such as activated sludge and aerobic biofilm reactors, are less dependent on temperature, as a result of the higher technological input and mechanisation level. The main purpose of this series of books is to present the technologies for urban wastewater treatment as applied to the specific condition of warm temperature, with the related implications in terms of design and operation. There is no strict definition for the range of temperatures that fall into this category, since the books always present how to correct parameters, rates and coefficients for different temperatures. In this sense, subtropical and even temperate climate are also indirectly covered, although most of the focus lies on the tropical climate.

Another important point is that most warm climate regions are situated in developing countries. Therefore, the books cast a special view on the reality of these countries, in which simple, economical and sustainable solutions are strongly demanded. All technologies presented in the books may be applied in developing countries, but of course they imply different requirements in terms of energy, equipment and operational skills. Whenever possible, simple solutions, approaches and technologies are presented and recommended.

Considering the difficulty in covering all different alternatives for wastewater collection, the books concentrate on off-site solutions, implying collection and transportation of the wastewater to treatment plants. No off-site solutions, such as latrines and septic tanks are analysed. Also, stronger focus is given to separate sewerage systems, although the basic concepts are still applicable to combined and mixed systems, especially under dry weather conditions. Furthermore, emphasis is given to urban wastewater, that is, mainly domestic sewage plus some additional small contribution from non-domestic sources, such as industries. Hence, the books are not directed specifically to industrial wastewater treatment, given the specificities of this type of effluent. Another specific view of the books is that they detail biological treatment processes. No physical-chemical wastewater treatment processes are covered, although some physical operations, such as sedimentation and aeration, are dealt with since they are an integral part of some biological treatment processes.

The books' proposal is to present in a balanced way theory and practice of wastewater treatment, so that a conscious selection, design and operation of the wastewater treatment process may be practised. Theory is considered essential for the understanding of the working principles of wastewater treatment. Practice is associated to the direct application of the concepts for conception, design and operation. In order to ensure the practical and didactic view of the series, 371 illustrations, 322 summary tables and 117 examples are included. All major wastewater treatment processes are covered by full and interlinked design examples which are built up throughout the series and the books, from the determination of the wastewater characteristics, the impact of the discharge into rivers and lakes, the design of several wastewater treatment processes and the design of the sludge treatment and disposal units.

The series is comprised by the following books, namely: *(1) Wastewater characteristics, treatment and disposal; (2) Basic principles of wastewater treatment; (3) Waste stabilisation ponds; (4) Anaerobic reactors; (5) Activated sludge and aerobic biofilm reactors; (6) Sludge treatment and disposal.*

Volume 1 (*Wastewater characteristics, treatment and disposal*) presents an integrated view of water quality and wastewater treatment, analysing wastewater characteristics (flow and major constituents), the impact of the discharge into receiving water bodies and a general overview of wastewater treatment and sludge treatment and disposal. Volume 1 is more introductory, and may be used as teaching material for undergraduate courses in Civil Engineering, Environmental Engineering, Environmental Sciences and related courses.

Volume 2 (*Basic principles of wastewater treatment*) is also introductory, but at a higher level of detailing. The core of this book is the unit operations and processes associated with biological wastewater treatment. The major topics covered are: microbiology and ecology of wastewater treatment; reaction kinetics and reactor hydraulics; conversion of organic and inorganic matter; sedimentation; aeration. Volume 2 may be used as part of postgraduate courses in Civil Engineering, Environmental Engineering, Environmental Sciences and related courses, either as part of disciplines on wastewater treatment or unit operations and processes.

Volumes 3 to 5 are the central part of the series, being structured according to the major wastewater treatment processes (*waste stabilisation ponds, anaerobic reactors, activated sludge and aerobic biofilm reactors*). In each volume, all major process technologies and variants are fully covered, including main concepts, working principles, expected removal efficiencies, design criteria, design examples, construction aspects and operational guidelines. Similarly to Volume 2, volumes 3 to 5 can be used in postgraduate courses in Civil Engineering, Environmental Engineering, Environmental Sciences and related courses.

Volume 6 (*Sludge treatment and disposal*) covers in detail sludge characteristics, production, treatment (thickening, dewatering, stabilisation, pathogens removal) and disposal (land application for agricultural purposes, sanitary landfills, landfarming and other methods). Environmental and public health issues are fully described. Possible academic uses for this part are same as those from volumes 3 to 5.

Besides being used as textbooks at academic institutions, it is believed that the series may be an important reference for practising professionals, such as engineers, biologists, chemists and environmental scientists, acting in consulting companies, water authorities and environmental agencies.

The present series is based on a consolidated, integrated and updated version of a series of six books written by the authors in Brazil, covering the topics presented in the current book, with the same concern for didactic approach and balance between theory and practice. The large success of the Brazilian books, used at most graduate and post-graduate courses at Brazilian universities, besides consulting companies and water and environmental agencies, was the driving force for the preparation of this international version.

In this version, the books aim at presenting consolidated technology based on worldwide experience available at the international literature. However, it should be recognised that a significant input comes from the Brazilian experience, considering the background and working practice of all authors. Brazil is a large country with many geographical, climatic, economical, social and cultural contrasts, reflecting well the reality encountered in many countries in the world. Besides, it should be mentioned that Brazil is currently one of the leading countries in the world on the application of anaerobic technology to domestic sewage treatment, and in the post-treatment of anaerobic effluents. Regarding this point, the authors would like to show their recognition for the Brazilian Research Programme on Basic Sanitation (PROSAB), which, through several years of intensive, applied, cooperative research has led to the consolidation of anaerobic treatment and aerobic/anaerobic post-treatment, which are currently widely applied in full-scale plants in Brazil. Consolidated results achieved by PROSAB are included in various parts of the book, representing invaluable and updated information applicable to warm climate regions.

Volumes 1 to 5 were written by the two main authors. Volume 6 counted with the invaluable participation of Cleverson Vitorio Andreoli and Fernando Fernandes, who acted as editors, and of several specialists, who acted as chapter authors: Aderlene Inês de Lara, Deize Dias Lopes, Dione Mari Morita, Eduardo Sabino Pegorini, Hilton Felício dos Santos, Marcelo Antonio Teixeira Pinto, Maurício Luduvice, Ricardo Franci Gonçalves, Sandra Márcia Cesário Pereira da Silva, Vanete Thomaz Soccol.

Many colleagues, students and professionals contributed with useful suggestions, reviews and incentives for the Brazilian books that were the seed for this international version. It would be impossible to list all of them here, but our heartfelt appreciation is acknowledged.

The authors would like to express their recognition for the support provided by the Department of Sanitary and Environmental Engineering at the Federal University of Minas Gerais, Brazil, at which the two authors work. The department provided institutional and financial support for this international version, which is in line with the university's view of expanding and disseminating knowledge to society.

Finally, the authors would like to show their appreciation to IWA Publishing, for their incentive and patience in following the development of this series throughout the years of hard work.

Marcos von Sperling
Carlos Augusto de Lemos Chernicharo

December 2006

The authors

Marcos von Sperling
PhD in Environmental Engineering (Imperial College, Univ. London, UK). Associate professor at the Department of Sanitary and Environmental Engineering, Federal University of Minas Gerais, Brazil. Consultant to governmental and private companies in the field of water pollution control and wastewater treatment.
marcos@desa.ufmg.br

ADDITIONAL CHAPTER AUTHORS

Carlos Augusto de Lemos Chernicharo, PhD. Federal University of Minas Gerais, Brazil.
calemos@desa.ufmg.br
Ricardo Franci Gonçalves, PhD. Federal University of Espírito Santo, Brazil.
franci@npd.ufes.br

1

Activated sludge process and main variants

1.1 INTRODUCTION

The activated sludge process is widely used around the world for the treatment of domestic and industrial wastewater, in situations where high effluent quality is necessary and space availability is limited. However, the activated sludge system is more heavily mechanised than the other treatment systems, involving a more sophisticated operation. Another disadvantage is the consumption of electrical energy for aeration.

To date, the largest application of the activated sludge system has been as a direct treatment of domestic or industrial effluents. More recently, the option of using the activated sludge system for the post-treatment of the effluent from anaerobic reactors is being investigated and used, by virtue of its various advantages. These are mainly associated with lower energy consumption and lower sludge production, while the effluent quality is comparable to that of the conventional activated sludge system.

The present chapter describes the main configurations of the activated sludge system, with its advantages, disadvantages and applicability.

The following units are integral parts and the essence of any continuous-flow activated sludge system (Figure 1.1):

- aeration tank (reactor)
- settling tank (secondary sedimentation tank)

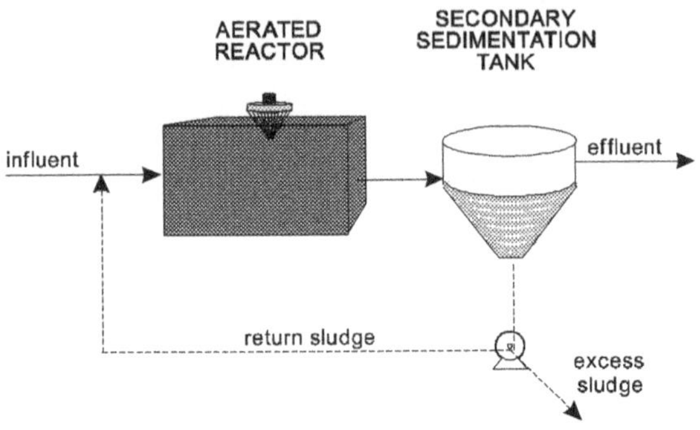

Figure 1.1.   Representation of the main units in the biological stage of the activated sludge system

- sludge recirculation
- excess sludge removal

The biochemical reactions associated with the removal of the organic matter and, under certain conditions the nitrogenous matter, take place in the biological reactor (also called aeration tank). The biomass develops by using the substrate present in the influent sewage. The settling of the solids (biomass), which leads to a clarified final effluent, occurs in the secondary sedimentation tank. A part of the solids that settle in the bottom of the secondary sedimentation tank is recirculated to the reactor (*return sludge*), to maintain a large biomass concentration in the reactor, which is responsible for the high efficiency of the system. The other part of the solids (*excess sludge*, also called *surplus sludge, secondary sludge, biological sludge* or *waste sludge*) is withdrawn from the system and is directed to the sludge treatment stage.

The biomass is separated in the secondary sedimentation tank due to its property of flocculating and settling. This is due to the production of a gelatinous matrix, which allows the agglutination of the bacteria, protozoa and other microorganisms responsible for the removal of the organic matter, into macroscopic flocs. The flocs individually are much larger than the microorganisms, which facilitates their sedimentation (Figure 1.2).

As a result of the recirculation of the sludge, the concentration of suspended solids in the aeration tank in the activated sludge systems is very high. In the activated sludge process, the detention time of the liquid (hydraulic detention time) is short, in the order of hours, which implies that the volume of the aeration tank is much reduced. However, the solids remain in the system for a longer period than the liquid, due to the recirculation. The retention time of the solids in the system is denominated *mean cell residence time (MCRT), solids retention time (SRT)* or *sludge age* (θ_c), and is defined as the ratio between the mass of biological sludge present in the reactor and the mass of biological sludge removed from (or produced in) the activated sludge system per day. It is this larger permanence of the solids

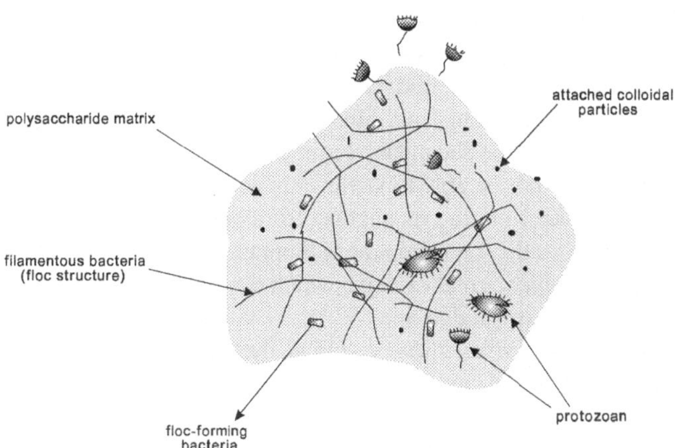

Figure 1.2.   Schematic representation of an activated sludge floc

in the system that guarantees the high efficiency of the activated sludge systems, since the biomass has enough time to metabolise practically all the organic matter in the sewage.

Another practical parameter used for the activated sludge process is the food/microorganism ratio (**F/M ratio**), which is defined as the load of food or substrate (BOD) supplied per day per unit biomass in the reactor (represented by MLVSS – mixed liquor volatile suspended solids), and expressed as kgBOD/kgMLVSS·d. Since the microorganisms have a limited capacity to consume the substrate (BOD) per unit time, a high F/M ratio can mean a larger offer of biodegradable organic matter than the consumption capacity of the biomass in the system, resulting in surplus substrate in the final effluent. On the other hand, low F/M values mean that the substrate offer is lower than the microorganisms' capacity to use it in the activated sludge system. As a consequence, they will consume practically all the organic matter from the influent wastewater, as well as their own organic cellular material. High sludge ages are associated with low F/M values, and vice versa.

For comparison purposes, anaerobic UASB (upflow anaerobic sludge blanket) reactors also have biomass retention in the reaction compartment, where a sludge blanket is developed, receiving the influent sewage and part of the recirculation of the biomass. This recirculation is attained through sedimentation of the solids in the settling compartment, followed by return by simple gravity to the reaction compartment. On the other hand, in the activated sludge system this recirculation of the solids is obtained by means of pumping (continuous-flow activated sludge) or by switching on and off the aerators (sequencing batch activated sludge reactors, described in Section 1.2.4). As a result, both in the UASB reactor and in the activated sludge system, the time of permanence of the biomass is longer than that of the liquid, guaranteeing the high compactness of the systems associated with their high efficiency.

Due to the continuous input of substrate (BOD from influent sewage) into the aeration tank, the microorganisms grow and continually reproduce. If the microorganisms were allowed to grow indefinitely, they would tend to reach excessive concentrations in the aeration tank, hindering the transfer of oxygen to all the bacterial cells. Besides, the secondary sedimentation tank would be overloaded, and the solids would not be able to settle satisfactorily and would be lost in the final effluent, thus deteriorating its quality. To maintain the system in balance, it is necessary to remove the same amount of biomass that is increased by reproduction. This is the *biological excess sludge,* which can be extracted directly from the reactor or from the return sludge line.

In the *conventional activated sludge system*, the excess sludge needs to undergo additional treatment in the sludge treatment line, usually comprising thickening, digestion and dewatering. The digestion is to decrease the amount of biodegradable bacterial mass (that is also organic matter) that could otherwise render the sludge septic in its final disposal. When activated sludge is used as *post-treatment for the effluent from anaerobic reactors*, due to the fact that a great part of the organic matter has already been removed in the anaerobic stage, the aerobic biomass growth in the activated sludge is lower (less substrate available). In this variant, the sludge production is, therefore, lower. The treatment of the sludge is also very simplified, since the aerobic excess sludge from the activated sludge can be returned to the UASB reactor, where it undergoes digestion and thickening.

The activated sludge system can be adapted to include the biological removal of *nitrogen* and *phosphorus*, now widely applied in several countries (see Chapters 6 and 7).

Regarding the removal of *coliforms* and pathogenic organisms, the efficiency is low and usually insufficient to meet the quality requirements of receiving water bodies, due to the reduced detention time in the units. This lower efficiency is also typical of other compact wastewater treatment processes. In case it is necessary, the effluent should be subjected to a subsequent disinfection stage. Due to the good quality of the effluent, the chlorine demand for disinfection is small: a concentration of a few mg/L of chlorine or its derivatives is enough for a substantial elimination of pathogens in a few minutes. As in every wastewater chlorination system, the possible need for dechlorination should be analysed for the reduction of the residual chlorine concentration, because of its toxicity to the receiving body biota. UV radiation is also attractive, due to the low level of suspended solids in the effluent from the activated sludge systems.

1.2 VARIANTS OF THE ACTIVATED SLUDGE PROCESS

1.2.1 Preliminaries

There are several variants of the activated sludge process. The present chapter focuses only on the main and more commonly used ones, which can be classified

Table 1.1. Classification of the activated sludge systems as a function of the sludge age and F/M ratio

Sludge age	Sludge age (day)	F/M ratio (kgBOD/kgMLVSS·day)	Usual designation
Low	4 to 10	0.25 to 0.50	Conventional activated sludge
High	18 to 30	0.07 to 0.15	Extended aeration

according to the following characteristics:

- division according to the sludge age (or F/M ratio)
  - conventional activated sludge (low sludge age, high F/M ratio)
  - extended aeration (high sludge age, low F/M ratio)
- division according to the flow
  - continuous flow
  - intermittent flow (sequencing batch reactors)
- division according to the influent to the biological stage of the activated sludge system
  - raw sewage
  - effluent from a primary sedimentation tank
  - effluent from an anaerobic reactor
  - effluent from another wastewater treatment process

There are other variants, related to the physical configuration of the aeration tank and the position of the inlets, but these are covered in Chapter 3.

The activated sludge systems can be classified in terms of the sludge age and the F/M ratio in one of the main categories listed in Table 1.1.

This classification according to the sludge age is applicable to both *continuous flow systems* (liquid entering and leaving the activated sludge reactor continuously) and *intermittent flow* or *sequencing batch systems* (intermittent input of the liquid in each activated sludge reactor). However, the extended aeration variant is more frequent for the intermittent flow systems. Regarding the activated sludge system acting as post-treatment for the effluent from anaerobic reactors, the most convenient option is the one with the reduced (conventional) sludge age.

Systems with very low sludge age (less than 4 days), also designated *modified aeration*, are less commonly used. Especially in warm-climate regions, the reactor volume would be very small, which could lead to some hydraulic instabilities in the system. In warm-climate areas, systems with *intermediate sludge ages* (between 10 and 18 days) do not present advantages for their use, since they do not enable a substantial increase in BOD removal, compared to the conventional sludge age, and they do not obtain the aerobic stabilisation of the sludge, which is a characteristic of the extended aeration. In temperate climate countries, the adoption of sludge ages of over 10 days can be necessary to reach complete nitrification throughout the year.

The biological stage of activated sludge (biological reactor and secondary sedimentation tank) can receive *raw wastewater* (usually in the extended aeration variant), *effluent from primary sedimentation tanks* (a classic conception of conventional activated sludge), *effluent from anaerobic reactors* (recent development) and *effluent from other wastewater treatment processes* (such as physical–chemical treatment or coarse trickling filters, for additional effluent polishing).

1.2.2 Conventional activated sludge (continuous flow)

To save energy for aeration and to reduce the volume of the biological reactor in the conventional system, part of the organic matter (suspended, settleable) from the wastewater is removed before the aeration tank, in the primary sedimentation tank. Thus, conventional activated sludge systems have primary treatment as an integral part of their flowsheet (Figure 1.3). In the figure, the top part corresponds to the treatment of the liquid phase (wastewater), while the bottom part exemplifies the stages involved in the treatment of the solid phase (sludge).

In the conventional system, the sludge age is usually of the order of 4 to 10 days, the F/M ratio is in the range of 0.25 to 0.50 kgBOD/kgMLVSS·d and the hydraulic detention time in the reactor is of the order of 6 to 8 hours. With this sludge age, the biomass removed from the system in the excess sludge still requires stabilisation

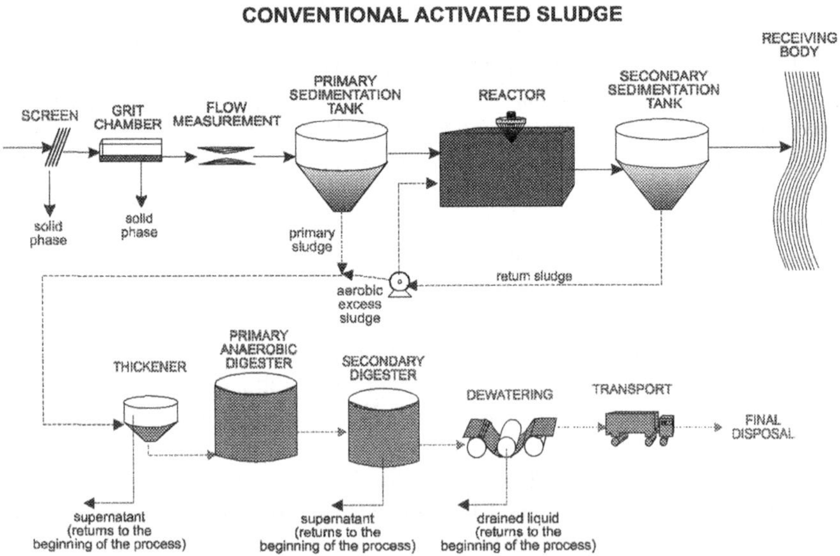

Figure 1.3. Typical flowsheet of the conventional activated sludge system

in the sludge treatment stage, since it still contains a high level of biodegradable organic matter in its cells. This stabilisation takes place in the digesters. To reduce the volume of the digesters, the sludge is previously subjected to a thickening stage, in which part of the water content is removed, thereby decreasing the sludge volume to be treated.

1.2.3 Extended aeration (continuous flow)

If the biomass stays in the system for longer periods, in the order of 18 to 30 days (hence the name *extended aeration*), and receives the same BOD load from the raw wastewater as in the conventional system, there will be less substrate available for the bacteria (F/M ratio of only 0.07 to 0.15 kgBOD/kgMLVSS·d). The amount of biomass (kgMLVSS) is larger than that in the conventional activated sludge system, the volume of the aerobic reactor is also higher and the detention time of the liquid is around 16 to 24 hours. Therefore, there is less organic matter per unit volume of the aeration tank and per unit biomass in the reactor. Consequently, to survive, the bacteria start to use in their metabolic processes their own biodegradable organic matter, which is a component of their cells. This cellular organic matter is transformed into carbon dioxide and water through respiration. This corresponds to an aerobic stabilisation of the biomass in the aeration tank. While in the conventional activated sludge system the stabilisation of the sludge is done separately (in the sludge digesters in the sludge treatment stage, usually in an anaerobic environment), in the extended aeration the sludge digestion is done jointly in the reactor, in an aerobic environment. The additional consumption of oxygen for the sludge stabilisation (endogenous respiration) is significant and it can be larger than the consumption for the assimilation of the organic matter from the influent (exogenous respiration).

Since there is no need to stabilise the excess biological sludge, the generation of another type of sludge is avoided in the extended aeration system, since this sludge would require subsequent separate stabilisation. For this reason, the extended aeration systems usually do not have primary sedimentation tanks, to avoid the need for a separate stabilisation of the primary sludge. With this, a great simplification in the process flowsheet is obtained: there are no primary sedimentation tanks or sludge digestion units (Figure 1.4).

The consequence of this simplification in the system is the increase in the energy consumption for aeration, since the sludge is stabilised aerobically in the aeration tank. On the other hand, the reduced substrate availability and its practically total assimilation by the biomass make the extended aeration variant one of the most efficient wastewater treatment processes for the removal of BOD.

However, it should be stressed that the efficiency of any variant of the activated sludge process is intimately associated with the performance of the secondary sedimentation tank. If there is a loss of solids in the final effluent, there will be a large deterioration in the effluent quality, independent of a good performance of the aeration tank in the BOD removal.

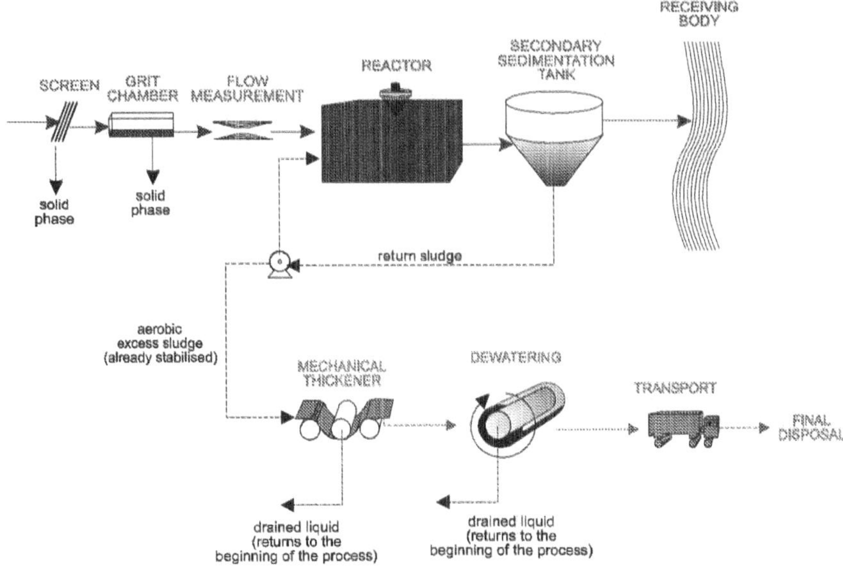

Figure 1.4.　Typical flowsheet of an extended aeration system

1.2.4 Intermittent operation (sequencing batch reactors)

The activated sludge systems described in Sections 1.2.2 and 1.2.3 are of *continuous flow* in relation to the wastewater, that is to say, the liquid is always entering and leaving the reactor. There is, however, a variant of the system with an *intermittent flow* operation, also called a *sequencing batch reactor.*

The principle of the activated sludge process with intermittent operation consists of the incorporation of all the units, processes and operations usually associated with the traditional activated sludge treatment, namely, primary settling, biological oxidation and secondary settling, in a single tank. In this tank, those processes and operations simply become sequences in time, and not separated units as in the conventional continuous-flow processes. The process of activated sludge with intermittent flow can be used both in the conventional and in the extended aeration modes, although the latter is more common, due to its greater operational simplicity. In the extended aeration mode, the single tank also incorporates a sludge digestion (aerobic) function. Figure 1.5 illustrates the flowsheet of a sequencing batch reactor system.

The process consists of a complete-mix reactor where all the treatment stages occur. That is obtained through the establishment of operational cycles and phases, each with a defined duration. The biomass is retained in the reactor during all phases, thus eliminating the need for separate settling tanks. A normal treatment

ACTIVATED SLUDGE-SEQUENCING BATCH REACTOR

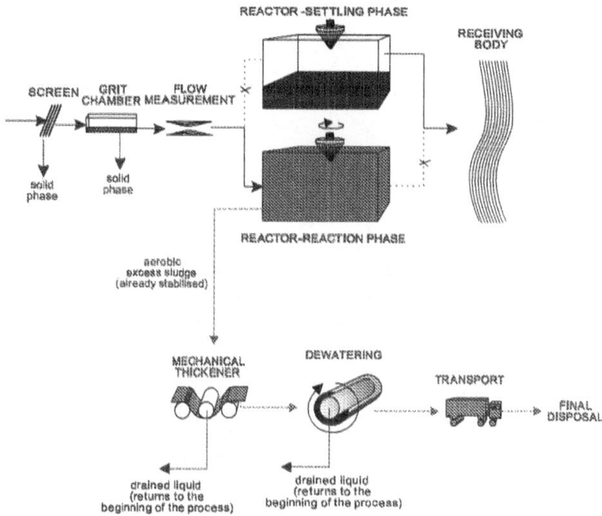

Figure 1.5. Typical flowsheet of an activated sludge system with intermittent operation (two reactors)

cycle is composed of the following phases:

- *filling* (input of raw or settled sewage to the reactor)
- *reaction* (aeration/mixing of the liquid contained in the reactor)
- *settling* (settling and separation of the suspended solids from the treated sewage)
- *withdrawal* (removal of the treated sewage from the reactor)
- *idle* (adjustment of cycles and removal of the excess sludge)

The usual duration of each phase and the overall cycle can be altered according to the influent flow variations, the treatment requirements, and the characteristics of the sewage and the biomass in the system.

Because sewage is continuously entering the treatment plant, more than one reactor is required: when one reactor is in the settling phase, no influent is allowed. Therefore, the influent is diverted to another reactor, which is in the fill stage.

The flowsheet of the process is largely simplified, due to the elimination of several units, compared to the continuous-flow activated sludge systems. In the extended aeration mode in the sequencing batch reactors, the only units of all the treatment processes (liquid and sludge) are: *screens, grit chambers, reactors, sludge thickeners (optional) and sludge dewatering units.*

There are some variants in the intermittent flow systems related to the operational procedure (continuous feeding and discontinuous emptying) and the sequence and duration of the cycles associated with each phase of the process. These variants can have additional simplifications in the process or incorporate the biological removal of nutrients, and are described in Chapter 8.

1.2.5 Activated sludge for the post-treatment of effluents from anaerobic reactors

A very promising alternative in warm-climate regions, which is the focus of recent research and is beginning to be implemented in full scale, is the one of activated sludge (with the conventional sludge age of 6 to 10 days) as a post-treatment of the effluent from anaerobic UASB-type reactors. In this case, there is an anaerobic reactor instead of a primary sedimentation tank. The excess aerobic sludge generated in the activated sludge system, not yet stabilised, is directed to the UASB reactor, where it undergoes thickening and digestion, together with the anaerobic sludge. As this aerobic excess sludge flow is very low, compared with the influent flow, there are no operational disturbances to the UASB reactor. The treatment of the sludge is largely simplified: thickeners and digesters are not needed, and there is only the dewatering stage. The mixed sludge withdrawn from the anaerobic reactor is digested and has similar concentrations to that of a thickened sludge, and also has excellent dewatering characteristics. Figure 1.6 shows the flowsheet for this configuration.

A comparison of this configuration with the traditional concept of the conventional activated sludge system is presented in Table 1.2.

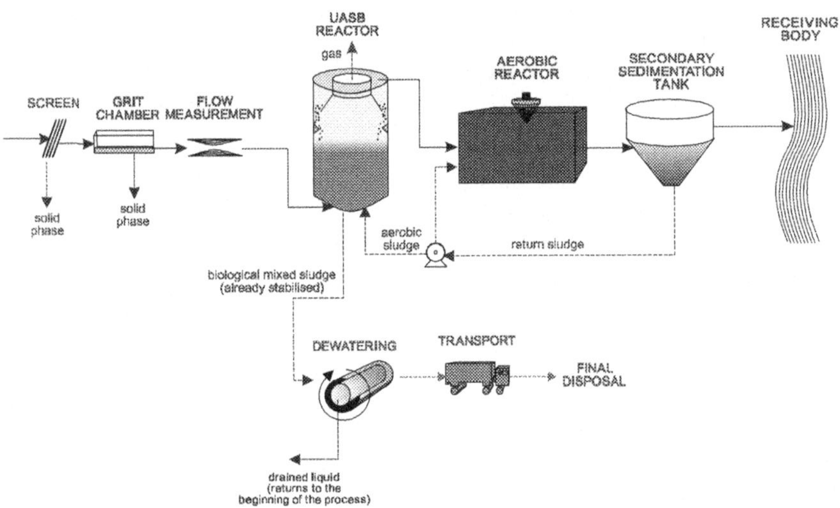

Figure 1.6.   Flowsheet of a system composed of a UASB reactor followed by an activated sludge system

Table 1.2. Main advantages, disadvantages and similarities of the UASB-activated sludge system with relation to the traditional concept of the conventional activated sludge system

Aspect	Item	Remark
Advantage	Reduction in the sludge production	• The mass of sludge produced and to be treated is of the order of 40 to 50% of the total value produced in the traditional concept of the conventional activated sludge, and 50 to 60% of the total produced in the extended aeration mode • The mass for final disposal is of the order of 60 to 70% of that from the traditional concepts • The reduction in the sludge volume is still larger, due to the fact that the mixed anaerobic sludge is more concentrated and has very good dewaterability
	Reduction in the energy consumption	• Since approximately 70% of the BOD is previously removed in the UASB reactor, the oxygen consumption is only for the remaining BOD and for nitrification, which in this case is the prevailing factor in the oxygen consumption (around 2/3 of the total consumption)
	Reduction in the total volume of the units	• The total volume of the units (UASB reactor, activated sludge reactor, secondary sedimentation tank and sludge dewatering) is lower than the total volume of the conventional activated sludge units (primary sedimentation tank, activated sludge reactor, secondary sedimentation tank, sludge thickener, sludge digester and sludge dewatering)
	Reduction in the consumption of chemical products for dewatering	• Reduction due to decreased sludge production and improved dewaterability
	Smaller number of different units to be installed	• There is no need for primary sedimentation tanks, thickeners and digesters, which are replaced by the UASB reactor
	Less equipment requirements	• The UASB reactor does not have electromechanical equipment, unlike the primary sedimentation tanks, thickeners and digesters of the conventional activated sludge systems

(Continued)

Table 1.2 (*Continued*)

Aspect	Item	Remark
	Greater operational simplicity	• Compared with the traditional concept of the conventional activated sludge system, there are less units to be operated and less electromechanical equipment to be maintained
Disadvantage	Lower capacity for biological nutrient removal (N and P)	• Nitrogen removal is only feasible if a minimum proportion between the concentration of nitrogenous matter (TKN) and the organic matter (COD) is reached • Similarly, there is also a minimum P/COD ratio for phosphorus removal • Once the UASB reactor removes a large portion of the organic carbon and hardly affects the nutrient concentration, in general the concentration of organic matter in the anaerobic effluent is smaller than the minimum necessary for good denitrification and phosphorus removal
Similarity	Similar efficiency to the traditional concept of conventional activated sludge	• The efficiency of the system in the removal of the main pollutants (except for N and P) is similar to that of the conventional activated sludge system

The operational experience with the new systems being built with this configuration will allow continuous progress in the knowledge of design criteria and parameters to be used. In this book, the same parameters usually adopted for conventional activated sludge systems have been used, based on the understanding that the main physical and biochemical phenomena involved are the same. However, it is possible that some coefficients of the mathematical model of the process are different, but this should not affect the design stage substantially.

1.2.6 Comparison among the main variants of the activated sludge process

In this section, the main variants of the activated sludge process are compared. The main dividing factor among the variables is the sludge age, characterising the extended aeration and conventional activated sludge modes, as well as the existence of pre-treatment (e.g., UASB reactor).

The following tables are presented to allow a comparison among the systems: (a) Table 1.3 – shows the main characteristics (efficiencies, requirements, production) of the systems; and (b) Table 1.4 gives comparison between several operational characteristics of the conventional activated sludge, extended aeration and UASB reactor-activated sludge systems.

Table 1.3. Main characteristics of the activated sludge systems used for the treatment of domestic sewage

		Type		
General item	Specific item	Conventional	Extended aeration	UASB-activated sludge
Sludge age	Sludge age (day)	4–10	18–30	6–10
F/M ratio	F/M ratio (kgBOD/kgMLVSS·d)	0.25–0.50	0.07–0.15	0.25–0.40
Removal efficiency	BOD (%)	85–95	93–98	85–95
	COD (%)	85–90	90–95	83–90
	Suspended solids (%)	85–95	85–95	85–95
	Ammonia (%)	85–95	90–95	75–90
	Nitrogen (%) [1]	25–30	15–25	15–25
	Phosphorus (%) [1]	25–30	10–20	10–20
	Coliforms (%)	60–90	70–95	70–95
Area required	Area (m^2/inhabitant) [2]	0.2–0.3	0.25–0.35	0.2–0.3
Total volume	Volume (m^3/inhabitant) [3]	0.10–0.12	0.10–0.12	0.10–0.12
Energy [4]	Installed power (W/inhabitant)	2.5–4.5	3.5–5.5	1.8–3.5
	Energy consumption (kW·hour/inhabitant·year)	18–26	20–35	14–20
Volume of sludge [5]	To be treated (L sludge/inhabitant·d)	3.5–8.0	3.5–5.5	0.5–1.0
	To be disposed of (L sludge/inhabitant·d)	0.10–0.25	0.10–0.25	0.05–0.15
Sludge mass	To be treated (gTS/inhabitant·d)	60–80	40–45	20–30
	To be disposed of (gTS/inhabitant·d)	30–45	40–45	20–30

Notes:
The values shown are typical, but may vary even outside the ranges depending on local circumstances.
[1]: Larger efficiencies can be reached in the removal of N (especially in conventional activated sludge and in the extended aeration) and P (especially in conventional activated sludge) through specific stages (denitrification and phosphorus removal). The UASB-activated sludge method is not efficient in the biological removal of N and P.
[2]: Smaller areas can be obtained by using mechanical dewatering. The area values represent the area of the whole WWTP, not just of the treatment units.
[3]: The total volume of the units includes UASB reactors, primary sedimentation tanks, aeration tanks, secondary sedimentation tanks, gravity thickeners and primary and secondary digesters. The dewatering process assumed in the computation of the volumes is mechanical. The need for each of the units depends on the variant of the activated sludge process.
[4]: The installed power should be enough to supply the O_2 demand in peak loads. The energy consumption requires a certain control of the O_2 supply, to be reduced at times of lower demand.
[5]: The sludge volume is a function of the concentration of total solids (TS), which depends on the processes used in the treatment of the liquid phase and the solid phase. The upper range of per capita volumes of sludge to be disposed of is associated with dewatering by centrifuges and belt presses (lower concentrations of TS in the dewatered sludge), while the lower range is associated with drying beds or filter presses (larger TS concentrations).
Source: von Sperling (1997), Alem Sobrinho and Kato (1999) and von Sperling *et al.* (2001)

Table 1.4. Comparison among the main variants of the activated sludge systems for the treatment of domestic sewage

Item	Conventional activated sludge	Extended aeration	UASB – activated sludge
Sludge age	⇩ 4 to 10 days	⇧ 18 to 30 days	⇩ 6 to 10 days
F/M ratio	⇧ • 0.25 to 0.50 kgBOD/kgMLVSS·d	⇩ • 0.07 to 0.15 kgBOD/kgMLVSS·d	⇧ • 0.25 to 0.4 kgBOD/kgMLVSS·d
Primary sedimentation tank	• Present	• Absent	• Absent
UASB reactor	• Absent	• Absent	• Present
Soluble effluent BOD	⇩ • Low • Can be practically ignored	⇩ • Very low • Can be practically ignored	⇩ • Low • Can be practically ignored
Particulate effluent BOD	⇕ • Depends on the settleability of the sludge and the performance of the secondary sedimentation tank • As nitrification is expected to happen, if there is no denitrification in the reactor, it can occur in the secondary sedimentation tank, causing rising sludge and solids loss • Secondary sedimentation tank is subject to problems with filamentous bacteria and other processes that deteriorate the settleability	⇕ • Depends on the settleability of the sludge and the performance of the secondary sedimentation tank • The larger load of influent solids in the secondary sedimentation tank requires unit sizes to be determined more conservatively • If there is no denitrification in the reactor, it can occur in the secondary sedimentation tank, causing rising sludge and solids loss • Secondary sedimentation tank is subject to problems with filamentous bacteria and other processes that deteriorate the settleability	⇕ • Depends on the settleability of the sludge and the performance of the secondary sedimentation tank • As nitrification is expected to happen, if there is no denitrification in the reactor, it can occur in the secondary sedimentation tank, causing rising sludge and solids loss • Secondary sedimentation tank is subject to problems with filamentous bacteria and other processes that deteriorate the settleability

Nitrification	⇕ • Very probable but subject to the instability in the lower range of the sludge ages, especially at lower temperatures • Totally consistent in the upper range of sludge ages, unless there are specific environmental problems (e.g., toxicity, lack of DO)	⇧ • Totally consistent in the upper range of sludge ages, unless there are specific environmental problems (e.g., toxicity, lack of DO)	⇕ • Likely to occur, unless there are environmental problems (e.g., toxicity, lack of DO) • The toxicity to the nitrifying bacteria by effluent sulphide from the UASB reactor is a topic that deserves investigation
Volume of the aerobic reactor (aeration tank)	⇩ • Low (hydraulic detention times in the order of 6 to 8 hours)	⇧ • High (hydraulic detention times in the order of 16 to 24 hours)	⇩ • Very reduced due to the previous removal of a large part of the organic matter (hydraulic detention times in the order of 3 to 5 hours)
Area of the secondary sedimentation tanks	⇩ • Low	⇧ • Higher, due to the large influent solids load and the settleability characteristics of the sludge	⇩ • Lower, due to the smaller influent solids load
Oxygen requirements	⇩ • Reduced, due to the lower respiration by the biomass and to the previous BOD removal in the primary settling	⇧ • High, due to the oxygen consumption for the respiration of the large amount of biomass present undergoing aerobic digestion and to the non-existence of primary settling	⇩ • Lower, due to the lower respiration by the biomass and to the high BOD removal in UASB reactor
Energy requirements	⇩ • Low, due to the low oxygen consumption	⇧ • High, due to the high oxygen consumption	⇩ • Lower, due to the lower oxygen consumption
Sludge production	⇧ • High, even though it decreases after anaerobic digestion, becoming reasonable	⇩ • Reasonable	⇩ • Low, since the anaerobic reactor produces a thick sludge in small quantities, and the aerobic sludge undergoes digestion and thickening in the anaerobic reactor

(*Continued*)

Table 1.4 (*Continued*)

Item		Conventional activated sludge		Extended aeration		UASB – activated sludge
Sludge stabilisation in the reactor	⇩ •	Lower and insufficient for the sludge to be directed to natural drying (generation of bad odours)	⇧ •	Sufficient and comparable to separate digestion processes, such as the anaerobic digesters	⇧ •	Sufficient and comparable to separate digestion processes, such as anaerobic digesters
Sludge thickening	•	Necessary (mainly for the secondary sludge)	•	Can be used, but thickening by gravity is not effective. Mechanised thickening is advised	•	Normally unnecessary
Separate digestion of the primary sludge	•	Necessary	•	No primary sludge	•	No primary sludge
Separate digestion of the aerobic sludge	•	Necessary	•	Unnecessary	•	The aerobic sludge is returned to the UASB reactor, where it undergoes digestion
Dewaterability of the sludge	⇧ •	Good dewaterability	⇩ •	Lower dewaterability	⇧ •	Excellent dewaterability
Stability of the process	⇩ •	Larger susceptibility to toxic discharges than extended aeration	⇧ •	High	⇧ •	Satisfactory, as it is made up of two stages in series (one anaerobic and one aerobic)
Operational simplicity	⇩ •	Low	⇧ •	Greater, due to the absence of primary sedimentation tanks and sludge digesters, and for being a more robust and stable system	⇕ •	Intermediate (larger complexity in the treatment of the liquid phase, but greater simplicity in the treatment of the solid phase)

Note: ⇩ = low or reduced   ⇕ = variable or intermediate   ⇧ = high or elevated
Source: von Sperling (1997) and von Sperling *et al.* (2001)

2

Principles of organic matter removal in continuous-flow activated sludge systems

2.1 PRELIMINARIES

In this chapter, the following items were discussed: influence of the solids recirculation, representation of the substrate and solids, solids production, hydraulic detention time, solids retention time, cell wash-out time, food/microorganism ratio, substrate utilisation rate and solids distribution in the wastewater treatment. All of these items are of fundamental importance for the activated sludge system and the reader must be familiar with them to understand the topics discussed below.

The present chapter covers the removal of the carbonaceous organic matter specifically in activated sludge systems and introduces new concepts that are applied to the system. The topics use the nomenclature shown in Figure 2.1.

It is known that there are very good and widely accepted models for the activated sludge process (e.g., IWA models), but these are at a higher level of sophistication and require the adoption of many parameters and input values. For these reasons, a more conventional approach of the activated sludge modelling is adopted in this book.

Two mass balances can be done, one for the substrate and the other for the biomass. These mass balances are essential for the sizing of the biological reactor and are detailed in the following sections.

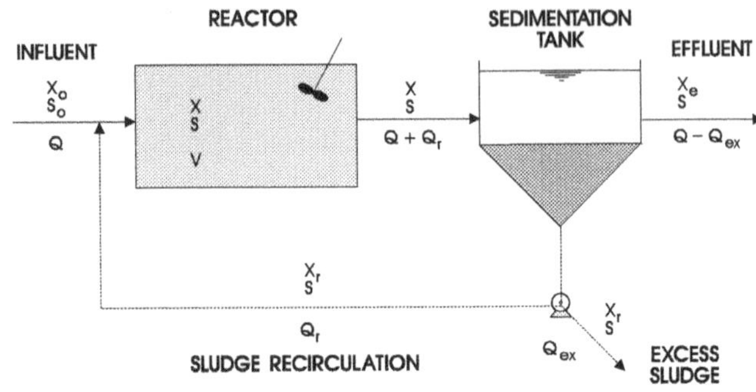

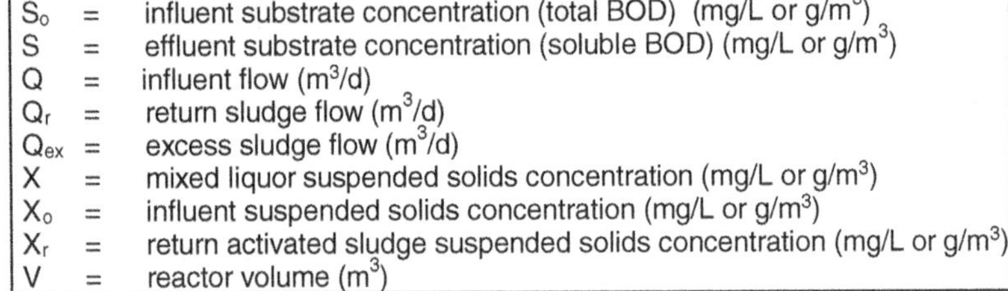

S_o = influent substrate concentration (total BOD) (mg/L or g/m^3)
S = effluent substrate concentration (soluble BOD) (mg/L or g/m^3)
Q = influent flow (m^3/d)
Q_r = return sludge flow (m^3/d)
Q_{ex} = excess sludge flow (m^3/d)
X = mixed liquor suspended solids concentration (mg/L or g/m^3)
X_o = influent suspended solids concentration (mg/L or g/m^3)
X_r = return activated sludge suspended solids concentration (mg/L or g/m^3)
V = reactor volume (m^3)

Figure 2.1.   Representation of the main variables in the activated sludge process

X is the concentration of solids. In the reactor these solids are typically biological solids and are represented by the *biomass* (microorganisms) produced in the reactor at the expense of the utilised substrate. These solids are called **mixed liquor suspended solids (MLSS)**.

In contrast, in the influent to the reactor, the solids are those originally present in the wastewater and, in many references they are neglected in the general mass balance. When appropriate, for the sake of clarity, these solids from the influent are not considered in some calculations in this chapter. However, it will still be shown in this chapter that in some formulae these solids have an influence on the sludge production of the system.

The solids can be represented as total suspended solids (**X**) or volatile suspended solids (**X_v**). When representing the biomass in the reactor, it is preferable to use X_v, while when analysing the behaviour of the solids in the secondary sedimentation tank, X is used. X_v is also called **mixed liquor volatile suspended solids (MLVSS)**.

The value of **X_r** is greater than X, that is, the **return sludge** X_r has a higher concentration of suspended solids, which leads to the maintenance of high SS concentrations in the reactor. The solids recycling can be done by pumping the sludge from the bottom of the secondary sedimentation tank (in continuous-flow systems) or through other operational procedures of solids retention in the reactor (in intermittent-flow systems).

In Figure 2.1 there is still another flow line, which corresponds to the **excess sludge** (Q_{ex}). This comes from the fact that, for the system to be maintained in equilibrium, the quantity of biomass production (bacterial growth) must be compensated by an equivalent wastage of solids. If solids are not wasted from the system, their concentration progressively increases in the reactor and the solids are transferred to the secondary sedimentation tanks, up to a point when they become overloaded. In this situation, the settling tank is not capable of transferring solids to its bottom anymore and the level of the sludge blanket starts to rise. Above a certain level, the solids start to leave with the final effluent, deteriorating its quality. Thus, in simplified terms, it can be said that the daily solids production must be counterbalanced by a withdrawal of an equivalent quantity (mass per unit time). The excess sludge can be wasted from the return sludge line (as shown in Figure 2.1) or directly from the reactor.

2.2 SLUDGE AGE IN ACTIVATED SLUDGE SYSTEMS

The sludge age is a fundamental parameter for the design and operation of the activated sludge process and is related to the reactor volume, production of solids, oxygen consumption and other operational variables of the process. Typical **sludge age** values in the activated sludge system are:

> - *Conventional activated sludge*: $\theta_c = 4$ to 10 days
> - *Extended aeration*: $\theta_c = 18$ to 30 days

The resultant **hydraulic detention time** in the reactor varies as follows:

> - *Conventional activated sludge*: $t = 6$ to 8 hours (<0.3 days)
> - *Extended aeration*: $t = 16$ to 24 hours (0.67 to 1.0 days)

The **F/M ratio** generally has the following values:

> - *Conventional activated sludge*: $F/M = 0.3$ to 0.8 kgBOD$_5$/kgVSS·d
> - *Extended aeration*: $F/M = 0.08$ to 0.15 kgBOD$_5$/kgVSS·d

2.3 SUSPENDED SOLIDS CONCENTRATION IN THE REACTOR

The design aspects related to the concept of X_v are examined in detail in this section.

To obtain the VSS concentration in the aeration tank, or MLVSS concentration in a system with solids recycling:

$$X_v = \frac{Y(S_o - S)}{1 + K_d \cdot f_b \cdot \theta_c} \left(\frac{\theta_c}{t} \right)$$

(2.1)

where:

θ_c = sludge age (d)

Y = yield coefficient (gVSS produced per gBOD removed) ($gX_v/gBOD_5$)

K_d = endogenous respiration coefficient (d^{-1})

f_b = biodegradable fraction of MLVSS (X_b/X_v)

Typical values of Y and K_d are:

$$Y = 0.5 \text{ to } 0.7 \text{ g VSS/g BOD}_5 \text{ removed}$$
$$K_d = 0.06 \text{ to } 0.10 \text{ gVSS/gVSS·d}$$

Equation 2.1 incorporates the concept of f_b, f_b is given by:

$$f_b = \frac{f_{b'}}{1 + (1 - f_{b'}) \cdot K_d \cdot \theta_c}$$

(2.2)

where:

f_b = biodegradable fraction of the VSS generated in a system subjected to a sludge age θ_c (X_b/X_v)

$f_{b'}$ = biodegradable fraction of the VSS immediately after its generation in the system, that is, with $\theta_c = 0$. This value is typically equal to 0.8 (= 80%).

Equation 2.1 is important in the estimation of the VSS concentration in a particular system once the other parameters and variables are known or have been estimated. The analysis of this equation also provides interesting considerations about the influence of the sludge recirculation on the VSS concentration in the reactor.

That $\theta_c = t$ in a *system without recirculation*. Under these conditions, Equation 2.1 is reduced to:

$$X_v = \frac{Y \cdot (S_o - S)}{1 + K_d \cdot f_b \cdot t}$$

(2.3)

It can be seen that the difference between both equations is the factor (θ_c/t), which has a multiplying effect on Equation 2.3, in that it increases the suspended solids concentration in the reactor. As mentioned in Section 2.4, any increase in

X_v, in a particular design, will result in a proportional decrease in the necessary volume for the reactor.

Typical values of X_v in an activated sludge system are:

> - conventional activated sludge = 1,500 to 3,500 mgVSS/L
> - extended aeration = 2,500 to 4,000 mgVSS/L

The maximum design concentration of MLSS in the reactor is generally limited to 4,500–5,000 mg/L. Extended aeration systems tend to have a higher MLSS concentration than the conventional activated sludge systems. Naturally, the larger the concentration of MLSS (or MLVSS) in the reactor, the greater the availability of the biomass to assimilate the influent substrate, resulting in the need for smaller reactor volumes (for a given removal efficiency). Some practical aspects, however, impose the mentioned upper limits:

- MLSS concentrations above a certain limit require larger secondary sedimentation tanks. Large surface areas for these units would become necessary for high SS loads flowing into them, which could offset the economic gain obtained with the reduced reactor volume.
- The transfer of oxygen to the entire biomass is adversely affected in the case of very high MLSS values.

The methodology for determining the VSS/SS ratio in an activated sludge reactor as a function of the sludge age is described. In general terms, the ranges of average VSS/SS values are as follows:

> - Conventional activated sludge: VSS/SS = 0.70 to 0.85
> - Extended aeration: VSS/SS = 0.60 to 0.75

2.4 CALCULATION OF THE REACTOR VOLUME

In Equation 2.1, replacing t with V/Q, and making V explicit leads to:

$$V = \frac{Y \cdot \theta_c \cdot Q \cdot (S_o - S)}{X_v \cdot (1 + K_d \cdot f_b \cdot \theta_c)} \qquad (2.4)$$

The volume of the reactor can be calculated by using this equation, provided that Q and S_o are known, a desired concentration for the soluble BOD effluent S is proposed, Y and K_d values are assumed, f_b is calculated and adequate values of the design parameters θ_c and X_v are adopted.

Equation 2.4 can be used for both the system with recirculation and the system without recirculation. In the latter case, when adopting $\theta_c = t$, the volume of the reactor can be calculated directly using the formula $V = t.Q$. However, the concentration of solids should be calculated using Equation 2.3.

Example 2.1

Calculate the volume of the reactor in the following systems:

- conventional activated sludge: $\theta_c = 6$ d; $X_v = 2,500$ mg/L
- extended aeration $\theta_c = 22$ d; $X_v = 3,000$ mg/L

General data:

- $Q = 1,500$ m^3/d (design data)
- $S_o = 300$ mg/L (design data, assuming that no primary sedimentation tanks are available in both systems, for comparison purposes)
- $S = 5$ mg/L (soluble BOD; desired value)
- $Y = 0.7$ (assumed)
- $K_d = 0.09$ d^{-1} (assumed)
- $f_b' = 0.8$ (adopted)

Solution:

(a) Conventional activated sludge

- Biodegradable fraction f_b (Equation 2.2)

$$f_b = \frac{f_{b'}}{1 + (1 - f_{b'}) \cdot K_d \cdot \theta_c} = \frac{0.8}{1 + (1 - 0.8) \times 0.09 \times 6} = 0.72$$

- Volume of the reactor (Equation 2.4)

$$V = \frac{Y \cdot \theta_c \cdot Q \cdot (S_o - S)}{X_v \cdot (1 + K_d \cdot f_b \cdot \theta_c)} = \frac{0.7 \times 6 \times 1,500 \times (300 - 5)}{2,500 \times (1 + 0.09 \times 0.72 \times 6)} = 535 \, \text{m}^3$$

- Hydraulic detention time

$$t = \frac{V}{Q} = \frac{535 \, \text{m}^3}{1,500 \, \text{m}^3/\text{d}} = 0.36 \, \text{d} = 8.6 \, \text{hours}$$

(b) Extended aeration

- Biodegradable fraction f_b (Equation 2.2)

$$f_b = \frac{f_{b'}}{1 + (1 - f_{b'}) \cdot K_d \cdot \theta_c} = \frac{0.8}{1 + (1 - 0.8) \times 0.09 \times 22} = 0.57$$

- Volume of the reactor (Equation 2.4)

$$V = \frac{Y \cdot \theta_c \cdot Q \cdot (S_o - S)}{X_v \cdot (1 + K_d \cdot f_b \cdot \theta_c)} = \frac{0.7 \times 22 \times 1500 \times (300 - 5)}{3,000 \times (1 + 0.09 \times 0.57 \times 22)} = 1067 \, \text{m}^3$$

Example 2.1 (Continued)

- Hydraulic detention time

$$t = \frac{V}{Q} = \frac{1{,}067 \text{ m}^3}{1{,}500 \text{ m}^3/\text{d}} = 0.71 \text{ d} = 17.1 \text{ hours}$$

It is observed that the extended aeration system requires larger reactor volumes compared to the conventional activated sludge system, due to the greater sludge age. However, the increase is not directly proportional to the relationship between the sludge ages.

An important aspect to be observed in Equation 2.4 is that *the calculation of the reactor volume is a function of the sludge age θ_c, and not of the hydraulic detention time t*. Because of this, t should not be used in the sizing of the reactor by means of the formula $V = t \cdot Q$, but only to evaluate the conditions of hydraulic stability and the resistance to shock loading. In case the system is without recirculation, naturally, the concept $\theta_c = t$ can be used.

The reason for using θ_c instead of t is as follows. A wastewater with a high flow, but a low BOD concentration, can require the same activated sludge reactor volume as a wastewater with a low flow but with a high BOD concentration, provided that the BOD loads are the same (load $=$ flow $\times$ concentration $= Q \cdot (S_o - S)$). However, once the same volumes are obtained, the hydraulic detention times will be essentially different, since the flow values differ from each other. Determining reactor volumes based only on the hydraulic detention time would, in this case, result in different values, which would induce under- or over-estimation, and in different treatment efficiencies. This is illustrated in Example 2.2.

Example 2.2

Calculate the reactor volume and the hydraulic detention time for an industrial wastewater in a conventional activated sludge system. Adopt the same parameters of Example 2.1 and compare the results with item "a" of the referred to example. The industrial wastewater data are:

- $Q = 300 \text{ m}^3/\text{d}$
- $S_o = 1{,}500 \text{ mg/L}$
- $S = 25 \text{ mg/L}$ (to keep the same removal efficiency as in Example 2.1)

Solution:

- Volume of the reactor (Equation 2.4)

$$V = \frac{Y \cdot \theta_c \cdot Q \cdot (S_o - S)}{X_v \cdot (1 + K_d \cdot f_b \cdot \theta_c)} = \frac{0.7 \times 6 \times 300 \times (1500 - 25)}{2{,}500 \times (1 + 0.09 \times 0.72 \times 6)} = 535 \text{ m}^3$$

Example 2.2 (Continued)

- Hydraulic detention time

$$t = \frac{V}{Q} = \frac{535\,\text{m}^3}{300\,\text{m}^3/\text{d}} = 1.78\,\text{d} = 42.8\ \text{hours}$$

When compared with the domestic sewage in Example 2.1, the volume of the reactor is the same (535 m^3), but the hydraulic detention time of Example 2.1 is five times lower (0.36 days). The reactor volumes are the same due to the fact that the BOD loads are the same (the industrial flow is five times smaller, but the concentration is five times larger). The detention times are different, since the industrial flow is five times smaller. For these reasons, it is important to size the system based on the sludge age instead of on the hydraulic detention time. For the calculation of the reactor volume, what ultimately matters is the BOD load, not the flow or the concentration itself.

2.5 SUBSTRATE REMOVAL

The bacterial growth rate, based on Monod's kinetics, is given by:

$$\frac{dX_v}{dt} = \mu_{max} \cdot \left(\frac{S}{K_s + S} \right) \cdot X_v - K_d \cdot f_b \cdot X_v \qquad (2.5)$$

where:

 μ_{max} = maximum specific growth rate (d^{-1})

  S = concentration of the limiting substrate (mg/L). In the case of treatment for BOD removal, the limiting nutrient is the organic matter itself

  K_s = half-saturation constant, which is defined as the concentration of the substrate for which $\mu = \mu_{max}/2$ (mg/L)

Dividing left- and right-hand sides of Equation 2.5 by X_v, and knowing that $\theta_c = X_v/(dX_v/dt)$:

$$\frac{1}{\theta_c} = \mu_{max} \cdot \left(\frac{S}{K_s + S} \right) - K_d \cdot f_b \qquad (2.6)$$

Rearranging this equation to make S (effluent soluble BOD) explicit:

$$S = \frac{K_s \cdot [(1/\theta_c) + K_d \cdot f_b]}{\mu_{max} - [(1/\theta_c) + K_d \cdot f_b]} \qquad (2.7)$$

This is the general equation to estimate the effluent soluble BOD from a complete-mix reactor. Since in complete-mix reactors S is generally much smaller than K_s in the denominator of Monod's equation, $(K_s + S)$ could be simply substituted by S. In these conditions, first-order kinetics would prevail. With such a replacement, Equation 2.7 can be presented in the following simplified way:

$$S = \frac{K_s}{\mu_{max}} \cdot \left(\frac{1}{\theta_c} + K_d \cdot f_b \right) \tag{2.8}$$

An interesting aspect in Equations 2.7 and 2.8 is that, in a complete-mix system in the steady state, the effluent BOD concentration (S) is independent of the influent concentration S_0 (Arceivala, 1981). This is justified by the fact that K_s, K_d and μ_{max} are constant and, therefore, S depends only on the sludge age θ_c. The larger the influent BOD load, the larger the production of biological solids and, consequently, the larger the biomass concentration X_v. Thus, the higher the substrate available, the greater the biomass availability for its assimilation. It should be emphasised that this consideration is applicable only to the steady state. In the dynamic state, any increase in the influent BOD load is not immediately followed by a corresponding increase in the biomass, since such an increase occurs slowly. Thus, until a new equilibrium is reached (if an equilibrium will ever be reached at all), the quality of the effluent in terms of BOD will deteriorate.

The value of S can also be obtained by rearranging Equation 2.4, used for the calculation of the volume of the reactor. When all of the terms are known, S can be made explicit. It should be noted that, for typically domestic sewage, S is usually low, especially in extended aeration systems. In these conditions, any deviation in the estimate of S can lead to significant relative errors. However, such errors are not expected to be substantial, since, ultimately, in a design the interest is mainly in the range of values of S, not in an exact estimate.

The minimum concentration of soluble substrate S that can be reached in a system is when the sludge age θ_c tends to be infinite. In these conditions, the term $1/\theta_c$ is equivalent to zero. By replacing $1/\theta_c$ with 0 in Equation 2.7, Equation 2.9 is obtained, defining the minimum reachable effluent soluble BOD (S_{min}). In a treatment system, in case one needs to obtain a value that is lower than S_{min}, this will not be possible with a single complete-mix reactor (Grady and Lim, 1980). S_{min} is independent of the presence of recirculation and is just a function of the kinetic coefficients.

$$S_{min} = \frac{K_s \cdot K_d \cdot f_b}{\mu_{max} - K_d \cdot f_b} \tag{2.9}$$

As already noted, for predominantly *domestic sewage*, the soluble effluent BOD is essentially small and could even be considered negligible (compared to the influent BOD). The exception is for systems with very small sludge ages ($\theta_c <$ 4 days), in which S can be representative.

Example 2.3

Calculate the soluble effluent BOD concentration from the systems described in Example 2.1:

- conventional activated sludge: $\theta_c = 6$ days
- extended aeration: $\theta_c = 22$ days

Adopt $\mu_{max} = 2.0\ d^{-1}$ and $K_s = 60$ mg/L.

Solution:

(a)   Conventional activated sludge

Using Equation 2.7:

$$S = \frac{K_s \cdot [(1/\theta_c) + K_d \cdot f_b]}{\mu_{max} - [(1/\theta_c) + K_d \cdot f_b]} = \frac{60 \times [(1/6) + 0.09 \times 0.72]}{2.0 - [(1/6) + 0.09 \times 0.72]} = 7.9\ \text{mg/L}$$

Note: if the simplified formula for first-order kinetics had been used (Equation 2.8), a value of $S = 6.9$ mg/L would have been obtained.

(b)   Extended aeration

Using Equation 2.7:

$$S = \frac{K_s \cdot [(1/\theta_c) + K_d \cdot f_b]}{\mu_{max} - [(1/\theta_c) + K_d \cdot f_b]} = \frac{60 \times [(1/22) + 0.09 \times 0.57]}{2,0 - [(1/22) + 0.09 \times 0.57]} = 3.1\ \text{mg/L}$$

Note: if the simplified formula for first-order kinetics had been used (Equation 2.8), a value of $S = 2.9$ mg/L would have been obtained.

(c)   Comments

- In both cases, the general and simplified formulae produce very similar values.
- The concentrations of soluble effluent BOD are low in both systems. In domestic sewage treatment by activated sludge, this is the most frequent situation.
- In the extended aeration system, due to the higher sludge age, the concentration of soluble effluent BOD is lower. It should be remembered that these values are for steady-state conditions, and that the conventional activated sludge system is more susceptible to variations in the influent load (which can cause the effluent to deteriorate during transients).
- It should also be noted that, in Example 2.1, it was estimated that the effluent BOD (S) would be equal to 5 mg/L in the two systems. In the present example, it is observed that there is a slight deviation from this estimate. The volume of the reactor can be recalculated with the new S values. Another option is to calculate the acceptable soluble BOD in the effluent, according to the desired total BOD_5 and SS values in the effluent

---

Example 2.3 (Continued)

(see Section 2.6). However, the difference between these two approaches in respect to the direct calculations of the volume is expected to be very small.

- The BOD values have been presented in this example with decimals only for the sake of clarity in the comparisons. In a real situation, there is no sensitivity in the BOD test to express its values with decimals.

2.6 SOLUBLE BOD AND TOTAL BOD IN THE EFFLUENT

All the calculations for the design of the reactor, or for the determination of the effluent BOD, were made by assuming that S was the effluent **soluble BOD**, that is to say, the biochemical oxygen demand caused by the organic matter dissolved in the liquid medium. This BOD could be considered the total effluent BOD from the system, if the final sedimentation tank were capable of removing 100% of the suspended solids flowing into it. However, it is worth remembering that the concentration of these solids that will reach the secondary sedimentation tank is in the order of 3,000 to 5,000 mg/L. Thus, it is expected that they will not be entirely removed, and that a residual fraction will leave with the final effluent. As these solids have a large fraction of organic matter (mainly represented by the biomass), they will still cause an oxygen demand when they reach the receiving body. This demand is named **suspended BOD** or **particulate BOD**. Thus, in the final effluent of an activated sludge plant, there are the following fractions:

$$\text{Total effluent BOD}_5 = \text{Soluble effluent BOD}_5 + \text{Particulate effluent BOD}_5$$

$$(2.10)$$

The soluble BOD can be estimated using Equation 2.7 or 2.8. For the estimation of the particulate BOD, some considerations should be made. The solids that generate oxygen demand are only the biodegradable solids, since the inorganic and the inert solids are not an organic substrate that can be assimilated by the bacteria and generate oxygen consumption. By using Equation 2.2, and knowing the process parameters, one can determine the parameter f_b, that is, estimate which fraction of the VSS present in the plant effluent is biodegradable and will, therefore, represent the BOD of the suspended solids. Once this biodegradable fraction is known, the oxygen consumption required to stabilise this fraction can be estimated. For this, Equation 2.11, relative to the stabilisation of the cellular material represented by the formula $C_5H_7NO_2$, can be used:

$$C_5H_7NO_2 + 5O_2 \rightarrow 5CO_2 + NH_3 + 2H_2 + \text{Energy} \qquad (2.11)$$

$$\underset{(X_b)}{\underset{MW=113}{}} \qquad MW=160$$

Thus, according to the stoichiometric relationship between the molecular weights (MW), 160 g of oxygen are required for the stabilisation of 113 g of biodegradable solids. Hence, this relationship is:

$$O_2/X_b = 160/113 = 1.42 \text{ gO}_2/\text{g biodegradable solids} \qquad (2.12)$$

The ultimate biochemical oxygen demand (BOD_u) of the biodegradable solids is equal to this O_2 consumption. Thus, expressed in other terms:

$$BOD_u \text{ of the biodegradable solids} = 1.42 \text{ mgBOD}_u/\text{mgX}_b \qquad (2.13)$$

In typical domestic sewage, the relationship between BOD_5 and BOD_u is approximately constant, and the ratio BOD_u/BOD_5 is usually adopted as 1.46. Thus, the ratio BOD_5/BOD_u is the same as the reciprocal of 1.46, that is, $BOD_5/BOD_u = 1/1.46 = 0.68 \text{ mgBOD}_5/\text{mgBOD}_u$. This means that when reaching the fifth day of the BOD test, 68% of the organic matter originally present has been stabilised, or else 68% of the total oxygen consumption takes place by the fifth day. Hence, Equation 2.13 can be expressed as:

$$BOD_5 \text{ of the SS}_{biodeg} = 0.68 \text{ mgBOD}_5/\text{mgBOD}_u \times 1.42 \text{ mgBOD}_u/\text{mgX}_b$$
$$BOD_5 \text{ of the SS}_{biodeg} \approx 1.0 \text{ mgBOD}_5/\text{mgX}_b \qquad (2.14)$$

To express this oxygen demand in terms of the volatile suspended solids, Equation 2.14 needs to be multiplied by f_b ($= X_b/X_v$). The f_b values can be obtained using Equation 2.2. Hence:

$$BOD_5 \text{ of VSS} \approx 1.0 \ (\text{mgBOD}_5/\text{mgX}_b) \times f_b(\text{mgX}_b/\text{mgVSS})$$
$$BOD_5 \text{ of VSS} \approx f_b \ (\text{mgBOD5/mgVSS}) \qquad (2.15)$$

To make this equation more realistic and yet practical, it is interesting to express the effluent solids not as volatile suspended solids, but as total suspended solids. This is because, in the operational control routine and in the determination of the performance of the treatment system, the usual procedure is to measure the performance of the secondary sedimentation tank based on the effluent *total suspended solids* concentration. In Section 2.3, the values of the VSS/TSS ratio were presented, and it was shown how to calculate the value of the ratio. For conventional activated sludge systems, VSS/TSS varies from 0.70 to 0.85, while for extended aeration systems, VSS/TSS varies from 0.60 to 0.75. The BOD_5 of the total suspended solids will then be:

$$\boxed{BOD_5 \text{ of the effluent SS } (\text{mgBOD}_5/\text{mgSS}) = (\text{VSS/TSS}) \cdot f_b} \qquad (2.16)$$

Based on the f_b values resulting from the application of Equation 2.2 and on the typical values of the relationship VSS/TSS described in the paragraph above, and by applying Equation 2.16, the following ranges of typical values of **particulate BOD** are obtained:

* *conventional activated sludge: 0.45 to 0.65 mgBOD$_5$/mgTSS*
* *extended aeration: 0.25 to 0.50 mgBOD$_5$/mgTSS*

Experimental studies by von Sperling (1990) and Fróes (1996) with two extended aeration systems led to a ratio in the range of 0.21 to 0.24 mgBOD$_5$ for each mgSS, close to the lower limit of the theoretical range.

The determination of the BOD$_5$ of the final effluent is, therefore, essentially dependent on the estimation of the suspended solids concentration in the effluent from the secondary sedimentation tank. Unfortunately, there are no widely accepted rational approaches that can be safely used to estimate the effluent solids concentration, since the number of variables involved in the clarification function of secondary sedimentation tanks is very high. There are some empirical criteria that correlate the solids loading rate in the settling tank and other variables with the effluent SS concentration, but these relationships are very site specific.

Designers usually assume a SS concentration to be adopted in the design (equal to or lower than the SS discharge standard), and through this value the particulate BOD$_5$ is estimated. Based on a desired value of *total BOD$_5$* in the effluent, and with the estimated *particulate BOD$_5$*, by difference, the required *soluble BOD$_5$* is obtained (simple rearrangement of Equation 2.10). With this value the biological stage of the treatment plant can be properly designed.

Example 2.4 illustrates the complete calculation of the effluent total BOD$_5$ of an activated sludge system.

Example 2.4

For the conventional activated sludge system described in Example 2.1, calculate the concentrations of particulate, soluble and total BOD in the effluent. Assume that the design value for the effluent SS concentration is 30 mg/L.

Data already obtained in Examples 2.1 and 2.3:

$$S = 8 \text{ mg/L}$$

$$f_b = 0.72$$

Solution:

(a) Particulate BOD$_5$ in the effluent from the secondary sedimentation tank

Adopt the VSS/SS ratio equal to 0.8 (see above). The particulate BOD$_5$ is calculated using Equation 2.16:

Particulate BOD$_5$ = (VSS/SS)·f$_b$ = 0.8 × 0.72 = 0.58 mgBOD$_5$/mgSS

For 30 mg/L of effluent suspended solids, the effluent particulate BOD$_5$ is:

30 mgSS/L × 0.58 mgBOD$_5$/mgSS = 17 mgBOD$_5$/L

Example 2.4 (Continued)

(b)   Summary of the effluent BOD_5 concentrations

Soluble BOD = 8 mg/L (calculated in Example 2.3)
Particulate BOD = 17 mg/L

Total BOD = 8 + 17 = 25 mg/L

If, for example, a better effluent quality, with a total effluent BOD_5 of 20 mg/L were desired, there would be two possibilities. The first would be to reduce the effluent SS concentration (effluent polishing), to decrease the particulate BOD_5. The second would be to allow a maximum value for the soluble BOD_5 of 3 mg/L ($= 20 - 17$ mg/L). In this case, the reactor should be redesigned.

(c)   Efficiency of the system in the BOD removal

The efficiency of the system in the BOD removal is given by:

$$E(\%) = \frac{BOD_5 \text{ influent} - BOD_5 \text{ effluent}}{DBO_5 \text{ influent}} \cdot 100$$

The biological removal efficiency (that considers only the soluble BOD in the effluent) is:

$$E = 100 \cdot (300 - 8)/300 = 97\%$$

The overall removal efficiency (considering total BOD in the effluent) is:

$$E = 100 \cdot (300 - 25) = 92\%$$

In the calculation of the reactor volume and of the BOD removal, **S** is considered as the **soluble effluent BOD**, and S_o is the **total influent BOD**. This is because the organic suspended solids, which are responsible for the influent particulate BOD, are adsorbed onto the activated sludge flocs, and subsequently undergo successive transformations into simpler substrate forms, until they become available for synthesis. Only after this transformation to soluble organic solids will they be removed by similar mechanisms to those that acted on the soluble BOD. Thus, the influent particulate BOD will also generate bacterial growth and oxygen demand, but with a time lag compared to soluble BOD. In dynamic models this time lag should be taken into account, but it has no influence in steady-state models. This is the reason why S_o is considered as the total influent BOD.

Another aspect to be remembered is that, if the treatment system is provided with primary sedimentation tanks, such as the conventional activated sludge system, part of the influent BOD is removed by sedimentation, corresponding to the settled fraction of the volatile suspended solids. These will undergo subsequent separate digestion processes in the sludge treatment line and will not enter the reactor. The BOD_5 removal efficiency of primary sedimentation tanks usually ranges from 25%

to 35%, that is, to say, the influent BOD to the reactor (S_o) is 65% to 75% of the raw sewage BOD.

2.7 SLUDGE DIGESTION IN THE REACTOR

Besides the removal of carbonaceous and nitrogenous matter, an additional purpose of the biological stage can be the stabilisation of the sludge in the reactor. This is the case of the extended aeration systems, which do not have separate digestion for the excess sludge. The high sludge ages are responsible, therefore, not just for the oxidation of BOD and ammonia, but also for the aerobic digestion of the biomass. The digestion of the biodegradable fraction can be partial or practically total, depending on the sludge age adopted. It was seen that the extended aeration system in question had a high removal efficiency of the biodegradable biological solids generated in the system (93%), which resulted in an efficiency of 53% in the removal of the volatile solids. This efficiency is comparable to that obtained through separate digestion of the sludge.

Theoretically, for a certain biomass type, the sludge age that leads to the total destruction of the biodegradable solids formed can be determined. This value of θ_c can be obtained through the sequence shown below.

The *gross production of volatile solids* in the reactor is:

$$Px_v \text{ gross} = Y \cdot Q \cdot (S_o - S) \tag{2.17}$$

The *gross production of volatile biodegradable solids* is obtained by multiplying the above equation by the biodegradability fraction f_b. Therefore:

$$Px_b \text{ gross} = f_b \cdot Y \cdot Q \cdot (S_o - S) \tag{2.18}$$

On the other hand, the *destruction of the biodegradable solids* is given by:

$$Px_b \text{ destroyed} = f_b \cdot Y \cdot Q \cdot (S_o - S) \cdot [K_d \cdot \theta_c / (1 + f_b \cdot K_d \cdot \theta_c)] \tag{2.19}$$

To achieve complete destruction of all the biodegradable biological solids generated in the system, the production of solids should equal their destruction. Thus:

$$X_b \text{ production} = X_b \text{ destruction}$$

$$f_b \cdot Y \cdot Q \cdot (S_o - S) = f_b \cdot Y \cdot Q \cdot (S_o - S) \cdot [K_d \cdot \theta_c / (1 + f_b \cdot K_d \cdot \theta_c)] \tag{2.20}$$

After making the necessary simplifications in Equation 2.20:

$$\theta_c = 1 / [K_d \cdot (1 - f_b)] \tag{2.21}$$

However, f_b is a function of θ_c. Using the formula

$$f_b = f_{b'} / [1 + (1 - f_{b'}) \cdot K_d \cdot \theta_c]$$

Table 2.1. Sludge age values (θ_c) to achieve total stabilisation of the biodegradable fraction of the generated suspended solids, as a function of the coefficient of endogenous respiration (K_d)

K_d (d^{-1})	0.05	0.07	0.09	0.11
θ_c (d)	45	32	25	20

(Equation 2.2), replacing f_b in Equation 2.21 and making rearrangements as required, the following equation is obtained:

$$\theta_c = \frac{1}{K_d \cdot \sqrt{1 - f_{b'}}} \qquad (2.22)$$

For values of $f_{b'}$ typically equal to 0.8, Equation 2.22 can still be rearranged into the following simplified form:

$$\theta_c = 2.24/K_d \qquad (2.23)$$

Equations 2.22 and 2.23 allow the theoretical determination of the *limit θ_c value, above which all the produced biodegradable biological solids are destroyed* through aerobic digestion in the reactor. Thus, in the volatile suspended solids only the non-biodegradable fraction (inert, or endogenous) will remain, and in the total suspended solids, only the inorganic fraction (fixed) and the non-biodegradable fraction will remain. In these conditions, the excess sludge requires no additional separate digestion. The oxidation of the organic carbonaceous matter from the wastewater will continue, because the active solids are present in higher concentrations than the biodegradable solids.

For typical values of K_d, Table 2.1 shows the limit θ_c values for complete digestion in the reactor of the biodegradable biological solids formed, according to the simplified Equation 2.23.

As expected, the larger the coefficient of bacterial decay K_d, the lower the sludge age required for the complete stabilisation of the biodegradable solids.

The above calculations can be confirmed, related to the solids distribution in the treatment. If a sludge age equal to the limit value is adopted, it can be seen that the destruction of the biodegradable solids will be the same as their production.

As an additional detail, the substrate utilisation rate (U) that leads to total stabilisation is given by:

$$U = K_d/Y \qquad (2.24)$$

For sludge ages under the limit value, the digestion of the produced biodegradable solids is incomplete, although it can be, in practical terms, sufficient (in the sense that no additional separate digestion is required). On the other hand, for sludge ages above the limit value, total destruction is achieved (in fact, in the calculation, the destruction component becomes larger than the production one).

For a given sludge age, the *removal percentage of biodegradable solids* is given by Equation 2.25, while the *removal percentage of the volatile solids* is obtained using Equation 2.26. When analysing the efficiency of a sludge digestion process, the concept of percentage destruction of volatile solids is normally used. For comparison purposes, the typical efficiencies in the reduction of volatile solids in the anaerobic sludge digestion vary from 45 to 60%, and in the aerobic (separate) digestion, they vary from 40 to 50% (Metcalf and Eddy, 1991). The formulae shown below do not take into account the solids present in the influent wastewater to the reactor:

$$\% \text{ destruction of } SS_b = \left(\frac{K_d \cdot \theta_c}{1 + f_b \cdot K_d \cdot \theta_c} \right) \cdot 100 \tag{2.25}$$

$$\% \text{ destruction of } VSS = \left(\frac{f_b \cdot K_d \cdot \theta_c}{1 + f_b \cdot K_d \cdot \theta_c} \right) \cdot 100 \tag{2.26}$$

Tables 2.2 and 2.3 show the calculated values of the percentage removal of SS_b and VSS, respectively, for different values of θ_c and K_d.

Table 2.2. Percentage removal of the volatile biodegradable suspended solids formed in the reactor

θ_c (day)	Percentage removal of the produced biodegradable volatile SS (SS_b) (%)			
	$K_d = 0.05$ d^{-1}	$K_d = 0.07$ d^{-1}	$K_d = 0.09$ d^{-1}	$K_d = 0.11$ d^{-1}
4	17	23	28	33
8	31	40	48	55
12	42	53	63	72
16	52	65	76	86
20	60	75	87	99
24	68	84	98	–
28	75	92	–	–
32	81	100	–	–

Table 2.3. Percentage removal of the volatile suspended solids formed in the reactor

θ_c (day)	Percentage removal of the produced VSS (%)			
	$K_d = 0.05$ d^{-1}	$K_d = 0.07$ d^{-1}	$K_d = 0.09$ d^{-1}	$K_d = 0.11$ d^{-1}
4	13	18	21	24
8	23	29	33	37
12	30	37	42	46
16	36	42	47	51
20	40	47	51	55
24	44	50	55	–
28	47	53	–	–
32	49	55	–	–

2.8 RECIRCULATION OF THE ACTIVATED SLUDGE

To achieve a high concentration of solids in the reactor and a sludge age greater than the hydraulic detention time ($\theta_c > t$), it is necessary to recirculate or retain the sludge in the system. The sludge *retention* processes can be adopted in systems with intermittent operation, such as batch systems. The sludge *recirculation* through pumping is the most commonly used and is typical of the continuous-flow conventional activated sludge and extended aeration processes.

The amount of sludge to be recirculated will depend fundamentally on the quality of the sludge settled in the secondary sedimentation tank: the more concentrated the sludge, the lower the recirculation flow needs to be to reach a certain solids concentration in the reactor. In other words, good sludge settleability and thickening properties in the secondary sedimentation tank, resulting in a return sludge with higher SS concentration, will lead to a reduction in the recirculation flow. However, this analysis is complex, since the flow at the bottom of the secondary sedimentation tank (usually equal to the return sludge flow plus the excess sludge flow) in itself affects the concentration of the settled sludge. The SS concentration in the return sludge is called **RASS (return activated sludge suspended solids**, also expressed as $\mathbf{X_r}$).

Figure 2.2 shows the items that integrate the solids mass balance in the biological stage of the activated sludge system.

The **return sludge ratio R** is defined as:

$$R = Q_r/Q \tag{2.27}$$

The mass balance in a complete-mix reactor operating in the steady state leads to:

$$\text{Accumulation} = \text{Input} - \text{Output} + \text{Production} - \text{Consumption}$$

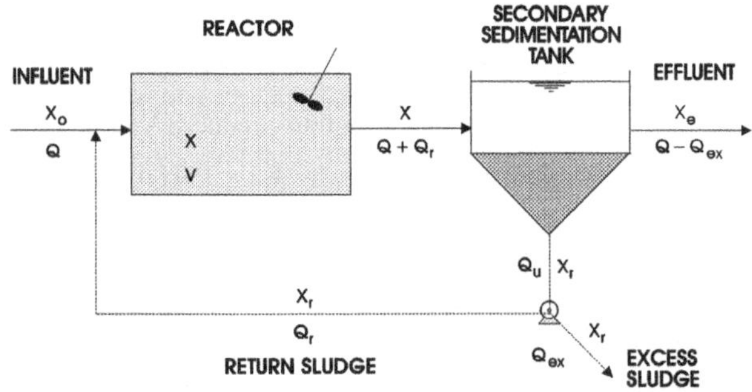

Figure 2.2. Suspended solids mass balance in the activated sludge system

In this mass balance, the following specific items are applicable:

- accumulation $= 0$ (there are no mass accumulations in the steady state)
- production $=$ consumption (bacterial growth equals the removal of excess sludge in the steady state)
- input $=$ raw sewage SS load $+$ return sludge SS load
- output $=$ MLSS load

The SS load in the raw sewage ($Q \cdot X_o$) is small, compared with the return sludge load ($Q_r \cdot X_r$). Neglecting the SS load in the raw sewage, one arrives at the following mass balance in the reactor:

$$\text{Input} = \text{Output}$$
$$Q_r \cdot X_r = (Q + Q_r) \cdot X \tag{2.28}$$

After rearrangement of Equation 2.28:

$$\boxed{R = \frac{Q_r}{Q} = \frac{X}{X_r - X}} \tag{2.29}$$

By rearranging Equation 2.29, the formula that expresses X_r as a function of X and R, in the steady state, can be obtained:

$$\boxed{X_r = X \cdot \frac{(R + 1)}{R}} \tag{2.30}$$

If the mass balance had been made in the secondary sedimentation tank, the results obtained would be the same:

$$\text{Input} = \text{Output}$$
$$(Q + Q_r) \cdot X = Q_r \cdot X_r \tag{2.31}$$

Equation 2.29 assumes that there are no biochemical mass production and consumption processes in the settling tanks, that the amount of solids leaving the settling tank through the final effluent (supernatant) is negligible and that $Q_r \approx Q_u$ (that is, the flow Q_{ex} is negligible compared to Q_r). Equation 2.31 is identical to Equation 2.28 and the values of R (Equation 2.29) and X_r (Equation 2.30) can be deduced from it.

Example 2.5

Calculate the required *return sludge ratio* to maintain a SS concentration in the reactor of Example 2.1 equal to 3,125 mg/L, knowing that the return sludge has an average SS concentration of 10,000 mg/L, as determined from measurements. Also calculate the *return sludge flow*, considering that the influent flow is 1,500 m³/d.

Example 2.5 (Continued)

Solution:

Using Equation 2.29:

$$R = \frac{X}{X_r - X} = \frac{3,125}{10,000 - 3,125} = 0.45$$

If the return sludge ratio is known, Q_r can be calculated through the rearrangement of Equation 2.27:

$$Q_r = R \cdot Q = 0.45 \times 1,500 \ m^3/d = 675 \ m^3/d$$

The reason for using SS instead of VSS in the example is that most frequently in the operational routine of the treatment plant the solids concentrations are determined as SS, for simplicity reasons. However, in the calculations that involve biological stages (reactor), it is interesting to consider VSS, for the sake of uniformity and greater coherence. In this example, the SS concentration of 3,125 mg/L in the reactor corresponds to a VSS of 2,500 mg/L (VSS/SS ratio of 0.8).

The concentration of suspended solids in the return sludge (RASS) depends on the settling and thickening characteristics of the sludge, the MLSS concentration and the underflow from the settling tank. Typical average values of RASS are around:

> SS in the return sludge (RASS): 8,000 to 12,000 mg/L

RASS can vary significantly along the day, outside the range given above, as a result of variations in the influent solids load to the settling tank.

The factors influencing the sludge quality are related to several design and operational parameters. Some important aspects are discussed here (Arceivala, 1981; Eckenfelder, 1980; Ramalho, 1977) and further examined in Chapters 10 and 12:

- Very low sludge ages can imply a bacterial growth with a tendency to be dispersed, instead of a flocculent growth.
- Very high sludge ages may result in a floc predominantly consisting of a highly mineralised residue of endogenous respiration, with a small flocculation capacity.
- Certain environmental conditions in the reactor such as low DO levels may lead to the predominance of filamentous microorganisms, which have a relatively high surface area per unit volume. These microorganisms, however, generate a poorly settling floc, giving rise to the so-called sludge bulking.
- A plug-flow reactor is capable of producing a sludge with a better settleability (predominance of the floc-forming bacteria over the filamentous ones) than a complete-mix reactor.

Usually, when maintaining the sludge age within the usual design ranges, the aeration is enough, and certain constituents of the raw sewage are within the acceptable limits, the sludge is expected to flocculate well and exhibit good settleability and compactness in the secondary sedimentation tank. As a result, the recirculation ratio can be lower.

A recirculation ratio around 0.5 is adopted in systems operating in temperate climates, in which good compaction of the sludge in the secondary sedimentation tank is aimed at. In warm-climate regions, however, the tendency is to use higher values of R. One reason is that in these regions nitrification is very likely to occur in the reactor, due to the high temperatures. Additionally, it is also probable that denitrification will occur in the secondary sedimentation tank. The denitrification corresponds to the transformation of the nitrate into gaseous nitrogen. The N_2 bubbles are released by the bottom sludge and, in their upward movement, they adhere to the sludge flocs, transporting them to the surface (*rising sludge*). The consequence is an increased solids concentration in the final effluent, which leads to its deterioration in terms of SS and particulate BOD. To avoid this effect, the sludge recirculation should be faster to minimise denitrification in the secondary settler and its effects (Marais and Ekama, 1976). As the sludge is more quickly recirculated and becomes less liable to thickening, the solids concentration in the underflow sludge is lower, which implies the need for a higher recirculation rate. Figure 2.3 schematically illustrates the influence of the return sludge flow (or more precisely the underflow Q_u) on the RASS concentration, on the level of the sludge blanket and on the sludge detention time in the secondary sedimentation tank.

Typical values adopted in the operational routine of treatment plants operating in warm-climate regions are:

> Return sludge ratio: R = 0.7 to 1.2

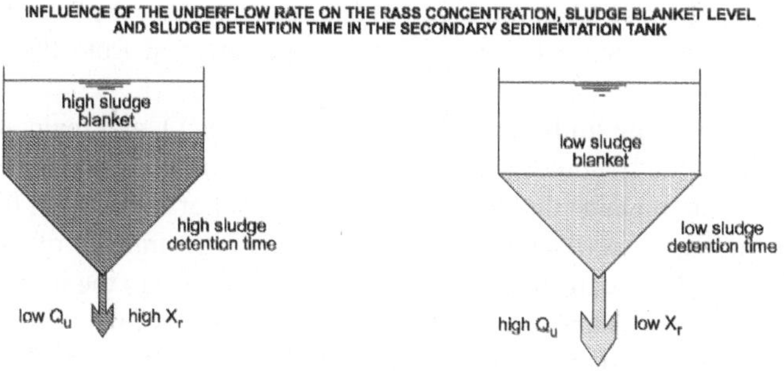

Figure 2.3. Influence of Q_u on the SS concentration in the return sludge, the sludge blanket level and the sludge detention time in the secondary sedimentation tank

However, the design should foresee a pumping capacity of around $R = 1.5$, for situations in which it becomes necessary to increase Q_r.

The increase in the capacity of an existing treatment plant can be obtained through an increase in the recirculation capacity, instead of expanding the reactor volume. This is true especially for systems having a low VSS concentration in the reactor. Through an increase in the value of R (and/or decrease in the excess sludge flow), an increase in VSS can be obtained, up to the practical limits discussed above (provided that the oxygenation capacity is enough for the new higher value of biomass respiration). The new value of R cannot be determined directly from Equation 2.29, using the new value of X. This is due to the fact that an increase in R may lead to a decrease in X_r.

It should be very clear that the existence of the sludge recirculation affects only the sludge age, with the hydraulic detention time remaining unaffected. The mass balance of the liquid remains constant (input = output), unlike the sludge, which is retained in the system. In a system with a return sludge ratio of $R = 1$ ($Q = Q_r$) each molecule of water has on average, probabilistically, the chance of passing twice through the reactor (an initial passage and another with the recirculation). As the influent flow is doubled ($Q + Q_r = 2Q$), the detention time in each passage is reduced to half ($t/2$). However, in the overall balance, after the two passages, the total time will be the same as t, therefore independent of the recirculation ratio.

Another aspect that should be very clear is the interaction between the return sludge flow Q_r, the excess sludge flow Q_{ex} and the sludge age. The flows Q_r and Q_{ex} are intimately connected, since both compose the underflow from the secondary sedimentation tank (Q_u). Thus:

$$Q_u = Q_r + Q_{ex} \qquad (2.32)$$

Irrespective of how the excess sludge is removed (directly from the reactor or from the return sludge line), the net contribution of the recirculated sludge will always be equal to $Q_r = Q_u - Q_{ex}$. Two situations can happen:

- **Fixed underflow Q_u.** In these conditions, increasing Q_r automatically decreases Q_{ex} (see Equation 2.32). When reducing Q_{ex}, the system's sludge age increases, since the amount of sludge removed from the system is reduced.
- **Increase of the underflow Q_u.** When increasing Q_r and maintaining Q_{ex} fixed, the underflow Q_u increases. However, the total mass of solids in the system remains the same, since the removal of solids from the system was not changed (fixed Q_{ex}). There is only a larger transfer of solids from the secondary sedimentation tank to the reactor, due to the increase in Q_r. Thus, the solids mass in the reactor increases, but the mass in the system (reactor+secondary sedimentation tank) remains the same. If the sludge age is computed in terms of only the *solids mass in the reactor*, there will be an apparent increase in the sludge age. On the other hand, if the sludge age

is computed in terms of the *solids mass in the system (reactor +secondary sedimentation tank)*, there will be no change in it.

In a simplified manner, the roles of Q_r and Q_{ex} in the activated sludge system can be understood as follows (Takase and Miura, 1985):

- The **return sludge flow** Q_r affects the *balance* of solids between the reactor and the secondary sedimentation tank.
- The **excess sludge flow** Q_{ex} affects the *total* mass of solids in the system (reactor+secondary sedimentation tank).

It is important to understand that the design and the operation of activated sludge systems require an integrated view of the reactor and the secondary sedimentation tank. The biological stage depends strongly on the solid–liquid removal stage. Therefore, it is fundamental to understand the settling and thickening phenomena.

2.9 PRODUCTION AND REMOVAL OF EXCESS SLUDGE

2.9.1 Sludge production

(a) Production of biological solids

As thoroughly discussed, an amount of sludge equivalent to the VSS produced daily, corresponding to the reproduction of the cells that feed on the substrate, should be removed from the system, so that it remains in balance (production of solids = removal of solids). A small part of this sludge leaves the system with the effluent (supernatant) of the secondary sedimentation tank, but most of it is extracted through the excess sludge (Q_{ex}). The excess sludge should be directed to the sludge treatment and final disposal stages.

An alternative way to present the net production is through the rearrangement of net production of VSS and Equation 2.4 (calculation of the reactor volume). Thus, P_{xv} can be expressed as:

$$P_{xv} \text{ net} = \text{Gross production of } X_v - \text{Destruction of } X_b$$

$$P_{xv} = Y \cdot Q \cdot (S_o - S) - K_d \cdot f_b \cdot X_v \cdot V \tag{2.33}$$

The same value can be arrived at using the concept of the *observed yield coefficient* (Y_{obs}), which directly reflects the net production of the sludge. Y_{obs} and P_{xv} can be obtained:

$$Y_{obs} = \frac{Y}{1 + f_b \cdot K_d \cdot \theta_c} \tag{2.34}$$

$$P_{xv} = Y_{obs} \cdot Q \cdot (S_o - S) \tag{2.35}$$

To obtain the production of biological solids in terms of TSS (P_x), P_{xv} should be divided by the VSS/TSS ratio.

Example 2.6

For the conventional activated sludge system described in Example 2.1, calculate the production of biological solids. Data from Examples 2.1, 2.3 and 2.4 include:

$$Q = 1{,}500 \text{ m}^3/\text{d} \qquad \theta_c = 6 \text{ days} \qquad Y = 0.7$$
$$S_o = 300 \text{ mg/L} \qquad X_v = 2{,}500 \text{ mg/L} \qquad K_d = 0.09 \text{ d}^{-1}$$
$$S = 8 \text{ mg/L} \qquad V = 535 \text{ m}^3 \qquad f_b = 0.72$$
$$\text{VSS/SS} = 0.80$$

Do not consider the solids in the raw sewage.

Solution:

(a) Calculation of the BOD load removed (information required in Example 2.7)

$$S_r = Q \cdot (S_o - S) = 1{,}500 \text{ m}^3/\text{d} \times (300 - 8) \text{ g/m}^3 \times 10^{-3} \text{ kg/g}$$
$$= 438 \text{ kgBOD/d}$$

(b) Calculation of the biological solids production according to Equation 2.33

$$P_{xv} = Y \cdot Q \cdot (S_o - S) - K_d \cdot f_b \cdot X_v \cdot V$$
$$P_{xv} = 0.7 \times 1{,}500 \text{ m}^3/\text{d} \times (300 - 8) \text{ g/m}^3 \times 10^{-3} \text{ kg/g}$$
$$- 0.09 \text{ d}^{-1} \times 0.72 \times 2{,}500 \text{ g/m}^3 \times 535 \text{ m}^3 \times 10^{-3} \text{ kg/g}$$
$$P_{xv} = 307 - 87 = 220 \text{ kgVSS/d}$$

In terms of TSS, the production is:

$$P_x = P_{xv}/(\text{VSS/SS}) = (220 \text{ kg/d})/(0.8) = 275 \text{ kgSS/d}$$

(c) Calculation of the biological solids production according to Equations 2.34 and 2.35

$$Y_{obs} = \frac{Y}{1 + f_b \cdot K_d \cdot \theta_c} = \frac{0.7}{1 + 0.72 \times 0.09 \text{ d}^{-1} \times 6 \text{ d}} = 0.50$$
$$P_{xv} = Y_{obs} \cdot Q \cdot (S_o - S) = 0.50 \times 1{,}500 \text{ m}^3/\text{d} \times (300 - 8) \text{ g/m}^3 \times 10^{-3} \text{ kg/g}$$
$$= 220 \text{ kgVSS/d}$$

> **Example 2.6 (*Continued*)**
>
> In terms of TSS, the production is:
>
> $$P_x = P_{xv}/(VSS/SS) = (220\,kg/d)/(0.8) = 275\,kgSS/d$$
>
> It is observed, therefore, that the values of P_{xv} and P_x obtained using Equations 2.33 and 2.35 are identical. Concerning the calculation of the distribution of the solids in the treatment, the daily production of VSS could have been obtained through the direct use of the simplified formula of Y_{obs}.

(b) Production of excess sludge

The *solids present in the raw sewage* (inorganic solids and non-biodegradable solids) also contribute to the production of excess sludge. The methodology for calculating the VSS/SS ratio and the production of secondary sludge including these solids is described in the general example in Chapter 5. If an initial approximation is desired, the values of Table 2.4 can be used, which were calculated for predominantly domestic sewage using the model described in this chapter, as well as the methodology exemplified in Section 2.13.

Table 2.4 includes the following alternatives of whether or not to consider the solids in the raw sewage and the presence of primary sedimentation tanks (in which approximately 60% of the suspended solids and 30% of BOD are removed):

- disregard the solids in the raw sewage (as is the case in most designs, but which leads to some distortions)
- consider the solids in the raw sewage in a system with primary sedimentation tanks
- consider the solids in the raw sewage in a system without primary sedimentation tanks

In Table 2.4, different combinations of the coefficients Y and K_d are presented (a high Y with a low K_d and vice versa). The VSS/SS ratio is relatively independent of the coefficients Y and K_d (in the range shown in the table) and is shown in Table 2.4 only as single intermediate values. The production of excess secondary sludge is more sensitive and is described according to three possible coefficient combinations (the first pair $Y - K_d$ results in the smallest sludge production, opposed to the last pair, in which the sludge production is the highest).

The utilisation of Table 2.4 is exemplified below. A conventional activated sludge plant that receives domestic sewage containing suspended solids, that includes a primary sedimentation tank and that has been designed for a sludge age of, say, 6 days, is expected to have a VSS/SS ratio of 0.76 and a sludge production between 0.75 to 0.95 kgSS/kgBOD$_5$ removed (depending on the coefficients Y and K_d adopted). An extended aeration plant that also contains solids in the influent, but does not include a primary sedimentation tank, and that has been designed for a sludge age of 26 days, is expected to have a VSS/SS of 0.68 and a sludge production between 0.88 and 1.01 kgSS/kgBOD$_5$ removed. Of course, in the design of the

Table 2.4. VSS/SS ratio in the reactor and production of excess secondary sludge per unit of BOD_5 removed from the reactor ($kgSS/kgBOD_5$ removed), as a function of the presence of solids in the influent, the existence of primary sedimentation tanks, the coefficients Y and K_d and the sludge age

Ratio	SS in the influent	Primary settling tank	Coefficients		Sludge age (day)							
			Y (g/g)	K_d (d^{-1})	2	6	10	14	18	22	26	30
VSS/SS (g/g)	No	No	0.5–0.7	0.07–0.09	0.89	0.87	0.85	0.84	0.83	0.82	0.81	0.81
	Yes	Yes	0.5–0.7	0.07–0.09	0.79	0.76	0.75	0.73	0.72	0.71	0.71	0.71
	Yes	No	0.5–0.7	0.07–0.09	0.75	0.73	0.71	0.70	0.69	0.69	0.68	0.68
SS/S$_r$ (kgSS / kgBOD$_5$rem)	No	No	0.5	0.09	0.50	0.42	0.37	0.33	0.31	0.29	0.28	0.28
			0.6	0.08	0.60	0.51	0.45	0.41	0.38	0.36	0.34	0.34
			0.7	0.07	0.71	0.61	0.55	0.50	0.47	0.44	0.42	0.40
	Yes	Yes	0.5	0.09	0.83	0.75	0.70	0.67	0.65	0.63	0.63	0.63
			0.6	0.08	0.96	0.87	0.81	0.78	0.75	0.73	0.71	0.71
			0.7	0.07	1.04	0.95	0.88	0.84	0.80	0.78	0.76	0.74
	Yes	No	0.5	0.09	1.08	1.00	0.95	0.92	0.90	0.88	0.88	0.88
			0.6	0.08	1.23	1.14	1.09	1.05	1.02	1.00	0.98	0.98
			0.7	0.07	1.29	1.20	1.13	1.08	1.06	1.03	1.01	0.99

Highlighted values: More usual values in activated sludge plants with typical flowsheets
• Per capita contributions: BOD = 50 g/inhabitant·day; SS = 60 g/inhabitant·day
• Removal efficiencies in the primary settling tank: BOD = 30%; SS = 60%

sludge treatment for the conventional activated sludge plant, the production of primary sludge also needs to be taken into account.

In this example, for a removed BOD_5 load of 100 $kgBOD_5/d$, a production of 43.2 kgVSS/d was estimated in the system with a sludge age of 6 days. Therefore, the calculated relation was $43.2/100 = 0.43$ $kgVSS/kgBOD_5$ removed. The VSS/SS ratio calculated in the example was 0.87. Thus, the specific production of SS can be expressed as $0.43/0.87 = 0.49$ $kgSS/kgBOD_5$ removed. This value is within the range expressed in Table 2.4 for systems without consideration of SS in the influent, without primary settling tanks and with a sludge age of 6 days (range from 0.42 to 0.61 $kgSS/kgBOD_5$ – the variation is due to the different values adopted for Y and K_d). The calculated VSS/SS value of 0.87 is identical to the value shown in Table 2.4.

Table 2.4 shows how important it is to consider the influent solids when calculating the production of excess secondary sludge. The sludge production values shown are quite similar to those in the German practice, related by Orhon and Artan (1994). According to this reference, conventional activated sludge systems, with an influent to the reactor with a SS/BOD_5 ratio of 0.7 (typical of systems with primary sedimentation tanks), have a sludge production in the range of 0.82 to 0.92 $kgSS/kgBOD_5$ applied (for sludge ages varying from 10 to 4 days, respectively). Extended aeration systems, with an influent to the reactor with a SS/BOD_5 ratio of 1.2 (typical of systems without primary settling tanks), result in a sludge production of around 1.00 $kgSS/kgBOD_5$ applied (sludge age of 25 days).

Example 2.11 included in Section 2.13 further illustrates how to use Table 2.4 for estimating the solids production in an activated sludge system taking into account the solids in the influent sewage.

2.9.2 Removal of the excess sludge

(a) Without consideration of the influent solids

In activated sludge systems, the excess sludge can be removed from two different locations: **reactor** or **return sludge line**. *If the solids in the influent are neglected* (unrealistic assumption for domestic sewage, but frequently adopted in the literature), the excess sludge concentration and flow, which vary with the removal place, can be determined as follows:

- **Withdrawal of the excess sludge directly from the reactor (or from the reactor effluent)**. This option is called *hydraulic control* of the system. The concentration of excess sludge is the same as the concentration of SS in the reactor (MLSS). If one wants to maintain the sludge age constant, the flow Q_{ex} can be obtained by:

$$Q_{ex}' = \frac{V}{\theta_c} \qquad (2.36)$$

$$SS\ concentration = MLSS\ (=X) \qquad (2.37)$$

where:

$Q_{ex}{}'$ = excess sludge flow removed from the reactor (m³/d)

V = reactor volume (m³)

θ_c = sludge age (d)

X = MLSS concentration (mg/L)

- **Withdrawal of the excess sludge from the return sludge line**. The concentration of excess sludge is the same as the concentration of SS in the return sludge (RASS). If one wants to maintain the sludge age constant, the flow Q_{ex} can be obtained by:

$$Q_{ex}{}'' = \frac{V}{\theta_c} \cdot \frac{X}{X_r} \tag{2.38}$$

$$\text{SS concentration} = \text{RASS}\,(= X_r) \tag{2.39}$$

where:

$Q_{ex}{}''$ = excess sludge flow removed from the return sludge line (m³/d)

X_r = RASS concentration (mg/L)

The removal of excess sludge from the return sludge line requires a flow Q_{ex} smaller than that required in the hydraulic control (X_r/X times smaller). Thus, the sludge *flow* to be treated is smaller, but the *load* of solids, which is equal to the product of concentration and flow, is the same. On the other hand, the hydraulic control is simpler, not requiring the determination of the SS concentration in the reactor and in the return line. In the hydraulic control, if one wants to maintain a sludge age of, for example, 20 days, it will suffice to remove 1/20 of the volume of the reactor per day as excess sludge.

(b) Considering the influent solids

It should be remembered that the methods shown in item (a) above do not take into account the influence of the solids in the influent wastewater (particularly the inert solids), and compute only the production and removal of the *biological solids produced in the system*. If the influent solids are considered, the calculations should be based on the total excess sludge production (P_x), as discussed in Section 2.9.1.b:

- **Withdrawal of the excess sludge directly from the reactor (or from the reactor effluent)**

$$Q_{ex}{}' = \frac{P_x \cdot 1{,}000}{X} \tag{2.40}$$

$$\text{SS concentration} = \text{MLSS}\,(= X) \tag{2.41}$$

where:

P_x = excess sludge production (kgSS/d)

Table 2.5. Items to be considered in the removal of the excess sludge from the activated sludge system

Process	Item
Conventional activated sludge	• The secondary excess sludge requires subsequent stabilisation, which is completed in the sludge treatment stage • The sludge can be removed directly from the reactor (smaller concentration of SS, larger Q_{ex}) or from the return sludge line (larger concentration of SS, smaller Q_{ex}) • The excess sludge can be removed continuously or intermittently • The excess sludge can be directed separately to the sludge treatment stage (including digestion) or returned to the primary settling tank, for sedimentation and treatment together with the primary sludge (smaller plants)
Extended aeration	• The secondary sludge is already largely stabilised and does not require a subsequent digestion stage • The sludge can be removed directly from the reactor or from the return sludge line • The sludge can be removed continuously or intermittently • The excess sludge is usually sent directly to the sludge-processing phase.

• **Withdrawal of the excess sludge from the return sludge line**

$$Q_{ex}'' = \frac{P_x \cdot 1,000}{X_r} \qquad (2.42)$$

$$\text{SS concentration} = \text{RASS}(= X_r) \qquad (2.43)$$

In the estimation of the excess sludge load to be removed, the *suspended solids load in the final effluent* can be discounted from the total value. The loss of solids in the final effluent is unintentional, but, in practice, it does occur. However, in most situations, this term is small, compared to the overall solids production.

A summary of additional aspects related to the removal of excess sludge is listed in Table 2.5.

Example 2.7

For the activated sludge system described in Example 2.6, determine the amount of excess sludge to be removed daily. Analyse the alternative methods of (a) removing the sludge directly from the reactor and (b) removing the sludge from the return sludge line. Make the calculations under two conditions: (i) without consideration of solids in the influent and effluent and (ii) with consideration of solids in the influent and in the effluent.

Example 2.7 (Continued)

Data from previous examples:

$Q = 1{,}500 \text{ m}^3/\text{d}$ $X = 3{,}125 \text{ mg/L (MLSS)}$

$V = 535 \text{ m}^3$ $X_r = 10{,}000 \text{ mg/L (RASS)}$

$\theta_c = 6 \text{ days}$ $X_e = 30 \text{ mg/L (suspended solids concentration in the final effluent)}$

Solution:

Without consideration of solids in the influent and effluent:

(a) Removal of the excess sludge directly from the reactor (hydraulic control)

- Daily flow to be wasted (Equation 2.36):

$$Q_{ex} = \frac{V}{\theta_c} = \frac{535 \text{ m}^3}{6 \text{ d}} = 89 \text{ m}^3/\text{d}$$

- SS concentration (Equation 2.37):

$$SS = X = 3{,}125 \text{ mg/L}$$

- Load to be wasted:

$$Q_{ex} \cdot X = 89 \text{ m}^3/\text{d} \times 3{,}125 \text{ g/m}^3 \times 10^{-3} \text{ kg/g} = 275 \text{ kgSS/d}$$

As expected, this value is equal to the production of biological excess sludge (as calculated in Example 2.6), since the system is in equilibrium in the steady state (production = removal).

(b) Removal of excess sludge from the sludge recirculation line

- Daily flow to be wasted (Equation 2.38):

$$Q_{ex} = \frac{V}{\theta_c} \cdot \frac{X}{X_r} = \frac{535 \text{ m}^3}{6 \text{ d}} \cdot \frac{3{,}125 \text{ g/m}^3}{10{,}000 \text{ g/m}^3} = 28 \text{ m}^3/\text{d}$$

Due to the larger concentration of the removed sludge ($=X_r$), the flow of the excess sludge Q_{ex} is much smaller than that in the alternative method of direct extraction from the reactor ($Q_{ex} = 89 \text{ m}^3/\text{d}$).

- SS concentration (Equation 2.39):

$$SS = X_r = 10{,}000 \text{ mg/L}$$

Example 2.7 (Continued)

- Load to be wasted:

$$Q_{ex} \cdot X_r = 27.8\,\text{m}^3/\text{d} \times 10{,}000\,\text{g/m}^3 \times 10^{-3}\,\text{kg/g} = 278\,\text{kgSS/d}$$

As expected, this value is equal to the production of excess sludge (as calculated in Example 2.6) and equal to the load to be extracted by the reactor in alternative "a" (any differences are due to rounding up).

With consideration of solids in the influent and effluent:

- BOD load removed (calculated in Example 2.6, item a):

$$S_r = 438\,\text{kgBOD/d}$$

- Load of SS produced:

From Table 2.4, sludge age of 6 days, considering solids in the influent and system with primary sedimentation tank: $P_x/S_r = 0.87$. Therefore, P_x is:

$$P_x = (P_x/S_r) \cdot \text{BOD load removed} = 0.87\,\text{kgSS/kgBOD} \times 438\,\text{kgBOD/d}$$

$$= 381\,\text{kgSS/d}$$

- Load of SS escaping with the final effluent:

$$\text{Load SS effluent} = Q \cdot X_e = 1{,}500\,\text{m}^3/\text{d} \times 30\,\text{mg/L} \times 10^{-3}\,\text{kg/g}$$

$$= 45\,\text{kgSS/d}$$

- Excess sludge load to be removed daily:

$$\text{Load excess sludge} = P_x - \text{load SS effluent} = 381 - 45 = 336\,\text{kgSS/d}$$

Note: the SS load to be removed (equal to the production of biological solids), calculated in the first part of this example, was 278 kgSS/d.

(a) Removal of the excess sludge directly from the reactor (hydraulic control)

- Daily flow to be wasted (adaptation of Equation 2.40, discounting the solids in the effluent):

$$Q_{ex}{}' = \frac{\text{load excess sludge} \times 1{,}000}{X} = \frac{336 \times 1{,}000}{3{,}125} = 108\,\text{m}^3/\text{d}$$

- SS concentration (Equation 2.41):

$$SS = X = 3{,}125\,\text{mg/L}$$

Example 2.7 (Continued)

(b) Removal of excess sludge from the sludge recirculation line

- Daily flow to be wasted (adaptation of Equation 2.42, discounting the solids in the effluent):

$$Q_{ex}' = \frac{\text{load excess sludge} \times 1,000}{X_r} = \frac{336 \times 1,000}{10,000} = 34\,\text{m}^3/\text{d}$$

- SS concentration (Equation 2.43):

$$SS = X_r = 10,000\,\text{mg/L}$$

The differences in loads and flows, compared to the calculations made in the first part of the example (without consideration of SS in the influent and effluent) should be noted.

2.10 OXYGEN REQUIREMENTS

2.10.1 Preliminaries

In aerobic biological treatment, oxygen should be supplied to satisfy the following demands:

- *oxidation of the carbonaceous organic matter*
  - oxidation of the organic carbon to supply energy for bacterial *synthesis*
  - *endogenous respiration* of the bacterial cells
- *oxidation of the nitrogenous matter (nitrification)*

In systems with biological denitrification, oxygen savings due to denitrification can be taken into consideration.

The present section is devoted to the analysis of aspects related to the oxygen **consumption**.

There are two ways to calculate the oxygen requirements for the satisfaction of the **carbonaceous** demand. Both are equivalent and interrelated, and naturally lead to the same values:

- method based on the total carbonaceous demand and on the removal of excess sludge
- method based on the oxygen demand for synthesis and for endogenous respiration

The oxygen demand for the **nitrification** is based on a stoichiometric relation with the oxidised ammonia. Although this chapter deals only with the removal of the carbonaceous matter, the consumption of oxygen for nitrification should also be taken into consideration, since in warm-climate regions nitrification takes place almost systematically in systems designed for the removal of BOD.

2.10.2 Carbonaceous oxygen demand

(a) Method based on the total carbonaceous demand and the removal of excess sludge

The supply of oxygen for the carbonaceous demand should be the same as the consumption of oxygen for the ultimate BOD (BOD_u) removed by the system. This demand corresponds to the total oxygen demand for the oxidation of the substrate and for the endogenous respiration of the biomass. The ultimate BOD, in turn, is the same as the BOD_5 multiplied by a conversion factor that is in the range of 1.2 to 1.6 for domestic sewage. A value usually adopted is BOD_u/BOD_5 equal to 1.46. Thus, the mass of oxygen required per day can be determined as a function of the removed BOD_5 load:

$$\text{OUR (kg/d)} = \frac{1.46 \cdot Q \cdot (S_o - S)}{10^3} \tag{2.44}$$

where:

$\quad$ OUR = oxygen utilisation rate, or oxygen requirement (kgO_2/d)
$\qquad$ Q = influent flow (m^3/d)
$\qquad$ S_o = influent BOD_5 concentration (total BOD) (g/m^3)
$\qquad$ S = effluent BOD_5 concentration (soluble BOD) (g/m^3)
$\quad$ 1.46 = conversion factor (BOD_u/BOD_5)
$\qquad$ 10^3 = conversion factor (g/kg)

However, in the activated sludge system, part of the influent organic matter is converted into new cells. A mass equivalent to that from the cells produced is wasted from the system (production = wastage in a system in the steady state). For this reason, the fraction corresponding to the oxygen consumed by these cells, which will not be completed inside the system, should be discounted from the total oxygen consumption. As demonstrated by Equations 2.11 and 2.12 (Section 2.6), each 1 g of cells consumes 1.42 g of oxygen for its stabilisation. Thus, Equation 2.44 can be expanded and written literally as:

$$\begin{array}{lll} \text{OUR} & = \text{Removed } BOD_u - [1.42 \times \text{(solids produced)}] \\ \text{(kg/d)} & \quad\text{(kg/d)} \qquad\qquad\qquad\qquad\text{(kg/d)} \end{array} \tag{2.45}$$

The mass of volatile suspended solids produced per day (P_{xv}) is given by Equation 2.33 or 2.35. Thus, the consumption of oxygen for the stabilisation of the carbonaceous organic matter can be expressed through (Metcalf and Eddy, 1991):

$$\boxed{\text{OUR (kg/d)} = 1.46 \cdot Q \cdot (S_o - S) - 1.42 \cdot P_{xv}} \tag{2.46}$$

By replacing P_{xv} in the above equation by the right-hand side of Equation 2.35 (equation that expresses P_{xv} in terms of Y_{obs}), another form of representing the consumption of oxygen is obtained, after some rearrangement:

$$\text{OUR (kg/d)} = Q \cdot (S_o - S) \cdot \left(1.46 - \frac{1.42 \cdot Y}{1 + K_d \cdot f_b \cdot \theta_c} \right) \qquad (2.47)$$

Example 2.8

Estimate the oxygen consumption for the oxidation of the carbonaceous matter in the conventional activated sludge ($\theta_c = 6$ days) and in the extended aeration ($\theta_c = 22$ days) systems.

Data:

Removed BOD load: $Q \cdot (S_o - S) = 100.0$ kg/d
SSV production: $P_{xv} = 43.2$ kg/d (conventional activated sludge)
SSV production: $P_{xv} = 28.2$ (extended aeration)

Solution:

- Conventional activated sludge (Equation 2.46)

$$\text{OUR} = 1.46 \cdot Q \cdot (S_o - S) - 1.42 \cdot P_{xv} = 1.46 \times 100.0 - 1.42 \times 43.2$$

$$= 84.7 \, \text{kgO}_2/\text{d}$$

- Extended aeration (Equation 2.46)

$$\text{OUR} = 1.46 \cdot Q \cdot (S_o - S) - 1.42 \cdot P_{xv} = 1.46 \times 100.0 - 1.42 \times 28.2$$

$$= 106.0 \, \text{kgO}_2/\text{d}$$

As expected, the extended aeration leads to a greater oxygen consumption, compared with the conventional activated sludge system. In this example, the difference is due to the lower removal of the excess sludge in the extended aeration plant. If the conventional activated sludge system had included a primary sedimentation tank (as is usual), the influent BOD_5 load to the biological treatment stage would have been smaller, resulting in an even smaller oxygen consumption.

(b) Method based on the oxygen demand for substrate oxidation and endogenous respiration

The oxygen demand for the oxidation of the carbonaceous organic matter can be divided into two main components:

- oxygen demand for synthesis
- oxygen demand for endogenous respiration

The equation for the O_2 consumption can be obtained by rearranging Equation 2.46. Thus, if P_{xv} is replaced by the right-hand-side of Equation 2.33, one will arrive at the following:

$$OUR\ (kg/d) = 1.46 \cdot Q \cdot (S_o - S) - 1.42 \cdot P_{xv} \tag{2.46}$$

$$OUR\ (kg/d) = 1.46 \cdot Q \cdot (S_o - S) - 1.42 \cdot [Y \cdot Q \cdot (S_o - S) - f_b \cdot K_d \cdot X_v \cdot V] \tag{2.48}$$

$$OUR\ (kg/d) = (1.46 - 1.42 \cdot Y) \cdot Q \cdot (S_o - S) + 1.42 \cdot f_b \cdot K_d \cdot X_v \cdot V] \tag{2.49}$$

The above equation can be expressed in the following simplified way:

$$\boxed{OUR\ (kg/d) = a' \cdot Q \cdot (S_o - S) + b' \cdot X_v \cdot V} \tag{2.50}$$

where:
$$a' = 1.46 - 1.42 \cdot Y$$
$$b' = 1.42 \cdot f_b \cdot K_d$$

This equation provides a very convenient way of expressing the oxygen consumption through its two main components: synthesis (first term on the right-hand side) and the biomass respiration (second term on the right-hand side). For example, in an existing system, the result of the manipulation of the concentration of the biomass (X_v) in the total oxygen consumption can be directly evaluated.

With respect to the coefficient values, it should be borne in mind that b' is a function of f_b, that is, indirectly, of θ_c. As a consequence, extended aeration systems should have smaller values of b'. However, as the volume V of the reactor is much larger in these systems, the term on the right-hand side (biomass respiration) is larger than that for the conventional activated sludge systems.

To allow expedited determinations of the average carbonaceous demand, Table 2.6 includes values of the $OUR/BOD_{removed}$, for different combinations of Y and K_d values.

With respect to Table 2.6, the following aspects are worth noting:

- The oxygen consumption for satisfaction of the carbonaceous demand increases with the sludge age.
- The lower sludge age range is more sensitive to the values of the coefficients Y and K_d. In the extended aeration range, the variation of the oxygen demand with the coefficients Y and K_d is smaller.

Table 2.6. Carbonaceous oxygen demand per unit of BOD_5 removed ($kgO_2/kgBOD_5$ rem), in domestic sewage, for different values of Y and K_d

Coefficients		Sludge age (day)							
Y (g/g)	$K_d(d^{-1})$	2	6	10	14	18	22	26	30
0.5	0.09	0.84	0.95	1.02	1.07	1.10	1.13	1.14	1.14
0.6	0.08	0.70	0.83	0.91	0.97	1.01	1.05	1.07	1.07
0.7	0.07	0.57	0.70	0.80	0.86	0.91	0.95	0.98	1.01

- The estimation of the oxygen consumption for the oxidation of the carbonaceous matter does not depend on whether solids are present in the influent sewage.
- The values included refer to average flow and load conditions, and do not take into account adjustments for peak conditions (see example in Chapter 5).

Example 2.9

Estimate the oxygen consumption for the oxidation of the carbonaceous matter in the conventional activated sludge ($\theta_c = 6$ days) and in the extended aeration ($\theta_c = 22$ days) systems. Data are:

$$Q \cdot (S_o - S) = 100.0 \text{ kg/d} \qquad f_b = 0.72 \text{ (conventional activated sludge)}$$
$$Y = 0.6 \qquad\qquad\qquad\quad f_b = 0.57 \text{ (extended aeration)}$$
$$K_d = 0.09 \text{d}^{-1}$$

Solution:

(a) Conventional activated sludge

- Calculation of a' (Equation 2.50)

$$a' = 1.46 - 1.42 \cdot Y = 1.46 - 1.42 \times 0.6 = 0.608 \text{ kgO}_2/\text{kgBOD}_5$$

- Calculation of b' (Equation 2.50)

$$b' = 1.42 \cdot f_b \cdot K_d = 1.42 \times 0.72 \times 0.09 = 0.092 \text{ kgO}_2/\text{kgVSS} \cdot \text{d}$$

- Calculation of $X_v \cdot V$ (Equation 2.4)

$$V \cdot X_v = \frac{Y \cdot \theta_c \cdot Q \cdot (S_o - S)}{1 + f_b \cdot K_d \cdot \theta_c} = \frac{0.6 \times 6 \times 100.0}{1 + 0.72 \times 0.09 \times 6} = 259.2 \text{ kgVSS}$$

- Calculation of the O_2 consumption
  - synthesis: $a' \cdot Q \cdot (S_o - S) = 0.608 \times 100.0 = 60.8 \text{ kgO}_2/\text{d}$
  - biomass respiration: $b' \cdot X_v \cdot V = 0.092 \times 259.2 = 23.8 \text{ kgO}_2/\text{d}$
  - total: $60.8 + 23.8 = 84.6 \text{ kgO}_2/\text{d}$
- $O_2/\text{BOD}_{5\,\text{removed}}$ ratio:

$$\text{OUR/BOD}_5 = 84.6/100.0 = 0.85 \text{ kgO}_2/\text{kgBOD}_5 \text{ rem}$$

(b) Extended aeration

- Calculation of a' (same as item (a))

$$a' = 0.608 \text{ kgO}_2/\text{kgBOD}_5$$

Example 2.9 (Continued)

- Calculation of b$'$ (Equation 2.50)

$$b' = 1.42 \cdot f_b \cdot K_d = 1.42 \times 0.57 \times 0.09 = 0.073 \text{ kgO}_2/\text{kgVSS.d}$$

- Calculation of $X_v \cdot V$ (Equation 2.4)

$$V \cdot X_v = \frac{Y \cdot \theta_c \cdot Q \cdot (S_o - S)}{1 + f_b \cdot K_d \cdot \theta_c} = \frac{0.6 \times 22 \times 100.0}{1 + 0.57 \times 0.09 \times 22} = 620.1 \text{ kgVSS}$$

- Calculation of the O_2 consumption
  - synthesis: 60.8 kgO$_2$/d (same as item (a))
  - biomass respiration: b$'\cdot X_v \cdot V = 0.073 \times 620.1 = 45.3$ kgO$_2$/d
  - total: $60.8 + 45.3 = 106.1$ kgO$_2$/d
- $O_2/BOD_{5\,\text{removed}}$ ratio:

OUR/BOD$_5 = 106.1/100.0 = 1.06$ kgO$_2$/kgBOD$_5$ rem (very similar to the value given in Table 2.6 – notice the difference in the values of Y and K_d)

(c) Summary

Variant	O$_2$ consumption (kgO$_2$/d)		
	Synthesis	Respiration	Total
Conventional	60.8	23.8	84.6
Extended aeration	60.8	45.3	106.1

Therefore, it is observed that the larger oxygen consumption in the extended aeration plant compared to the conventional activated sludge is due to the biomass respiration. It can also be noticed that the total values of oxygen consumption are the same ones obtained in Example 2.8 (any differences are due to rounding up).

The O_2 consumption for biomass respiration can also be calculated by multiplying the load of destroyed biodegradable solids by the factor 1.42.

The present example assumed, for comparison purposes, that the conventional activated sludge plant had no primary sedimentation tank. In most real situations, primary clarifiers are included, leading to a reduction in the influent BOD load to the biological stage and, therefore, an even lower oxygen consumption.

The oxygen consumption calculated following the methods described above refers to the average steady-state conditions. During **peak hours**, the maximum influent flow usually coincides with the maximum concentration of influent BOD$_5$

(Metcalf and Eddy, 1991; von Sperling, 1994c). Thus, if both peaks are coincident, the maximum influent load of BOD_5 is $(Q_{max}/Q_{average}) \times (BOD_{max}/BOD_{average})$ times greater than the average load. However, the peak oxygen consumption does not necessarily coincide with the peak BOD_5 load, being dampened and lagged in some hours. The reason for this is that the soluble BOD is assimilated rapidly, while the particulate BOD takes some time to be hydrolysed (without oxygen consumption) and later assimilated (Clifft and Andrews, 1981). When calculating the total oxygen consumption, a **safety factor** should be included, which is associated with the influent peak load or with the maximum flow.

2.10.3 Oxygen demand for nitrification

Nitrification corresponds to the oxidation of ammonia to nitrite and, subsequently, to nitrate. This oxidation implies an oxygen consumption, which should be included in the total oxygen requirements. The organic nitrogen, also present in the raw sewage, does not directly undergo nitrification, but is initially converted into ammonia, which then results in its subsequent nitrification. Thus, it is assumed that the organic nitrogen and ammonia are capable of generating oxygen consumption in the nitrification process. The sum of the organic nitrogen and the ammonia nitrogen is represented by TKN (total Kjeldahl nitrogen).

The principles of nitrification, as well as the conditions for its occurrence, are discussed in Chapters 6 and 7. For the purpose of the current section, it is sufficient to know that, stoichiometrically:

1 g TKN requires 4.57 gO_2 for conversion to NO_3^{-}

Thus:

$$\boxed{OUR\ (kg/d) = 4.57 \cdot Q \cdot TKN/10^3}$$
(2.51)

where:
  TKN = total Kjeldahl nitrogen, equal to the organic nitrogen and the ammonia
          nitrogen (mgN/L)

In fact, it can be considered that in the raw sewage, TKN represents the total influent nitrogen, since nitrite and nitrate concentrations in the influent are normally negligible. Thus, TKN is the nitrogen potentially oxidisable to nitrate.

The bacteria responsible for nitrification have a very slow growth rate, besides being very sensitive to changes in the environmental conditions. Consequently, nitrification is subject to the compliance to some minimum criteria. In the conventional activated sludge system, in **warm-climate** countries, the chances of occurrence of nitrification are very high, even in activated sludge systems with low sludge ages, because of the high temperatures that accelerate the growth rate of the nitrifying bacteria. Therefore, even if only for safety reasons, it is recommended that the consumption of oxygen for nitrification should be added to the total oxygen requirements. In the extended aeration process, in view of the higher sludge ages that allow comfortably the growth of the nitrifying bacteria, it can be considered that nitrification takes place systematically, unless some environmental restrictions (such as low dissolved oxygen) are present.

Denitrification implies decreased oxygen requirements. However, to obtain significant savings, denitrification should be included as a specific goal in the design of the plant. The presence of anoxic conditions is essential for the occurrence of denitrification.

2.11 NUTRIENT REQUIREMENTS

The microorganisms responsible for the oxidation of the organic matter require other nutrients, besides carbon, for their metabolic activities. The main nutrients are usually **nitrogen** and **phosphorus**, besides other elements in trace concentrations.

For the treatment system to remove BOD, organic carbon must be the limiting nutrient in the medium and the other nutrients must be present in concentrations above the minimum level required by the microorganisms. For domestic sewage this requirement is usually satisfied, while for certain industrial wastewaters there may be a lack of some nutrients, leading to a decrease in the biomass growth rate. In several situations, it is advantageous to combine domestic and industrial wastewaters in the public sewerage network, so that, after mixing and dilution, the influent to the treatment plant will be self-sufficient in terms of nutrient requirements.

The amount of N and P required depends on the composition of the biomass. When expressing the typical composition of a bacterial cell in terms of the empirical formulae $C_5H_7O_2N$ or $C_{60}H_{87}O_{23}N_{12}P$ (Metcalf and Eddy, 1991), the biomass synthesised in the treatment plant contains approximately 12.3% of nitrogen and 2.6% of phosphorus. The cellular residue after endogenous respiration has around 7% of nitrogen and 1% of phosphorus (Eckenfelder, 1980, 1989).

According to Eckenfelder (1980, 1989), the amount of **nitrogen** required is equivalent to the nitrogen removed from the system through the excess sludge. The main fractions are the nitrogen present in the active biomass that leaves the system in the form of excess sludge, and the nitrogen present in the non-active residue from the endogenous respiration. Based on the above mentioned percentages of the cellular composition, the nitrogen requirement can be estimated:

$$\text{N required} = \text{N in the active cells from excess sludge}$$

$$+ \text{N in the non-active cells of the excess sludge} \qquad (2.52)$$

$$\boxed{N_{req} = 0.123 \cdot \left(\frac{f_b}{f_b{}'}\right) \cdot P_{xv} + 0.07 \cdot \left(1 - \frac{f_b}{f_b{}'}\right) \cdot P_{xv}} \qquad (2.53)$$

where:

N_{req} = required nitrogen load (kgN/d)

f_b = biodegradable fraction of the volatile suspended solids (SS_b/VSS)

$f_b{}'$ = biodegradable fraction of the volatile suspended solids immediately after its generation, usually adopted as 0.8.

P_{xv} = net production of volatile suspended solids (kgVSS/d) = $X_v \cdot V / (10^3 \cdot \theta_c) = Y_{obs} \cdot Q \cdot (S_o - S)$

Table 2.7. Minimum nutrient requirements

Activated sludge	θ_c (day)	Ratio between nutrients (in mass)		
		BOD_5	N	P
Conventional	4–10	100	4.0–6.0	0.9–1.2
Extended aeration	20–30	100	2.5–3.5	0.5–0.6

Similarly, for **phosphorus**, one has:

$$P_{req} = 0.026 \cdot \left(\frac{f_b}{f_{b'}} \right) \cdot P_{xv} + 0.01 \cdot \left(1 - \frac{f_b}{f_{b'}} \right) \cdot P_{xv} \qquad (2.54)$$

To be used by the microorganisms, the nitrogen needs to be in a form that can be assimilated, such as *ammonia* and *nitrate*. The organic nitrogen first needs to undergo hydrolysis to become available for the biomass.

It can be seen from Equations 2.53 and 2.54 that systems with a high sludge age, such as extended aeration, imply lower nutrient requirements, due to the lower production of excess sludge. Table 2.7 presents the ranges of N and P requirements for conventional activated sludge and extended aeration systems.

Values usually mentioned in literature are a BOD_5:N:P ratio of 100:5:1. However, it should be borne in mind that these values will apply only to the **conventional** activated sludge, as shown in Table 2.7.

Example 2.10

Calculate the nitrogen requirement for the two activated sludge systems described in Example 2.1. Important data from this and subsequent examples are:

- Conventional activated sludge:

$$\theta_c = 6 \text{ days} \qquad f_b = 0.72$$
$$X_v = 2{,}500 \text{ mg/L} \qquad S = 8 \text{ mg/L}$$
$$V = 535 \text{ m}^3$$

- Extended aeration:

$$\theta_c = 22 \text{ days} \qquad f_b = 0.57$$
$$X_v = 3{,}000 \text{ mg/L} \qquad S = 3 \text{ mg/L}$$
$$V = 1{,}067 \text{ m}^3$$

- General data:

$$S_o = 300 \text{ mg/L}$$
$$TKN = 45 \text{ mg/L}$$
$$Q = 1{,}500 \text{ m}^3/\text{d}$$

Example 2.10 (Continued)

Solution:

(a)   Conventional activated sludge

The production of biological solids P_{xv} is given by:

$$P_{xv} = \frac{X_v \cdot V}{10^3 \cdot \theta_c} = \frac{2{,}500 \times 535}{1000 \times 6} = 229 \text{ kgVSS/d}$$

According to Equation 2.53, the required daily nitrogen load is:

$$N_{req} = 0.123 \cdot \left(\frac{f_b}{f_b'}\right) \cdot P_{xv} + 0.07 \cdot \left(1 - \frac{f_b}{f_b'}\right) \cdot P_{xv}$$

$$= 0.123 \times \left(\frac{0.72}{0.80}\right) \times 229 + 0.07 \times \left(1 - \frac{0.72}{0.80}\right) \times 229$$

$$N_{req} = 24.7 + 1.6 = 26.3 \text{ kgN/d}$$

(b)   Extended aeration

The biological solids production P_{xv} is given by:

$$P_{xv} = \frac{X_v \cdot V}{10^3 \cdot \theta_c} = \frac{3{,}000 \times 1.067}{1000 \times 22} = 146 \text{ kgVSS/d}$$

According to Equation 2.53, the required daily nitrogen load is:

$$N_{req} = 0.123 \cdot \left(\frac{f_b}{f_b'}\right) \cdot P_{xv} + 0.07 \cdot \left(1 - \frac{f_b}{f_b'}\right) \cdot P_{xv}$$

$$= 0.123 \times \left(\frac{0.57}{0.80}\right) \times 146 + 0.07 \times \left(1 - \frac{0.57}{0.80}\right) \times 146$$

$$N_{req} = 12.8 + 2.9 = 15.7 \text{ kgN/d}$$

(c)   Available nitrogen

For comparison purposes, the influent nitrogen load (TKN) is:

$$\text{Influent TKN load} = \frac{Q \cdot TKN}{1{,}000} = 1{,}500 \frac{m^3}{d} \cdot 45 \frac{g}{m^3} \cdot \frac{1 \text{ kg}}{1{,}000 \text{ g}} = 67.5 \text{ kgTKN/d}$$

The influent nitrogen load expressed in terms of TKN is thus higher than the required load, in both activated sludge process variants.

(d)   BOD:N ratio

The BOD consumed in the two systems is:

Example 2.10 (Continued)

- conventional activated sludge:

$$BOD_{rem} = \frac{Q \cdot (S_o - S)}{1,000} = 1,500 \frac{m^3}{d} \cdot (300 - 8) \frac{g}{m^3} \cdot \frac{1\ kg}{1,000\ g} = 438\ kgBOD/d$$

- extended aeration:

$$BOD_{rem} = \frac{Q \cdot (S_o - S)}{1,000} = 1,500 \frac{m^3}{d} \cdot (300 - 3) \frac{g}{m^3} \cdot \frac{1\ kg}{1,000\ g} = 446\ kgBOD/d$$

Thus, the required BOD:N ratio is:

- conventional activated sludge: BOD:N = 438:26.3 or **100:6.0**
- extended aeration: BOD:N = 446:15.7 or **100:3.5**

As can be seen, systems with higher sludge ages have lower nutrient requirements.

2.12 INFLUENCE OF THE TEMPERATURE

The temperature has a great influence on the microbial metabolism, thereby affecting the oxidation rates for the carbonaceous and nitrogenous matters.

In general terms and within certain limits, the rates of most chemical and biological reactions increase with temperature. In some chemical reactions, an approximate rule of thumb is that the reaction rate doubles for each increase of $10\,^{\circ}C$ in the medium temperature, resulting from the increased contact between the chemical molecules. In biological reactions, the tendency to increase the rates with the temperature will remain approximately valid up to a given optimum temperature. Above this temperature, the rate will decrease, due probably to the destruction of enzymes in the higher temperatures (Sawyer and Mc Carthy, 1978).

The relation between the temperature and the reaction coefficient can be expressed in the following manner:

$$\boxed{K_T = K_{20} \cdot \theta^{(T-20)}} \tag{2.55}$$

where:

K_T = reaction coefficient at a temperature T (d^{-1})
K_{20} = reaction coefficient at a standard temperature of $20\,^{\circ}C$ (d^{-1})
θ = temperature coefficient (–)
T = temperature of the medium $(^{\circ}C)$

Equation 2.55 is usually valid in the temperature range from 4 to $30\,^{\circ}C$, defined as the *mesophilic* range, in which most of the aerobic systems are included. The biological activity can also take place in the *thermophilic* range, at higher temperatures, found for example, in some anaerobic systems and aerobic digestion systems.

The interpretation of the coefficient θ is made in the sense that, if θ is equal to, say, 1.02, the value of the reaction rate increases by 2% ($= 1.02 - 1.00 = 0.02$) for each increment of $1\,^{\circ}C$ in the temperature.

The influence of the temperature decreases with the increase of the sludge age (Eckenfelder, 1980) and is not of great significance in systems with high sludge ages (Ekama and Marais, 1977; Cook, 1983; Matsui and Kimata, 1986; Markantonatos, 1988; von Sperling and Lumbers, 1989), such as extended aeration. Additionally, compared with other treatment processes, the activated sludge system is less sensitive to temperature. According to Eckenfelder (1980), this is due to the fact that a great part of the BOD, present in the form of particulate BOD, is removed physically by adsorption in the floc, which is independent of temperature. For example, in aerated lagoons, with low solids concentrations, each organism is more directly affected by temperature changes, which justify the large value of θ.

The adaptation of the microorganisms to abrupt temperature changes seems to be much slower at higher temperatures. For example, it was observed that several months would be needed for the acclimatisation of the biomass to a change of $5\,^{\circ}C$ in the temperature range of $30\,^{\circ}C$, while only 2 weeks were necessary for a similar adaptation in the range of $15\,^{\circ}C$ (Winkler, 1981).

Between 10 and $30\,^{\circ}C$, μ_{max} and K_d increase with temperature. K_s decreases slightly between 10 and $20\,^{\circ}C$ and increases substantially up to $30\,^{\circ}C$. Y increases between 10 and $20\,^{\circ}C$, but it decreases after that. Thus, the effect of the temperature on substrate removal depends on the combined effect of μ_{max}, K_s and Y. Similarly, the effect on the production of solids depends on the combined effect on K_d and Y (Arceivala, 1981).

The Task Group for the IWA models (IAWPRC, 1987) recognises the difficulty in obtaining temperature correction rates for the model parameters (especially the K_s-type half-saturation constants), and suggests that the parameters are determined in operational conditions considered to be more critical. This aspect is particularly important in countries with a temperate climate, where the amplitude of temperature between winter and summer is significant. However, in many warm-climate countries the temperatures of the liquid are not substantially far from $20\,^{\circ}C$, for which the kinetic parameters and stoichiometric coefficients are usually reported.

2.13 FUNCTIONAL RELATIONS WITH THE SLUDGE AGE

This section analyses the influence of the sludge age on selected important process parameters of the activated sludge system. All values have been calculated applying the model presented in this chapter.

To broaden the results, the values are given for three different combinations of the parameters Y and K_d, selected to reflect conditions of lower biomass production (smaller Y and larger K_d) and of larger biomass production (larger Y and smaller K_d).

The main relations presented in Table 2.8 and in Figure 2.4 are (von Sperling, 1996d):

- *Production of suspended solids (SS) per unit of BOD_5 (S_r) removed.* Used for the estimation of the production of secondary excess sludge
- *Volatile suspended solids (VSS) to total suspended solids (SS) ratio.* Used in several design stages
- *Oxygen consumption (O_2) needed to satisfy the carbonaceous demand per unit of BOD_5 (S_r) removed.* Used for the design of the aeration system
- *Mass of mixed liquor volatile suspended solids required ($X_v \cdot V$) per unit of BOD_5 (S_r) removed.* With the product $X_v \cdot V$, for a given adopted value of the mixed liquor volatile suspended solids (X_v), the required volume for the reactor (V) may be determined

The following comments can be made with respect to Table 2.8 and Figure 2.4:

- The P_{SS}/S_r and VSS/SS ratios were presented in Section 2.9.1.
- The O_2/S_r ratio was presented in Section 2.10.2.
- The influence of the consideration of the influent solids to the reactor and of the presence of primary settling on the production of secondary excess sludge and on the VSS/SS ratio in the reactor can be seen clearly.
- The relations O_2/S_r and $X_v \cdot V/S_r$ are not affected by the presence of primary settling or solids in the influent. Obviously, in a system with primary settling the BOD load to the reactor will be lower, but the values of O_2 and $X_v \cdot V$ *per unit of BOD removed in the reactor* will be the same.
- The VSS/SS ratio is little affected by the values of the coefficients Y and K_d.
- The relations O_2/S_r and $X_v \cdot V/S_r$ are highly influenced by the values of the coefficients Y and K_d.

When using the data from Table 2.8 for a **quick design**, the following points should be taken into consideration (further details are given in Chapter 5):

- If *nitrification* is desired to be included in the computation of the average oxygen consumption (which is always advisable), the values of the O_2/S_r ratio in the above table starting from the sludge age of 4 days (in warm-climate regions) should be increased by around 50 to 60% (for typical values of influent TKN and assuming full nitrification, oxygen savings through the removal of nitrogen with the excess sludge and absence of intentional denitrification).
- To estimate the oxygenation capacity to be added to the system, the average oxygen consumption needs to be multiplied by a factor, such as the ratio between the maximum flow and the average flow (approximately 1.5 in medium to large plants, and 2.0 in smaller plants). This is the value of the oxygen demand in the *field*.

Table 2.8. Functional relations in the activated sludge system as a function of the presence of solids in the influent, existence of primary settling, coefficients Y and K_d and the sludge age

Item	Ratio and unit	SS in the influent	Primary settling	Coefficients		Sludge age (day)							
				Y (g/g)	K_d (d^{-1})	2	6	10	14	18	22	26	30
Production of solids	**SS/S$_r$** (kgSS/ kgBOD$_5$ rem)	*No*	*No*	0.5	0.09	0.50	0.42	0.37	0.33	0.31	0.29	0.28	0.28
				0.6	0.08	0.60	0.51	0.45	0.41	0.38	0.36	0.34	0.34
				0.7	0.07	0.71	0.61	0.55	0.50	0.47	0.44	0.42	0.40
		Yes	*Yes*	0.5	0.09	0.83	0.75	0.70	0.67	0.65	0.63	0.63	0.63
				0.6	0.08	0.96	0.87	0.81	0.78	0.75	0.73	0.71	0.71
				0.7	0.07	1.04	0.95	0.88	0.84	0.80	0.78	0.76	0.74
		Yes	*No*	0.5	0.09	1.08	1.00	0.95	0.92	0.90	0.88	0.88	0.88
				0.6	0.08	1.23	1.14	1.09	1.05	1.02	1.00	0.98	0.98
				0.7	0.07	1.29	1.20	1.13	1.08	1.06	1.03	1.01	0.99
VSS/SS ratio in reactor	**VSS/SS** (g/g)	*No*	*No*	0.5–0.7	0.07–0.09	0.89	0.87	0.85	0.84	0.83	0.82	0.81	0.81
		Yes	*Yes*	0.5–0.7	0.07–0.09	0.79	0.76	0.75	0.73	0.72	0.71	0.71	0.71
		Yes	*No*	0.5–0.7	0.07–0.09	0.75	0.73	0.71	0.70	0.69	0.69	0.68	0.68
Carbonaceous oxygen demand	**O$_2$/S$_r$** (kgO$_2$/ kgBOD$_5$ rem)	–	–	0.5	0.09	0.84	0.95	1.02	1.07	1.10	1.13	1.14	1.14
		–	–	0.6	0.08	0.70	0.83	0.91	0.97	1.01	1.05	1.07	1.07
		–	–	0.7	0.07	0.57	0.70	0.80	0.86	0.91	0.95	0.98	1.01
Volume of the reactor	**X$_v$·V/S$_r$** [kgVSS/ (kgBOD$_5$/d)]	–	–	0.5	0.09	0.88	2.16	3.11	3.88	4.55	5.15	5.71	6.24
		–	–	0.6	0.08	1.07	2.67	3.87	4.85	5.70	6.47	7.17	7.84
		–	–	0.7	0.07	1.26	3.21	4.69	5.92	6.98	7.93	8.80	9.62

Notes:
- Highlighted values: More usual values in activated sludge plants with typical flowsheets
- Per capita contributions: BOD$_5$ = 50 g/inhabitant·d; SS = 60 g/inhabitant·day
- Removal efficiencies in the primary sedimentation tank: BOD = 30%; SS = 60%
- S$_r$: Removed BOD$_5$ load (kgBOD$_5$/d)

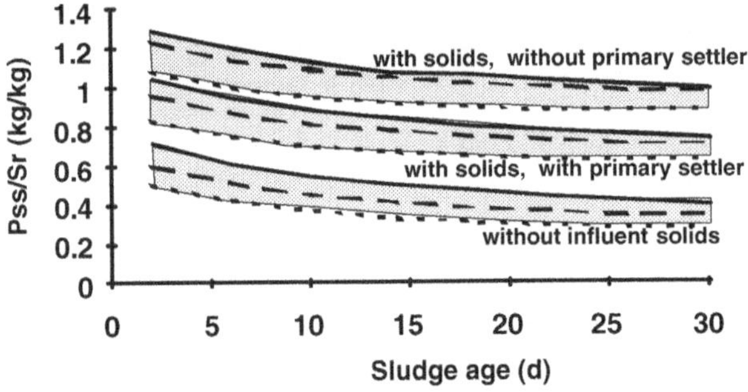

(in each range: lower curve: $Y = 0.5$ and $K_d = 0.09$ d^{-1}; intermediate curve: $Y = 0.6$ and $K_d = 0.08$ d^{-1}; upper curve: $Y = 0.7$ and $K_d = 0.07$ d^{-1})

Conventional activated sludge systems ($\theta_c \leq 10$ d) usually include primary settling, while extended aeration systems ($\theta_c \geq 18$ d) do not usually include primary settling

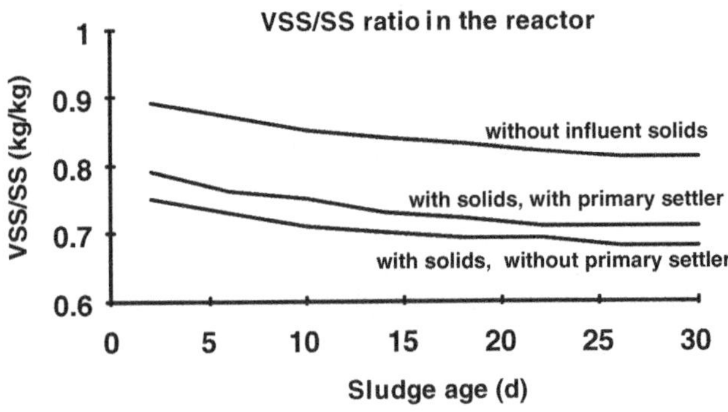

(variation of the coefficients Y and K_d: small influence; not considered)

Conventional activated sludge systems ($\theta_c \leq 10$ d) usually include primary settling, while extended aeration systems ($\theta_c \geq 18$ d) do not usually include primary settling

Figure 2.4.   Functional relations of process variables with the sludge age

- If one wants to express the oxygen demand under *standard conditions* (20 °C, clean water, sea level), the field demand has to be divided by a factor between 0.55 and 0.65.
- If one wants to express the relations in terms of the BOD load *applied* to the reactor, instead of the *removed* load, the corresponding values must be multiplied by 0.93 to 0.98, which correspond to the typical BOD removal efficiencies [$(S_o{-}S)/S_o$].

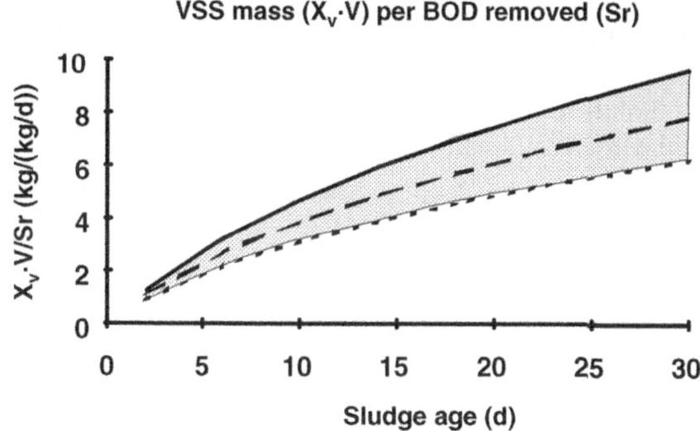

(in the range: lower curve: $Y = 0.5$ and $K_d = 0.09$ d^{-1}; intermediate curve: $Y = 0.6$ and $K_d = 0.08$ d^{-1}; upper curve: $Y = 0.7$ and $K_d = 0.07$ d^{-1})

To obtain the required reactor volume, simply divide the value of $X_v \cdot V$ (kg) by the adopted value of X_v (kg/m³)

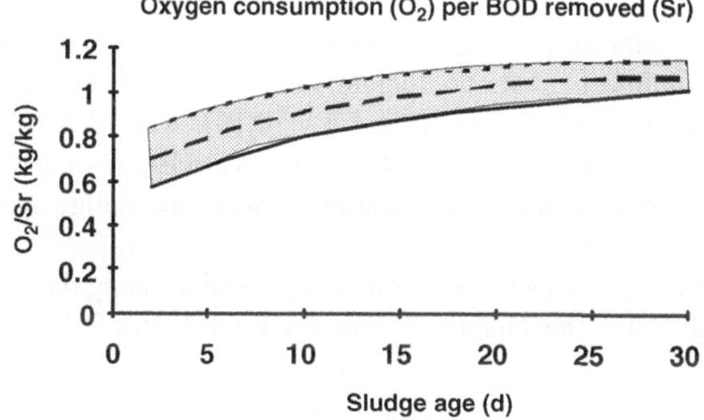

(in the range: lower curve: $Y = 0.7$ and $K_d = 0.07$ d^{-1}; intermediate curve: $Y = 0.6$ and $K_d = 0.08$ d^{-1}; upper curve: $Y = 0.5$ and $K_d = 0.09$ d^{-1})

This consumption is for the average carbonaceous demand. To obtain the total demand, the consumption for nitrification and the provision of oxygen for the maximum flow must be considered

Figure 2.4   (*Continued*)

To facilitate the implementation of a quick automated design tool in the computer, thus avoiding the need to refer to Table 2.8 and allowing a continuous solution for any sludge age within the range (not only those given in Table 2.8), von Sperling (1996d) made regression analyses which correlated the various variables and ratios in Table 2.8 with the sludge age. The structure adopted for the regression equation was the multiplicative ($y = a \cdot x^b$). The results are included in Table 2.9.

In all the regressions the fitting was excellent (coefficients of determination R^2 greater than 0.98). The utilisation of the equations is as follows. For example,

Table 2.9. Regression analysis comparing several relations included in Table 2.8 and the sludge age (influent solids considered)

Relation (y)	Solids in the influent	Primary settling	Coefficients		Equation $y = a \cdot (\theta_c)^b$	
			Y (g/g)	K_d (d^{-1})	a	b
SS/S$_r$ (kgSS/ kgBOD$_5$ rem)	*Yes*	*Yes*	0.5	0.09	0.900	−0.110
			0.6	0.08	1.053	−0.115
			0.7	0.07	1.158	−0.126
	Yes	*No*	0.5	0.09	1.145	−0.081
			0.6	0.08	1.318	−0.087
			0.7	0.07	1.401	−0.098
VSS/SS (g/g)	*Yes*	*Yes*	0.5–0.7	0.07–0.09	0.817	−0.043
	Yes	*No*	0.5–0.7	0.07–0.09	0.774	−0.038
O$_2$/S$_r$ (kgO$_2$/ kgBOD$_5$ rem)	–	–	0.5	0.09	0.777	0.118
	–	–	0.6	0.08	0.630	0.161
	–	–	0.7	0.07	0.483	0.218
X$_v$·V/Sr [kgVSS/ (kgBOD$_5$/d)]	–	–	0.5	0.09	0.662	0.663
	–	–	0.6	0.08	0.809	0.671
	–	–	0.7	0.07	0.959	0.682

the oxygen consumption per unit of BOD removed (O_2/S_r) for the sludge age of 8 days and the intermediate coefficient values ($Y = 0.6$ and $K_d = 0.08$ d^{-1}) will be (from Table 2.9) $a = 0.630$ and $b = 0.161$. The equation is: $O_2/S_r = 0.630 \cdot (\theta_c)^{0.161} = 0.630 \times (8)^{0.161} = 0.88$ kgO$_2$/kgBOD$_5$ removed. This value is consistent with Table 2.8, after interpolating between the sludge ages of 6 and 10 days.

The *detailed* design sequence of the activated sludge, using the various process formulae introduced in this chapter, is presented in Chapter 5.

Example 2.11

Undertake a **quick design** of the biological reactor, based on the data included in Table 2.8 and the associated remarks. Determine the volume of the reactor, the oxygen consumption, the power of the aerators and the production and removal of the excess sludge. Consider an extended aeration plant, with a sludge age of 25 days and a MLVSS concentration of 3,500 mg/L. Use the influent load of 3,350 kgBOD$_5$/d (adopted in the detailed design in Chapter 5). Take into account the solids in the influent and assume that the system will not have primary settling.

Solution:

(a) Estimation of the removed BOD load

The removed BOD load can be estimated as 95% of the applied BOD. Thus:

$$S_r = 0.95 \times 3{,}350 \text{ kgBOD}_5/d = 3{,}183 \text{ kgBOD}_5/d$$

Example 2.11 (Continued)

(b) Estimation of the VSS/SS ratio in the reactor and the resulting MLSS concentration

According to Table 2.8 (considering the influent solids and not using primary settling), after a linear interpolation between the sludge ages of 22 and 26 days for the sludge age of 25 days, one gets: **VSS/SS = 0.68**.

For $X_v = 3,500$ mgMLVSS/L, the resulting MLSS concentration is:

$$MLSS = 3,500/0.68 = \textbf{5,147 mg/L} = 5.147 \text{ kg/m}^3$$

(c) Estimation of the reactor volume

According to Table 2.8, by adopting the intermediate coefficient values ($Y = 0.6$; $K_d = 0.08$ d^{-1}) and interpolating between the sludge ages of 22 and 26 days, one gets: $X_v \cdot V/S_r = 7.0$ d^{-1}.

For $X_v = 3.5$ kg/m^3 ($= 3,500$ mgMLVSS/L) and $S_r = 3,183$ kgBOD$_5$/d, the resulting reactor volume is:

$$V = 7.0 \times 3,183/3.5 = \textbf{6,366 m}^3$$

(d) Estimation of the production and removal of excess sludge

According to Table 2.8, by considering the influent solids and not adopting primary settling, one gets: $P_x/S_r = 0.98$ (interpolating between the sludge ages of 22 and 26 days).

For $S_r = 3,183$ kgBOD$_5$/d, the sludge production is calculated as:

$$P_x = 0.98 \times 3,183 = \textbf{3,119 kgSS/d}$$

If the sludge is removed directly from the reactor, its concentration is the same as MLSS (X). Thus, the excess sludge flow is (disregarding the loss of solids in the final effluent) is:

$$Q_{ex} \text{ reactor} = P_x/X = 3,119/5.147 = \textbf{606 m}^3\textbf{/d}$$

Note that this value is different from the value V/θ_c ($= 6,366/25 = 255$ m^3/d), usually adopted when controlling the system by the sludge age (hydraulic control), disregarding the influent solids.

If the sludge is removed from the return sludge line, its concentration is the same as RASS (X_r). For a recirculation ratio R ($=Q_r/Q$) equal to 1.0 (adopted), one has:

$$X_r = X \cdot (R + 1)/R = 5.147 \times (1 + 1)/1 = 10.294 \text{ kg/m}^3 (=10,294 \text{ mg/L})$$

$$Q_{ex} \text{ return sludge line} = P_x/X_r = 3,119/10.294 = \textbf{303 m}^3\textbf{/d}$$

Example 2.11 (Continued)

The excess sludge flow removed from the return sludge line (303 m^3/d) is half of the flow removed from the reactor (606 m^3/d), due to the fact that the solids concentration in the return sludge line (10,294 mg/L) is twice the concentration in the reactor (5,147 mg/L).

(e)   Calculation of the oxygen consumption and aerator power requirements

According to Table 2.8, $O_2/S_r = 1.06$ kgO$_2$/kgBOD$_5$ (interpolating between the sludge ages of 22 and 26 days).

For $S_r = 3,183$ kgBOD$_5$/d:

$$O_2 \text{ carbonaceous} = 1.06 \times 3,183$$

$$= 3,374 \text{ kgO}_2/\text{d (average carbonaceous demand)}$$

To take into account the nitrification in the total O_2 consumption, the carbonaceous demand value must be increased by 50 to 60%. By adopting a value of 55%, the total average demand (disregarding denitrification) is:

$$O_2 \text{ total} = 1.55 \times 3,374 = 5,230 \text{ kgO}_2/\text{d}$$

To take into account the demand under maximum load conditions, the average oxygen demand must be multiplied by a correction factor. This factor may be adopted varying between 1.5 and 2.0 ($\approx Q_{max}/Q_{average}$), depending on the size of the plant. Adopting a factor of 2.0, one has:

$$\text{Total maximum } O_2 = 2.0 \times 5,230$$

$$= 10,460 \text{ kgO}_2/\text{d (in the field, under operational conditions)}$$

To express it in standard conditions, the field value should be divided by a factor that varies between 0.55 and 0.65. By adopting the value of 0.60, one has:

$$O_2 \text{ standard} = 10,460/0.60 = 17,433 \text{ kgO}_2/\text{d} = 726 \text{ kgO}_2/\text{hour}$$

By adopting a standard oxygenation efficiency of 1.8 kgO$_2$/kW·hour for mechanical aeration, the power requirement is:

$$\text{Power required} = 726/1.8 = 403 \text{ kW} = 537 \text{ HP}$$

(f)   Comments

*   If the detailed design sequence presented in Chapter 5 had been followed, it could be verified that the values found for the volume of the reactor,

---

Example 2.11 (Continued)

production of excess sludge and oxygen requirements are very similar to those obtained in the present quick design (differences of less than 2.5%).

- This example could have been also undertaken based on the equations presented in Table 2.9. The results obtained should be very similar.
- The design did not foresee the intentional denitrification in the reactor. Although still little used in most developing countries, its implementation in a more systematic way should be encouraged, especially in warm-climate countries.
- To complete the plant, the designs of the secondary sedimentation tanks and the preliminary treatment units (screen and grit chamber) and sludge processing units (thickening and dewatering) are still needed. The design of these units is simpler than the design of the reactor and associated variables.

3

Design of continuous-flow activated sludge reactors for organic matter removal

3.1 SELECTION OF THE SLUDGE AGE

The selection of the sludge age is the main step in the design of an activated sludge plant. As shown in Section 2.19, several process variables are directly associated with the sludge age. The first decision concerns the selection of θ_c values that place the plant within one of the main operational ranges presented in Table 1.1 (Chapter 1). Tables 1.3 and 1.4 in the same chapter present a comparison among the main activated sludge variants (conventional activated sludge, extended aeration), focusing on several important aspects.

The advantages of incorporating the UASB reactor upstream the activated sludge system were presented in Section 1.2.5. The main design parameters of this configuration are detailed in Chapter 9. Due to the large number of advantages in warm-climate regions, it is recommended that this alternative be carefully analysed by the designer, prior to making a decision for the classical conceptions (without UASB reactor).

With relation to the classical conceptions, the first decision to be taken by the designer is the adoption of either conventional activated sludge or extended aeration. Although there are no fixed rules, the following approximate applicability ranges could be mentioned as an initial guideline in preliminary studies:

- smaller plants (less than $\approx$50,000 inhabitants): extended aeration
- intermediate plants (between $\approx$50,000 and $\approx$150,000 inhabitants): technical–economical studies
- larger plants (more than $\approx$150,000 inhabitants): conventional activated sludge

In smaller plants, operational simplicity has a strong weight, which leads to the extended aeration alternative. In larger plants, the economy in power

Table 3.1. Design parameters for carbon removal in the biological reactor

Category	Parameter	Conventional activated sludge	Extended aeration
Parameter to be initially assumed	Sludge age (d)	4–10	18–30
	MLVSS concentration (mg/L)	1,500–3,500	2,500–4,000
	Effluent SS (mg/L)	10–30	10–30
	Return sludge ratio (Q_r/Q)	0.6–1.0	0.8–1.2
	Average DO concentration in the reactor (mg/L)	1.5–2.0	1.5–2.0
Data resulting from the design or parameter to be used in quick designs	F/M ratio (kgBOD$_5$/kgMLVSS·d)	0.3–0.8	0.08–0.15
	Hydraulic detention time (hour)	6–8	16–24
	MLSS concentration (mg/L)	2,000–4,000	3,500–5,000
	VSS/SS ratio in the reactor (–)	0.70–0.85	0.60–0.75
	Biodegradable fraction of MLVSS(f_b) (–)	0.55–0.70	0.40–0.65
	BOD removal efficiency (%)	85–93	90–98
	Effluent soluble BOD$_5$ (mg/L)	5–20	1–4
	BOD$_5$ of the effluent SS (mgBOD$_5$/mgSS)	0.45–0.65	0.20–0.50
	VSS production per BOD$_5$ removed (kgVSS/kgBOD$_5$)	0.5–1.0	0.5–0.7
	Excess sludge production per BOD$_5$ rem. (kgSS/kgBOD$_5$)	0.7–1.0	0.9–1.1
	Average O$_2$ requirements (without nitrification) (kgO$_2$/kgBOD$_5$)	0.7–1.0	–
	Average O$_2$ requirements with nitrification (kgO$_2$/kgBOD$_5$)	1.1–1.5	1.5–1.8
	Nutrient requirements – nitrogen (kgN/100 kgBOD$_5$)	4.3–5.6	2.6–3.2
	Nutrient requirements – phosphorus (kgP/100 kgBOD$_5$)	0.9–1.2	0.5–0.6
	N removed per BOD$_5$ removed (kgN/100 kgBOD$_5$)	0.4–1.0	0.1–0.4
	P removed per BOD$_5$ removed (kgP/100 kgBOD$_5$)	4–5	2.4

Sources: Arceivala (1981), Orhon and Artan (1994) and the author's adaptations

consumption assumes great importance, and the operational issue is no longer critical, leading to the conventional activated sludge system. In intermediate plants these items overlap, and more detailed technical and economical assessments are necessary.

After the selection concerning the sludge age range, a refinement should be performed, and the ideal sludge age should be selected for the system at issue. Depending on the degree of detail desired, the selection can be made based on either economic assessments, a simple comparison between volumes, areas, and required powers, obtained from a preliminary design, or even be based on the designer's experience. The figures, tables and equations presented in Section 2.13 can aid in this selection process.

3.2 DESIGN PARAMETERS

The main parameters for the design of a reactor aiming at the removal of organic carbon (BOD) are listed in Table 3.1.

The kinetic and stoichiometric coefficients necessary for the design of the BOD removal stage are summarised in Table 3.2.

Table 3.2. Kinetic and stoichiometric coefficients and basic relations for the calculation of the BOD removal in activated sludge systems

Coefficient	Description	Unit	Range	Typical value
Y	Yield coefficient (cellular production)	$gVSS/gBOD_5$	0.4–0.8	0.6
K_d	Endogenous respiration coefficient	$gVSS/gVSS\cdot d$	0.06–0.10	0.08–0.09
θ	Temperature coefficient for K_d	–	1.05–1.09	1.07
f_b'	Biodegradable fraction when generating solids (X_b/X_v)	$gSS_b/gVSS$	–	0.80
VSS/SS	SSV/SS in the raw sewage	$gVSS/gSS$	0.70–0.85	0.80
SS_b/SS	$SS_{biodegradable}/SS$ in the raw sewage	gSS_b/SS	–	0.60
VSS/SS	SSV/SS when generating solids	$gVSS/gSS$	–	0.90
O_2/SS_b	Oxygen per biodegradable solids destroyed	$gBOD_u/gSS_b$	–	1.42
BOD_u/BOD_5	$BOD_{ultimate}/BOD_5$ ratio	$gBOD_u/gBOD_5$	1.2–1.5	1.46

Base: BOD_5 and VSS; temperature = 20 °C
Sources: Eckenfelder (1989), Metcalf and Eddy (1991), WEF/ASCE (1992), Orhon and Artan (1994), and von Sperling (1996d)

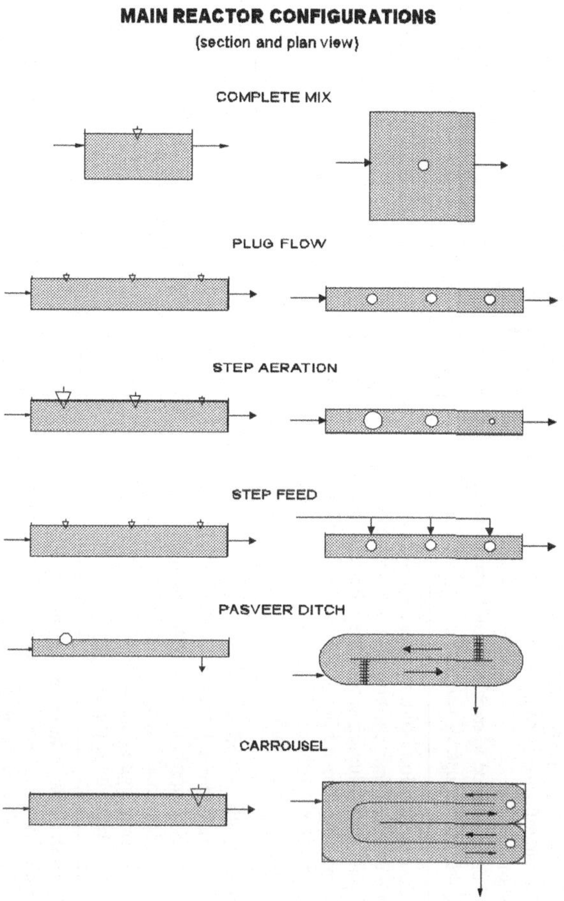

Figure 3.1. Simplified schematics of the main physical configurations of activated sludge reactors (section and plan view)

3.3 PHYSICAL CONFIGURATION OF THE REACTOR

There are several variants in the design of continuous-flow activated sludge systems regarding the physical configuration of the biological reactor. Table 3.3 presents a summary of the main variants, and Figure 3.1 shows schematic sections and plan views.

The Pasveer- and Carrousel-type **oxidation ditches** deserve some additional considerations. Regarding the mixing regime, the oxidation ditches have the following characteristics (Johnstone *et al.*, 1983):

- *complete-mix* behaviour for most of the variables (such as BOD and suspended solids)

Table 3.3. Description of the main physical configurations of activated sludge reactors

Variant	Scheme	Description	Longitudinal BOD profile	Longitudinal DO profile
Complete mix		• Predominantly square dimensions • The concentrations are the same at any point of the reactor • Higher resistance to overloads and toxic substances • Suitable for industrial wastes		
Plug flow		• Predominantly longitudinal dimensions • More efficient than the complete-mix reactor • Capable of producing sludge with better settleability • The oxygen demand decreases along the reactor • The DO concentration increases along the reactor • Less resistant to shock loads		
Step aeration		• Similar to the plug flow • The oxygen demand decreases along the reactor • The oxygen supply is equal to the demand (the aeration decreases along the reactor) • Savings in oxygen supply • Easier control by diffused air		
Step feeding		• Inlet to the reactor at several points • The reactor behaves as a complete-mix reactor • The oxygen consumption is homogeneous • Operationally flexible • Possibility of temporary solids storage in the initial part, if the initial inlet is closed		

Pasveer-type oxidation ditch	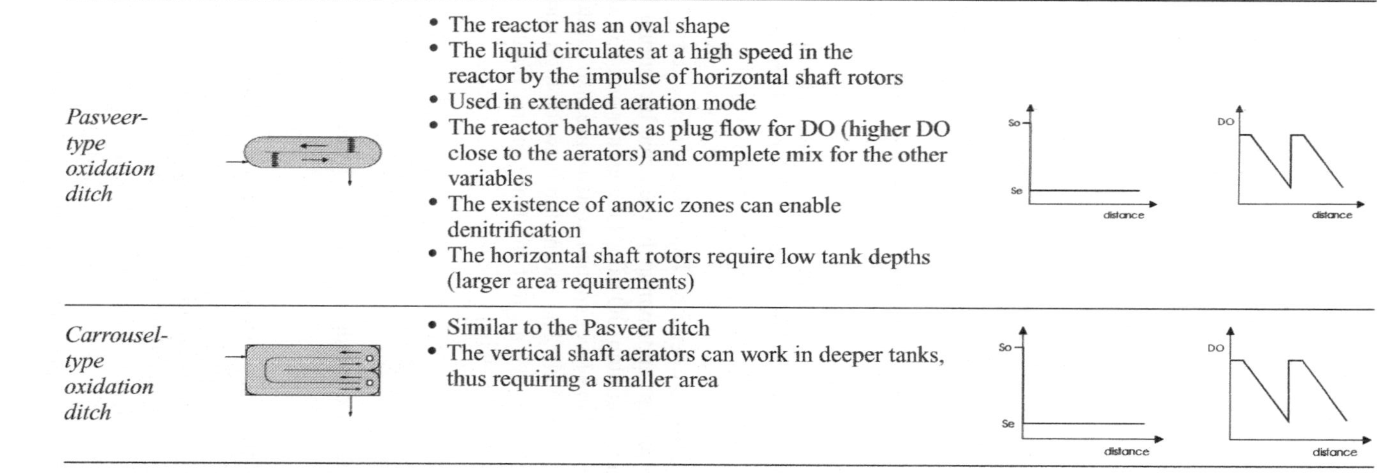	• The reactor has an oval shape • The liquid circulates at a high speed in the reactor by the impulse of horizontal shaft rotors • Used in extended aeration mode • The reactor behaves as plug flow for DO (higher DO close to the aerators) and complete mix for the other variables • The existence of anoxic zones can enable denitrification • The horizontal shaft rotors require low tank depths (larger area requirements)	
Carrousel-type oxidation ditch		• Similar to the Pasveer ditch • The vertical shaft aerators can work in deeper tanks, thus requiring a smaller area	

- *plug-flow* behaviour for variables with fast dynamics (such as dissolved oxygen)

This double behaviour results from the interrelation between the hydraulic dynamics of the reactor and the dynamics of the reaction rates of the variables. The dilution capacity of a ditch is high, due to the high horizontal velocity, which allows a complete circuit around the ditch in 15 to 20 minutes. Thus, variables with relatively slow dynamics, such as most of the variables involved in wastewater treatment, present approximately the same concentration at any point of the reactor, characterising a hydraulic regime approaching complete mix. However, variables with fast dynamics, such as dissolved oxygen, present a concentration gradient along the reactor, approaching the plug-flow regime. In spite of the fast flow velocities, the oxygen reaction rate (production and consumption) velocity is higher. The liquid has high DO concentrations soon after leaving the aerator. As the liquid flows downstream, the oxygen is consumed to satisfy various oxygen demands, and its concentration decreases until it reaches another aerator (or the same one, if there is only one).

The existence of this oxygen gradient in oxidation ditches affects all the variables that depend, either directly or indirectly, on dissolved oxygen. The DO values are always closely related to the place where they are measured. Monod's relations, which include oxygen as a substrate, are specific to each ditch, in view of its geometry, travel time of the liquid, oxygen utilisation rate and, above all, the DO measurement location. The comparison between DO values from ditch to ditch only makes sense when they represent approximately the same measuring location. Generalisations in relation to optimum DO values for aerator on/off or outlet weir level setting in automated aeration level control systems are also limited by the existing variations from ditch to ditch. The alternation between high and low DO values along the ditch can also have a great influence on nitrification and denitrification processes, as described in Chapter 7.

3.4 DESIGN DETAILS

Some aspects of the design of activated sludge reactors are listed below.

General aspects:

- The length and width of the reactor should allow a homogeneous distribution of the aerators on the surface of the tank.
- The liquid depth of the reactor is within the following range: 3.5 to 4.5 m (mechanical aeration) and 4.5 to 6.0 m (diffused air).
- The depth of the reactor should be established in accordance with the aerator to be adopted (consult the manufacturer's catalogue).
- The freeboard of the tank is approximately 0.5 m.
- The plan dimensions should be established according to the hydraulic regime selected, and should be compatible with the areas of influence of the aerators.

- In plants with a high flow (say, >200 L/s), more than one reactor should be adopted.
- The tanks are usually made of reinforced concrete with vertical walls but, whenever possible, the alternative of sloped tanks should be analysed (lighter wall structure and foundations).
- Should there be more than one unit, common walls can be used between them.
- Low-speed fixed mechanical aerators should be supported by catwalks and pillars (designed to resist torsion). High-speed floating mechanical aerators are anchored to the borders.
- Mechanical aerators may have their oxygenation capacity controlled by means of a variable submergence of the aerators (variation in the level of the outlet weir or in the aerator shaft), by a variable speed of the aerators or by switching on/off the aerators.
- The diffused-air aeration can have its oxygenation capacity controlled by means of adjustment in the outlet valves from the blowers or in the inlet valves in the reactors.
- A submerged inlet avoids the release of hydrogen sulphide present in the raw sewage.
- The outlet from the tank is generally by weirs at the opposite end to the inlet.
- If there is more than one tank, the inlet and outlet arrangements should allow the isolation of one unit for occasional maintenance.
- The scum occasionally formed should be broken by means of hoses or sprinklers, and be either removed to scum boxes or directed to secondary sedimentation tanks.
- The possible drainage of the tank (occasional emptying for maintenance purposes) by means of submersible pumps (simpler and more reliable) or by bottom discharge valves should be considered.
- In case of interference with the groundwater level, there should be a means to relieve the sub-pressure when the tank is empty.

Oxidation ditches:

- The oxidation ditches are designed using the same principles as those used for the design of other extended aeration reactors, resulting in the same reactor volumes and oxygen demands.
- The aerators have horizontal shaft in the Pasveer-type ditches and vertical shaft in the Carrousel-type ditches.
- The aerators should ensure a horizontal velocity between 0.30 and 0.50 m/s, to avoid the sedimentation of solids in the reactor.
- Pasveer-type oxidation ditches usually have a maximum depth of approximately 1.2 m, in view of the limited capacity of the horizontal-shaft rotors to transfer oxygen and maintain the liquid in movement at higher depths.
- Carrousel oxidation ditches are deeper as a result of the type of aerator employed (vertical shaft), which leads to a lower land requirement, compared

with the Pasveer ditch. Depths can be up to 5 m in the aeration zone, and approximately 3.5 m in the non-aerated zone.

- Smaller ditches can have their walls sloped at 45°, while larger ditches have vertical walls.
- There is no maximum number of curves between each aerator. However, it is suggested that the sum of the curves does not exceed 360° between one aerator and another.
- To avoid stagnation and solids sedimentation zones in the internal face of the dividing wall, downstream the curve, additional inner semi-circular walls following the curve can be adopted. These walls are slightly decentralised in relation to the internal wall: the largest opening receives most of the liquid, while the smallest opening discharges this larger portion of the liquid at a higher speed, in the downstream zone, internally to the central wall, thus avoiding the sedimentation of solids.

4

Design of activated sludge sedimentation tanks

4.1 TYPES OF SEDIMENTATION TANKS

This chapter mostly deals with *secondary* sedimentation tanks, in view of their fundamental importance in the biological stage of the activated sludge process. However, the design of *primary* sedimentation tanks is briefly covered in Section 4.4.

The most used shapes for the secondary sedimentation tanks are the *horizontal-flow rectangular tank* and the *central-feeding circular tank*. Both tanks require the continuous removal of sludge by scrapers or bottom suction. A schematic view of both tanks can be seen in Figures 4.1 and 4.2. The circular tank allows an easier continuous removal of sludge, besides the greater structural advantage of the ring effect. On the other hand, the rectangular tank allows a larger economy of area (absence of dead areas between tanks) and the possibility of using common walls between adjoining tanks. Both tanks are used in medium- and large-sized plants.

In small-sized plants, sludge removal mechanisms will not be necessary if the bottom has a high slope (approximately 60° with relation to the horizontal line), assuming the shape of an inverted pyramid. The sludge is thus directed to the bottom sludge hoppers, from where it is removed by hydraulic pressure. These tanks are named *Dortmund* tanks. Their use is restricted to smaller plants, due to the fact that high bottom slopes require very large depths in case of large surface areas. Figure 4.3 exemplifies one possible shape for this tank, rectangular in the upper plan, but divided into three equal chambers.

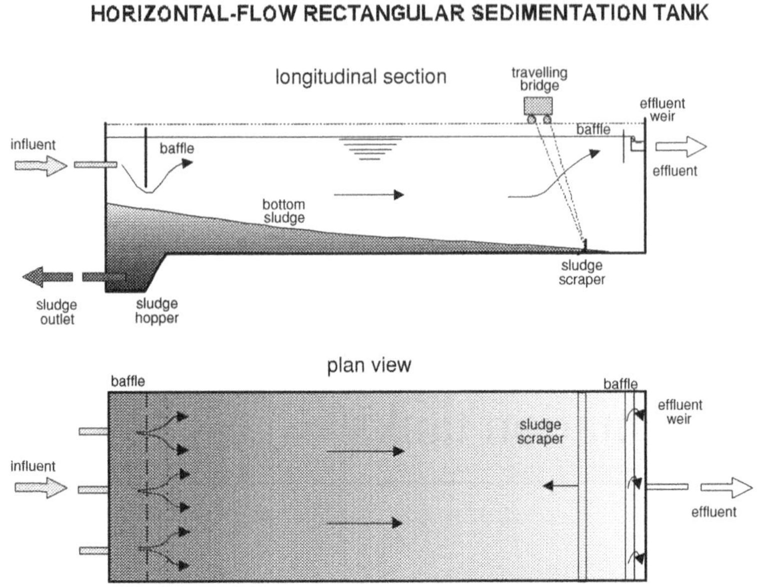

Figure 4.1.   Schematics of a rectangular secondary sedimentation tank (section and plan view)

Section 4.2 deals with the determination of the main design aspect (surface area of the sedimentation tanks), while Section 4.3 presents several design details for the three types of sedimentation tanks mentioned.

4.2 DETERMINATION OF THE SURFACE AREA REQUIRED FOR SECONDARY SEDIMENTATION TANKS

4.2.1 Determination of the surface area according to conventional hydraulic loading rates and solids loading rates

The calculation of the required surface area is the main aspect in the design of a sedimentation tank. The area is usually determined by considering the following design parameters:

- **Hydraulic loading rate: (Q/A).** It corresponds to the quotient between the influent flow to the plant (Q) and the surface area of the sedimentation tanks (A).
- **Solids loading rate: $(Q + Q_r) \cdot X/A$.** It corresponds to the quotient between the applied solids load $(Q + Q_r) \cdot X$ and the surface area of the sedimentation tanks (A).

CIRCULAR SEDIMENTATION TANK

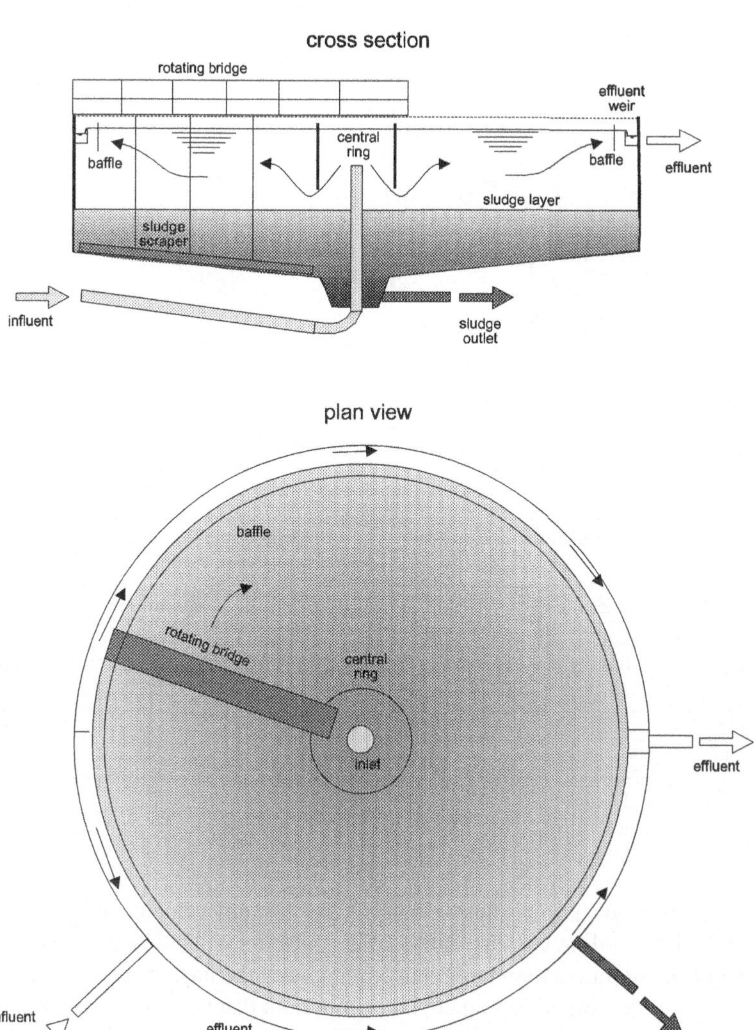

Figure 4.2.   Schematics of a circular secondary sedimentation tank (section and plan view)

It is important to note that the hydraulic loading rate (HLR) is based only on the influent flow to the plant (Q), and not on the total influent flow to the sedimentation tank ($Q + Q_r$). This is because only the Q flow has an upward component, leaving through the weirs on the top (the return sludge flow Q_r has a downward direction, leaving through the bottom of the sedimentation tank). The

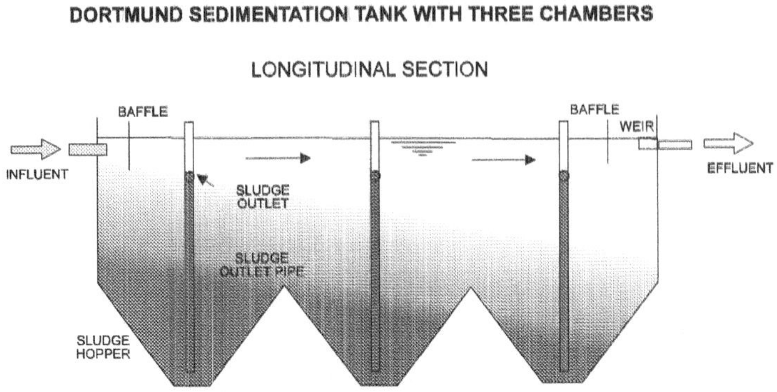

DORTMUND SEDIMENTATION TANK WITH THREE CHAMBERS

LONGITUDINAL SECTION

BAFFLE

BAFFLE WEIR

INFLUENT

EFFLUENT

SLUDGE OUTLET

SLUDGE OUTLET PIPE

SLUDGE HOPPER

Figure 4.3.   Dortmund-type tank, with high bottom slope and no sludge removal mechanism

Table 4.1. Hydraulic and solids loading rates for secondary sedimentation tanks

System	Hydraulic loading rate ($m^3/m^2 \cdot$ hour)		Solids loading rate ($kg/m^2 \cdot$ hour)		Reference
	Average Q	Maximum Q	Average Q	Maximum Q	
Conventional	0.67–1.33	1.70–2.00	4.0–6.0	10.0	(1)
activated	0.67–1.20	1.70–2.70	4.0–6.0	–	(2)
sludge					
Extended	0.33–0.67	1.00–1.33	1.0–5.0	7.0	(1)
aeration					

Ref: (1) Metcalf and Eddy (1991); (2) WEF/ASCE (1992)

upflow component is important because if the upward velocity of the liquid is higher than the settling velocity of the solids, the latter will not be able to go to the bottom of the sedimentation tank, thus leaving with the final effluent.

In terms of the solids loading rate, it is important that the load of solids applied per unit area is not higher than the limiting solid flux. In this case, the load applied is the actual influent load to the sedimentation tank, that is, $(Q + Q_r) \cdot X$.

These aspects are covered by the limiting solids flux theory. If data on the settleability of the sludge under study are available, the limiting flux theory can be used for the design of the secondary sedimentation tanks, as demonstrated in Chapter 10. In this case, the solids loading rate (SLR) to be used is equal to the limiting solids flux. A simplified approach using concepts of the solids flux theory is described in Section 4.2.2.

Surface loading rates based on the designers' experience are usually used. Table 4.1 presents typical loading rate values according to some traditional references (Metcalf and Eddy, 1991; WEF/ASCE, 1992).

Example 4.1 presents the calculation of the area required for the secondary sedimentation tanks, based on the concepts of the hydraulic and solids loading rates. The complete design of the secondary sedimentation tank is covered in the example given in Chapter 5.

Example 4.1

Calculate the area required for the secondary sedimentation tanks of a conventional activated sludge plant, which has the following data (same data as those for the design example of Chapter 5, for conventional activated sludge):

- average influent flow: $Q = 9{,}820$ m^3/d
- maximum influent flow: $Q_{max} = 19{,}212$ m^3/d
- average return sludge flow: $Q_r = 9{,}820$ m^3/d
- concentration of suspended solids in the reactor: MLSS $= 3{,}896$ g/m^3

Solution:

(a) Express the flows in m^3/hour

$Q = 9{,}820/24 = 409$ m^3/hour
$Q_{max} = 19{,}212/24 = 801$ m^3/hour
$Q_r = 9{,}820/24 = 409$ m^3/hour

(b) Calculate the surface area based on the hydraulic loading rate

From Table 4.1, adopt the following hydraulic loading rate values:

For $Q_{av} \rightarrow$ HLR $= 0.80$ m^3/m^2·hour
For $Q_{max} \rightarrow$ HLR $= 1.80$ m^3/m^2·hour

The required surface area is given by:

For $Q_{av} \rightarrow$ A $= Q/$HLR $= 409/0.8 = 511$ m^2
For $Q_{max} \rightarrow$ A $= Q_{max}/$HLR $= 801/1.80 = 445$ m^2

(c) Calculate the surface area based on the solids loading rate

The influent suspended solids load to the secondary sedimentation tank is:

For $Q_{av} \rightarrow (Q + Q_r) \cdot X = (409 + 409) \times 3{,}896/1{,}000 = 3{,}187$ kgSS/hour
For $Q_{max} \rightarrow (Q_{max} + Q_r) \cdot X = (801 + 409) \times 3{,}896/1{,}000 =$
   $4{,}714$ kgSS/hour

From Table 4.1, adopt the following solids loading rate values:

For $Q_{av} \rightarrow$ SLR $= 5.0$ kgSS/m^2·hour
For $Q_{max} \rightarrow$ SLR $= 10.0$ kgSS/m^2·hour

The required surface area is given by:

For $Q_{av} \rightarrow$ A $= (Q + Q_r) \cdot X/(1{,}000 \cdot$ SLR$) = 3{,}187/5.0 = 637$ m^2
For $Q_{max} \rightarrow$ A $= (Q_{max} + Q_r) \cdot X/(1{,}000 \cdot$ SLR$) = 4{,}714/10.0 = 471$ m^2

Example 4.1 (Continued)

(d) Surface area to be adopted

The surface area to be adopted for the secondary sedimentation tanks should correspond to the largest value among the four values obtained ($511\,\mathrm{m}^2$, $445\,\mathrm{m}^2$, $637\,\mathrm{m}^2$ and $471\,\mathrm{m}^2$). Therefore, the area required for the secondary sedimentation tanks is:

$$A = 637\,\mathrm{m}^2$$

It is to be noted that in this example the most restrictive criterion, that is, the criterion that led to the largest required area, was that of the solids loading rate for the average flow. This conclusion reinforces the notion that the secondary sedimentation tanks should be designed by taking into consideration the solids loading rate, and not just the hydraulic loading rate, as is usual in the design of other sedimentation tanks.

Apparently conservative loading rate values were adopted in this example, but the importance of the adequate performance of the secondary sedimentation tanks justifies that.

4.2.2 Determination of the surface area according to loading rates based on a simplified approach to the solids flux theory

The solids flux theory is an important tool for the design and control of secondary sedimentation tanks. However, its use is often difficult due to *a priori* lack of knowledge of the parameters characterising the sludge settleability (v_0 and K), unless the design is intended for expansion of an already existing plant, with an already known sludge.

Aiming at expanding the use of the limiting solids flux theory, several authors (White, 1976; Johnstone *et al.*, 1979; Tuntoolavest and Grady, 1982; Koopman and Cadee, 1983; Pitman, 1984; Daigger and Roper, 1985; Ekama and Marais, 1986; Wahlberg and Keinath, 1988, 1995; van Haandel *et al.*, 1988; von Sperling, 1990; Daigger, 1995) tried to express the interface settling velocity according to easily determinable or assumable variables, such as the sludge volume index (SVI and its variants). Once the sludge settling velocity is estimated, the limiting flux theory can be easily employed for design and operation. However, each author used a different form to express the sludge volume index, thus making the calculation of unified values more difficult.

By using the methodology proposed by von Sperling (1994b), Fróes (1996) conjugated the data obtained by the authors above and presented a unique formulation,

Table 4.2. Typical SVI ranges and values (average), according to its four variants and five settleability ranges

| | Range of sludge volume index values (mL/g) | | | | | | | |
| | SVI | | DSVI | | SSVI | | SSVI$_{3.5}$ | |
Settleability	Range	Typical	Range	Typical	Range	Typical	Range	Typical
Very good	0–50	45	0–45	40	0–50	45	0–40	35
Good	50–100	75	45–95	70	50–80	65	40–80	60
Fair	100–200	150	95–165	130	80–140	110	80–100	90
Poor	200–300	250	165–215	190	140–200	170	100–120	110
Very poor	300–400	350	215–305	260	200–260	230	120–160	140

Notes: The ranges were established based on the analysis of the various references mentioned
The typical values were considered, in most of the cases, as the mean value in the range

Table 4.3. Values of the coefficients v_0, K, m and n, for each settleability range

| | Settling velocity (m/hour) $v = v_0 \cdot e^{-K \cdot C}$ | | Limiting flux (kg/m^2·hour) $G_L = m \cdot (Q_r/A)^n$ | |
Settleability	v_0(m/hour)	K (m^3/kg)	m	n
Very good	10.0	0.27	14.79	0.64
Good	9.0	0.35	11.77	0.70
Fair	8.6	0.50	8.41	0.72
Poor	6.2	0.67	6.26	0.69
Very poor	5.6	0.73	5.37	0.69

v: Interface settling velocity (m/hour)
C: Influent SS concentration to the sedimentation tank (MLSS) (kgSS/m^3)
G_L: Limiting solid flux (kg/m^2·hour)
Q_r: Return sludge flow (approximately equal to the underflow from the sedimentation tank) (m^3/hour)
A: Surface area of the sedimentation tanks (m^2)
v_0, K, m, n: Coefficients

based on settleability ranges. The proposition made by von Sperling and Fróes (1998, 1999) is described in this section.

Table 4.2 presents the typical settleability ranges, according to the several variants of the sludge volume index test. The interpretation of Table 4.2 is that the average or *fair* settleability can be characterised by a sludge with a SVI of 150 mL/g, a DSVI of 130 mL/g, a SSVI of 110 mL/g and a SSVI$_{3.5}$ of 90 mL/g. This unification, according to the settleability ranges, forms a common base, from which the values obtained by the several authors mentioned above can be integrated.

The average v_0 and K values obtained by the various authors were calculated for each settleability range (*very good* to *very poor*) (see Table 4.3). As a whole, data presented in 17 publications were used, representing dozens of activated sludge plants operating in real scale. Based on the v_0 and K values obtained by each author, the limiting solids flux as a function of Q_r/A was calculated. After that, the relation between the limiting solids flux (GL) and Q_r/A was determined by regression analysis, for each author and for each settleability range, according to the multiplicative equation $G_L = m \cdot (Q_r/A)^n$. After that, the average m and n values obtained by the various authors were calculated for each settleability range, which are presented in Table 4.3.

Knowing the v_o, K, m and n values for each settleability range, the design can be done according to the criteria presented below. From the unification proposed, there is no need to work with the various authors' values and with different sludge volume indices, but only with the settleability ranges estimated for the sludge. Thus, the area required for the sedimentation tanks can be calculated using the principles of the solids flux theory just by knowing the concentration of MLSS in the reactor (C), the influent flow (Q) and the return sludge flow (Q_r), and by assuming the sludge settleability (settleability range).

(a) Design principles

In order not to lose solids in the effluent, the secondary sedimentation tank should not be overloaded in terms of clarification and thickening (Keinath, 1981). This means that the following two criteria need to be met:

- *sedimentation tank should not be overloaded in terms of clarification:* the hydraulic loading rate should not exceed the sludge settling velocity

$$Q/A \leq v \tag{4.1}$$

$$Q/A \leq v_o \cdot e^{-K \cdot C} \tag{4.2}$$

- *sedimentation tank should not be overloaded in terms of thickening:* the applied solids flux should not exceed the limiting solids flux

$$G_a \leq G_L \tag{4.3}$$

$$(Q + Q_r) \cdot C_o/A \leq m \cdot (Q_r/A)^n \tag{4.4}$$

where:
     Q = influent flow to the plant (m³/hour)
     Q_r = return sludge flow (m³/hour)
     v = settling velocity of the interface at the concentration C (m/hour)
     v_o = coefficient, expressing the settling velocity of the interface at a concentration C = 0 (m/hour)
     K = sedimentation coefficient (m³/kg)
     G_a = applied solids flux (kg/m²·hour)
     G_L = limiting solids flux (= maximum allowable solids loading rate) (kg/m²·hour)

C_0 = concentration of influent solids to the secondary sedimentation tank (=MLSS) (kg/m³)

A = surface area required for the sedimentation tanks (m²)

(b) Design for clarification

Considering the concentration of influent suspended solids to the sedimentation tank (C_0) as equal to MLSS, the settling velocity can be obtained from the equation $v = v_0 \cdot e^{-K \cdot C}$, with v_0 and K values obtained from Table 4.3. Thus, the hydraulic loading rate should be equal to or lower than the value of v calculated (Equation 4.2). Figure 4.4 presents the resulting curves of hydraulic loading rates for the different sludge settleabilities and for different MLSS concentrations.

(c) Design for thickening

After meeting the clarification criteria, in which an adequate value was adopted for the hydraulic loading rate (Q/A), the SLR can be established. The first step is to select a value for the return sludge ratio $R(Q_r/Q)$. In other words, Q_r/A should be equal to R·Q/A. Using the coefficient values given in Table 4.3, and adopting the concept of Equation 4.4, the allowable maximum solids rate should be equal to:

$$SLR = m \cdot (R \cdot Q/A)^n \qquad (4.5)$$

In terms of design, the SLR is adopted as being equal to the limiting flux G_L. Since the Q/A value is itself a function of the MLSS concentration (C_0), to meet

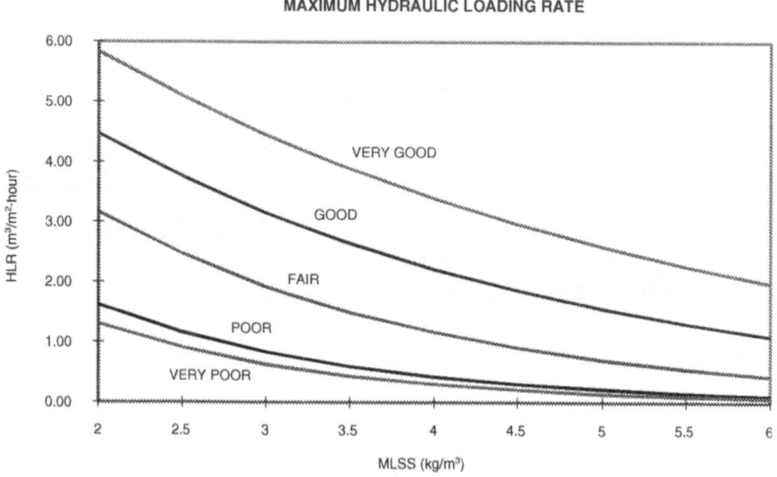

Figure 4.4.   Hydraulic loading rates (HLR) for the design of secondary sedimentation tanks, as a function of different sludge settleabilities and MLSS concentrations

the clarification criteria, the maximum solid loading rate can also be expressed as presented in Equation 4.6, where Q/A was replaced by the settling velocity given on the right – hand side of Equation 4.2.

$$SLR = m \cdot \left[R \cdot v_0 e^{(-K \cdot C_0)}\right]^n \qquad (4.6)$$

By knowing the intervening coefficients (v_0, K, m, n), the maximum solid loading rate can be easily determined (Equation 4.6). In these conditions, the **clarification and thickening criteria are simultaneously met**. It should be remembered that the four coefficients are functions of the sludge settleability (*very good, good, fair, poor* and *very poor*), as expressed in Table 4.3.

Figure 4.5 presents the curves of the maximum solid loading rates for the different sludge settleabilities (*very good, good, fair, poor* and *very poor*), MLSS concentrations (C_0), and return sludge ratios ($= Q_r/Q$). The graph clearly shows the essential relations: the better the sludge settleability, or the lower the MLSS concentration, or still the higher the return sludge ratio, then the higher the allowable solid loading rate and, consequently, the smaller the required surface area.

After obtaining the value for the maximum allowable G_L, the required area A can be finally calculated using

$$A = \frac{(Q+Q_r) \cdot C_0}{SLR} \qquad (4.7)$$

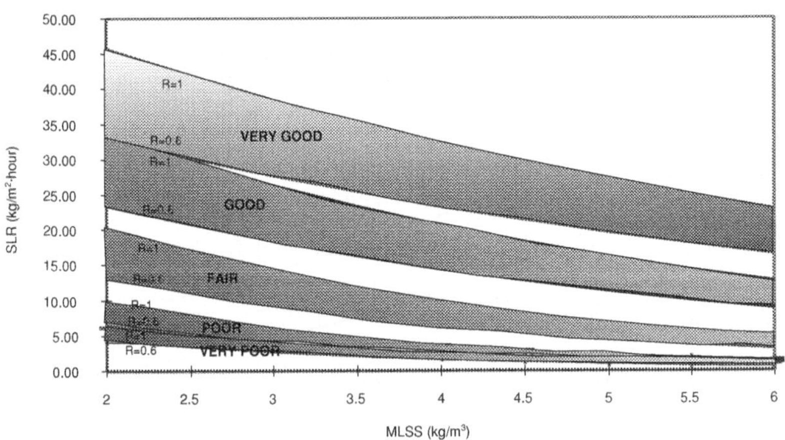

Figure 4.5. Solid loading rate (SLR) for the design of secondary sedimentation tanks, as a function of different sludge settleabilities, return sludge ratios R and MLSS concentrations. In each range, the highest value corresponds to R = 1.0, while the lowest value corresponds to R = 0.6

For design purposes, the sludge settleability should be considered as **fair** or **poor**, depending on the desirable safety degree. An intermediate **fair–poor** range can also be adopted, for which the curve and coefficient values can then be interpolated.

(d)   Summary of the proposed approach

The main advantage of the proposed methodology is that of working with the integrated data from 17 publications, representing a database of dozens of activated sludge plants operating in full scale worldwide.

The proposed methodology can also be used by means of Table 4.4, which represents a synthesis of Figure 4.5 and Equation 4.6, meeting simultaneously the clarification and thickening criteria. It should be emphasised that Table 4.4 can be used for both design and control of secondary sedimentation tanks (under the steady-state assumption).

Figures 4.6 and 4.7 present the curves from Figures 4.4 and 4.5, but only for the fair and poor settleability ranges, usually of higher interest for design and operation. Both figures also present the design ranges according to the conventional criteria covered in Section 4.2.1. For the two main activated sludge variants, the following MLSS ranges are considered typical: (a) conventional activated sludge: MLSS from 2.5 to 4.5 g/L and (b) extended aeration: MLSS from 3.5 to 5.5 g/L.

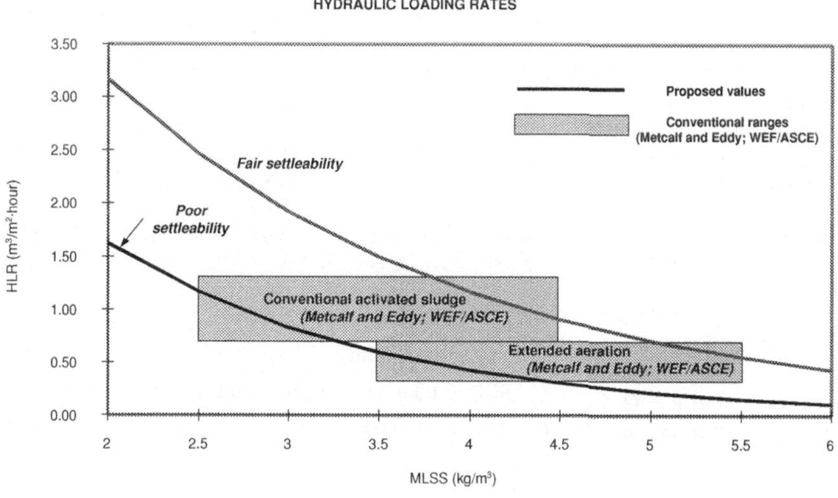

Figure 4.6.   Comparison between the proposed HLR and those from traditional methods (Metcalf and Eddy, 1991; WEF/ASCE, 1992)

Table 4.4. Maximum values for the HLR and SLR in secondary sedimentation tanks, as a function of MLSS and R ($=Q_r/Q$)

		Very good settleability				
MLSS (mg/L)	HLR $= Q/A$ ($m^3/m^2 \cdot$hour)					
		SLR $= (Q+Q_r) \cdot$MLSS$/(1,000 \cdot A)$(kgSS/m$^2 \cdot$hour)				
		$R=0.4$	$R=0.6$	$R=0.8$	$R=1.0$	$R=1.2$
2,000	5.83	25.42	32.95	39.61	45.70	51.35
2,500	5.09	23.32	30.23	36.34	41.91	47.10
3,000	4.45	21.39	27.72	33.33	38.44	43.20
3,500	3.89	19.62	25.43	30.57	35.26	39.63
4,000	3.40	17.99	23.32	28.04	32.34	36.35
4,500	2.97	16.50	21.39	25.72	29.67	33.34
5,000	2.59	15.14	19.62	23.59	27.21	30.58
5,500	2.27	13.88	18.00	21.64	24.96	28.05
6,000	1.98	12.74	16.51	19.85	22.89	25.73

		Good settleability				
MLSS (mg/L)	HLR $= Q/A$ ($m^3/m^2 \cdot$hour)	SLR $= (Q+Q_r) \cdot$MLSS$/(1,000 \cdot A)$(kgSS/m$^2 \cdot$hour)				
		$R=0.4$	$R=0.6$	$R=0.8$	$R=1.0$	$R=1.2$
2,000	4.47	17.68	23.48	28.71	33.57	38.14
2,500	3.75	15.64	20.77	25.40	29.70	33.74
3,000	3.15	13.84	18.38	22.48	26.27	29.85
3,500	2.64	12.24	16.26	19.88	23.25	26.41
4,000	2.22	10.83	14.38	17.59	20.57	23.36
4,500	1.86	9.58	12.72	15.56	18.19	20.67
5,000	1.56	8.48	11.26	13.77	16.10	18.29
5,500	1.31	7.50	9.96	12.18	14.24	16.18
6,000	1.10	6.63	8.81	10.78	12.60	14.31

		Fair settleability				
MLSS (mg/L)	HLR $= Q/A$ ($m^3/m^2 \cdot$hour)	SLR $= (Q+Q_r) \cdot$MLSS$/(1,000 \cdot A)$(kgSS/m$^2 \cdot$hour)				
		$R=0.4$	$R=0.6$	$R=0.8$	$R=1.0$	$R=1.2$
2,000	3.16	9.96	13.34	16.41	19.27	21.98
2,500	2.46	8.32	11.14	13.71	16.10	18.36
3,000	1.92	6.95	9.31	11.45	13.45	15.33
3,500	1.49	5.81	7.77	9.56	11.23	12.81
4,000	1.16	4.85	6.49	7.99	9.38	10.70
4,500	0.91	4.05	5.42	6.67	7.84	8.93
5,000	0.71	3.38	4.53	5.57	6.54	7.46
5,500	0.55	2.83	3.78	4.66	5.47	6.23
6,000	0.43	2.36	3.16	3.89	4.57	5.21

		Poor settleability				
MLSS (mg/L)	HLR $= Q/A$ ($m^3/m^2 \cdot$hour)	SLR $= (Q+Q_r) \cdot$MLSS$/(1,000 \cdot A)$(kgSS/m$^2 \cdot$hour)				
		$R=0.4$	$R=0.6$	$R=0.8$	$R=1.0$	$R=1.2$
2,000	1.62	4.65	6.15	7.50	8.75	9.92
2,500	1.16	3.69	4.88	5.95	6.94	7.87
3,000	0.83	2.93	3.87	4.72	5.51	6.25
3,500	0.59	2.32	3.07	3.75	4.37	4.96
4,000	0.43	1.84	2.44	2.97	3.47	3.93
4,500	0.30	1.46	1.94	2.36	2.75	3.12
5,000	0.22	1.16	1.54	1.87	2.19	2.48
5,500	0.16	0.92	1.22	1.49	1.73	1.97
6,000	0.11	0.73	0.97	1.18	1.38	1.56

Table 4.4 (*Continued*)

		Very poor settleability				
MLSS	HLR = Q/A	$SLR = (Q + Q_r) \cdot MLSS/(1{,}000 \cdot A)(kgSS/m^2 \cdot hour)$				
(mg/L)	(m³/m²·hour)	R = 0.4	R = 0.6	R = 0.8	R = 1.0	R = 1.2
2,000	1.30	3.42	4.53	5.52	6.44	7.30
2,500	0.90	2.66	3.52	4.29	5.00	5.68
3,000	0.63	2.07	2.73	3.33	3.89	4.41
3,500	0.44	1.61	2.13	2.59	3.02	3.43
4,000	0.30	1.25	1.65	2.02	2.35	2.67
4,500	0.21	0.97	1.28	1.57	1.83	2.07
5,000	0.15	0.75	1.00	1.22	1.42	1.61
5,500	0.10	0.59	0.78	0.95	1.10	1.25
6,000	0.07	0.46	0.60	0.74	0.86	0.97

MLSS (mg/L); Q (m³/hour); Q_r (m³/hour); A (m²); $HLR = v_0 \cdot e^{(-KMLSS/1{,}000)}$; $SLR = m \cdot (HLR.R)^n$
Classification of the settleability: see Table 4.2; v_0, K, m, n values: see Table 4.3

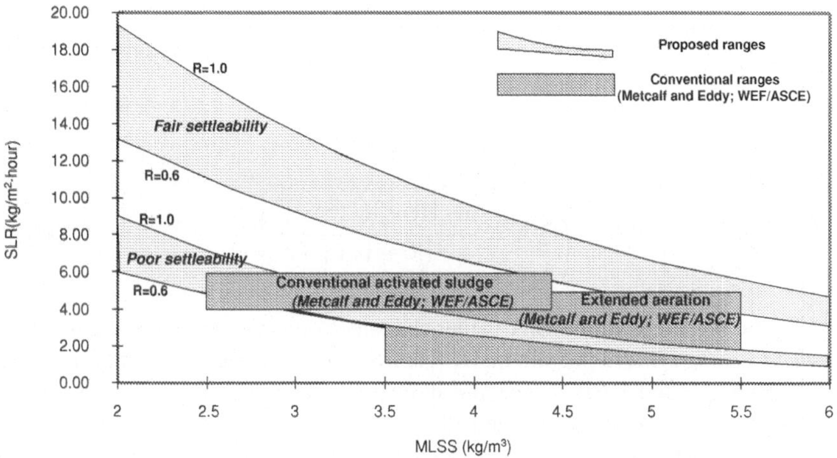

Figure 4.7. Comparison between the proposed SLR and those from traditional methods (Metcalf and Eddy, 1991; WEF/ASCE, 1992)

The analysis of the figures leads to the following points:

- The conventional loading rates are situated between the *fair* and *poor* settleability ranges.
- The extended aeration, which operates with a higher solid concentration, requires more conservative loading rates.
- The influence of the return sludge ratio, not taken into consideration in the traditional approach, can be clearly noticed.
- The proposed method allows a continuous solution for any MLSS and R values and settleability ranges, not leading to stepwise solutions, such as those from the conventional loading-rate methods.

Table 4.5 presents a more detailed version of Table 4.4, including more values of the MLSS concentration and return sludge ratio R. Its presentation form is different, clearer and simpler for design and operation (see Examples 4.2 and 4.3). The table presents, for each MLSS and R pair, the maximum **Q/A** values that meet the clarification and thickening criteria (hydraulic and solids loading rates). The lowest Q/A value found should be adopted for the design. Only the *fair, poor* and *very poor* settleability ranges are presented, since they are more important for design and operational purposes.

The maximum Q/A values required to meet the **clarification criteria** are given by Equation 4.2, already presented:

$$\boxed{Q/A = v_o \cdot e^{-K \cdot C_o}} \tag{4.2}$$

The maximum Q/A values required to meet the **thickening criteria** are obtained from the following calculation:

$$SLR = \frac{(Q + Q_r) \cdot C_o}{A} = \frac{(Q + R \cdot Q) \cdot C_o}{A} = \frac{Q}{A} \cdot (R + 1) \cdot C_o \tag{4.8}$$

$$Q/A = \frac{SLR}{(R + 1) \cdot C_o} \tag{4.9}$$

$$\boxed{Q/A = \frac{m \cdot \left[R \cdot v_o \cdot e^{(-K \cdot C_o)}\right]^n}{(R + 1) \cdot C_o}} \tag{4.10}$$

Figure 4.8 presents the **Q/A** values that *meet simultaneously the clarification and thickening criteria* (maximum HLR and SLR), according to the concepts above (Equations 4.2 and 4.10), for the *fair* and *poor* settleability ranges and for several MLSS (2,000 to 6,000 mg/L) and R (0.6, 0.8 and 1.0) values. There are three curves for each settleability range, each one representing a return sludge ratio R. Where the three curves are merged, the clarification criteria (Equation 4.2) are more restrictive, controlling the process, and the return sludge ratio has no influence. Where the three curves are separated, the thickening criteria are more demanding, controlling the process. It is noticed that the HLR values mentioned in literature (Table 4.1) are located in a range between the *fair* and *poor* settleabilities.

In terms of design, by knowing the influent flow and by adopting reasonable MLSS and R values and settleability characteristics (*fair* or *poor*), the required surface area can be calculated by means of Equations 4.2 and 4.10, Table 4.5, or even Figure 4.8. For a long-term operational control (not for daily control, which requires a dynamic model of the sedimentation tanks), different combinations of MLSS and R can be tried, by adopting the appropriate settleability range, to obtain the Q/A value corresponding to that existing in the plant.

Table 4.5. Q/A values ($m^3/m^2 \cdot$ hour) that meet the clarification and thickening criteria (HLR and SLR), for several MLSS and R values, and for different settleability ranges

MLSS (mg/L)	Q/A for clarification ($m^3/m^2 \cdot$hour)	Fair settleability								
		Q/A for thickening ($m^3/m^2 \cdot$hour)								
		R = 0.4	R = 0.5	R = 0.6	R = 0.7	R = 0.8	R = 0.9	R = 1.0	R = 1.1	R = 1.2
2,000	3.16	3.56	3.90	4.17	4.38	4.56	4.70	4.82	4.91	4.99
2,200	2.86	3.01	3.30	3.53	3.71	3.86	3.98	4.08	4.16	4.23
2,400	2.59	2.57	2.81	3.01	3.16	3.29	3.39	3.48	3.55	3.60
2,600	2.34	2.21	2.42	2.58	2.72	2.83	2.91	2.99	3.05	3.10
2,800	2.12	1.91	2.09	2.23	2.35	2.44	2.52	2.58	2.63	2.67
3,000	1.92	1.66	1.81	1.94	2.04	2.12	2.19	2.24	2.29	2.32
3,200	1.74	1.44	1.58	1.69	1.78	1.85	1.91	1.96	1.99	2.03
3,400	1.57	1.26	1.39	1.48	1.56	1.62	1.67	1.71	1.75	1.77
3,600	1.42	1.11	1.22	1.30	1.37	1.42	1.47	1.50	1.53	1.56
3,800	1.29	0.98	1.07	1.15	1.21	1.26	1.29	1.33	1.35	1.38
4,000	1.16	0.87	0.95	1.01	1.07	1.11	1.14	1.17	1.20	1.22
4,200	1.05	0.77	0.84	0.90	0.95	0.98	1.01	1.04	1.06	1.08
4,400	0.95	0.68	0.75	0.80	0.84	0.87	0.90	0.92	0.94	0.96
4,600	0.86	0.61	0.67	0.71	0.75	0.78	0.80	0.82	0.84	0.85
4,800	0.78	0.54	0.59	0.63	0.67	0.69	0.71	0.73	0.75	0.76
5,000	0.71	0.48	0.53	0.57	0.60	0.62	0.64	0.65	0.67	0.68
5,200	0.64	0.43	0.47	0.51	0.53	0.55	0.57	0.59	0.60	0.61
5,400	0.58	0.39	0.42	0.45	0.48	0.50	0.51	0.52	0.54	0.54
5,600	0.52	0.35	0.38	0.41	0.43	0.45	0.46	0.47	0.48	0.49
5,800	0.47	0.31	0.34	0.37	0.38	0.40	0.41	0.42	0.43	0.44
6,000	0.43	0.28	0.31	0.33	0.35	0.36	0.37	0.38	0.39	0.39

(Continiued)

Table 4.5 (*Continued*)

		Poor settleability								
MLSS (mg/L)	Q/A for clarification (m³/m²·hour)	Q/A for thickening (m³/m²·hour)								
		R = 0.4	R = 0.5	R = 0.6	R = 0.7	R = 0.8	R = 0.9	R = 1.0	R = 1.1	R = 1.2
2,000	1.62	1.66	1.81	1.92	2.01	2.08	2.14	2.19	2.22	2.25
2,200	1.42	1.38	1.50	1.59	1.67	1.73	1.77	1.81	1.84	1.87
2,400	1.24	1.15	1.25	1.33	1.39	1.44	1.48	1.51	1.54	1.56
2,600	1.09	0.97	1.05	1.12	1.17	1.21	1.25	1.27	1.30	1.31
2,800	0.95	0.82	0.89	0.95	0.99	1.03	1.06	1.08	1.10	1.11
3,000	0.83	0.70	0.76	0.81	0.84	0.87	0.90	0.92	0.93	0.95
3,200	0.73	0.60	0.65	0.69	0.72	0.75	0.77	0.78	0.80	0.81
3,400	0.64	0.51	0.56	0.59	0.62	0.64	0.66	0.67	0.68	0.69
3,600	0.56	0.44	0.48	0.51	0.53	0.55	0.57	0.58	0.59	0.60
3,800	0.49	0.38	0.41	0.44	0.46	0.48	0.49	0.50	0.51	0.52
4,000	0.43	0.33	0.36	0.38	0.40	0.41	0.42	0.43	0.44	0.45
4,200	0.37	0.29	0.31	0.33	0.35	0.36	0.37	0.38	0.38	0.39
4,400	0.33	0.25	0.27	0.29	0.30	0.31	0.32	0.33	0.33	0.34
4,600	0.28	0.22	0.24	0.25	0.26	0.27	0.28	0.29	0.29	0.29
4,800	0.25	0.19	0.21	0.22	0.23	0.24	0.24	0.25	0.25	0.26
5,000	0.22	0.17	0.18	0.19	0.20	0.21	0.21	0.22	0.22	0.23
5,200	0.19	0.15	0.16	0.17	0.18	0.18	0.19	0.19	0.19	0.20
5,400	0.17	0.13	0.14	0.15	0.15	0.16	0.16	0.17	0.17	0.17
5,600	0.15	0.11	0.12	0.13	0.14	0.14	0.14	0.15	0.15	0.15
5,800	0.13	0.10	0.11	0.11	0.12	0.12	0.13	0.13	0.13	0.13
6,000	0.11	0.09	0.09	0.10	0.11	0.11	0.11	0.11	0.12	0.12

Very poor settleability

MLSS (mg/L)	Q/A for clarification (m³/m²·hour)	Q/A for thickening (m³/m²·hour)								
		R = 0.4	R = 0.5	R = 0.6	R = 0.7	R = 0.8	R = 0.9	R = 1.0	R = 1.1	R = 1.2
2,000	1.30	1.22	1.33	1.41	1.48	1.53	1.58	1.61	1.64	1.66
2,200	1.12	1.00	1.09	1.16	1.22	1.26	1.29	1.32	1.35	1.36
2,400	0.97	0.83	0.91	0.96	1.01	1.04	1.07	1.10	1.12	1.13
2,600	0.84	0.69	0.76	0.80	0.84	0.87	0.90	0.92	0.93	0.94
2,800	0.73	0.58	0.63	0.68	0.71	0.73	0.75	0.77	0.78	0.79
3,000	0.63	0.49	0.54	0.57	0.60	0.62	0.63	0.65	0.66	0.67
3,200	0.54	0.42	0.45	0.48	0.51	0.52	0.54	0.55	0.56	0.57
3,400	0.47	0.36	0.39	0.41	0.43	0.45	0.46	0.47	0.48	0.48
3,600	0.40	0.30	0.33	0.35	0.37	0.38	0.39	0.40	0.41	0.41
3,800	0.35	0.26	0.28	0.30	0.31	0.33	0.33	0.34	0.35	0.35
4,000	0.30	0.22	0.24	0.26	0.27	0.28	0.29	0.29	0.30	0.30
4,200	0.26	0.19	0.21	0.22	0.23	0.24	0.25	0.25	0.26	0.26
4,400	0.23	0.17	0.18	0.19	0.20	0.21	0.21	0.22	0.22	0.23
4,600	0.19	0.14	0.16	0.17	0.17	0.18	0.18	0.19	0.19	0.19
4,800	0.17	0.12	0.14	0.14	0.15	0.16	0.16	0.16	0.17	0.17
5,000	0.15	0.11	0.12	0.12	0.13	0.14	0.14	0.14	0.14	0.15
5,200	0.13	0.09	0.10	0.11	0.11	0.12	0.12	0.12	0.13	0.13
5,400	0.11	0.08	0.09	0.09	0.10	0.10	0.11	0.11	0.11	0.11
5,600	0.09	0.07	0.08	0.08	0.09	0.09	0.09	0.09	0.10	0.10
5,800	0.08	0.06	0.07	0.07	0.08	0.08	0.08	0.08	0.08	0.08
6,000	0.07	0.05	0.06	0.06	0.07	00.07	0.07	0.07	0.07	0.07

For each MLSS and R values, compare Q/A for clarification with Q/A for thickening, and adopt the lowest value
Calculation of the Q/A values: Clarification (Equation 4.2); thickening (Equation 4.10)
Classification of the settleability: See Table 4.2

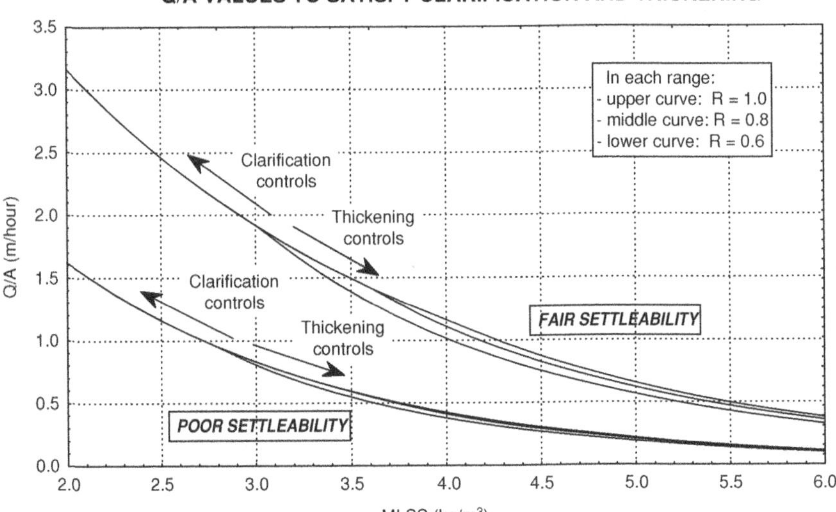

Figure 4.8.   Q/A values to meet simultaneously the clarification and thickening criteria (HLR and SLR), for the *fair* and *poor* settleability ranges and for several MLSS and R values

---

Example 4.2

Using the methodology presented in this section, calculate the surface area required for the secondary sedimentation tanks of a conventional activated sludge plant. Use the same data as those in Example 4.1 and the design example in Chapter 5:

- average influent flow: $Q = 9{,}820$ m³/d
- average return sludge flow: $Q_r = 9{,}820$ m³/d
- mixed liquor suspended solids concentration: MLSS = 3,896 g/m³

Solution:

(a)   Express the flows in m³/hour

$$Q = 9{,}820/24 = 409 \text{ m}^3/\text{hour}$$

$$Q_r = 9{,}820/24 = 409 \text{ m}^3/\text{hour}$$

The return sludge ratio R is equal to $Q_r/Q = 409/409 = 1.0$

(b)   Calculation of the surface area based on the hydraulic and solids loading rates

Assume a *poor* sludge settleability.

$$\text{With MLSS} = 3{,}896 \text{ mg/L} \rightarrow C_o = 3.896 \text{ kg/m}^3.$$

Example 4.2 (Continued)

Using the v_o, K, m and n values of Table 4.3: $v_o = 6.2$ m/hour; $K = 0.67$ m^3/kg; $m = 6.26$; $n = 0.69$

Clarification requirements (Equation 4.2):

$$HLR = v_o \cdot e^{-K \cdot C} = 6.2 \cdot e^{-0.67 \times 3.896} = 0.456 \text{ m/hour}$$

The required area is given by:

$$A = \frac{Q}{HLR} = \frac{409}{0.456} = 897 \text{ m}^2$$

Note: the hydraulic loading rate value of 0.456 m/hour can approximately be also obtained by interpolation in Tables 4.4 and 4.5 (*poor* settleability and MLSS = 3,896 mg/L).

Thickening requirements (Equation 4.6):

$$SLR = m \cdot \left[R \cdot v_o \cdot e^{(-K \cdot C_o)} \right]^n = 6.26 \cdot \left[1.0 \times 6.2 \cdot e^{(-0.67 \times 3.896)} \right]^{0.69}$$

$$= 3.640 \text{ kgSS/m}^2 \cdot \text{hour}$$

The required area is given by:

$$A = \frac{(Q + Q_r) \cdot C_o}{SLR} = \frac{(409 + 409) \times 3.896}{3.640} = 876 \text{ m}^2$$

Note: the solids loading rate value of 3.640 kgSS/m$^2 \cdot$hour can approximately be also obtained by interpolation in Table 4.4 (*poor* settleability, $R = 1.0$, and MLSS = 3,896 mg/L).

According to the clarification criteria, the required area is 897 m^2, and according to the thickening criteria, the required area is 876 m^2. The higher value should be adopted, that is, **897 m^2**. In this case, the clarification is controlling the process, as it is more restrictive in terms of loading rates and required area.

(c)   Calculation of the surface area based on the direct equation for Q/A

For the *clarification* criteria, the Q/A value was calculated in the above item as being 0.456 m/hour, and the required area was equal to 897 m^2.

For the *thickening* criteria, the Q/A value is directly obtained from Equation 4.10:

$$Q/A = \frac{m \cdot \left[R \cdot v_o \cdot e^{(-K \cdot C_o)} \right]^n}{(R + 1) \cdot C_o} = \frac{6.26 \cdot \left[1.0 \times 6.2 \cdot e^{(-0.67 \times 3.896)} \right]^{0.69}}{(1.0 + 1) \times 3.896}$$

$$= 0.467 \text{ m/hour}$$

Since $Q = 409$ m^3/hour $\rightarrow A = 409/0.467 = 876$ m^2

Example 4.2 (Continued)

This value is, as expected, identical to that obtained in item (b), for thickening. Once again, the largest area (or the lowest Q/A value) should be adopted. In this sense, Q/A = 0.456 m/hour is adopted, and the area is equal to **897 m^2**. The Q/A value could have also been approximately obtained from Figure 4.8.

(d)   Calculation of the surface area using Table 4.5
(HLR and SLR criteria)

With R = 1.0 and *poor* settleability, according to Table 4.5:

For MLSS = 3,800 mg/L: clarification: Q/A = 0.90 m/hour; thickening: Q/A = 0.50 m/hour
For MLSS = 4,000 mg/L: clarification: Q/A = 0.43 m/hour; thickening: Q/A = 0.43 m/hour

Since MLSS = 3,896 mg/L, linearly interpolating the values, the following is obtained:

Clarification: Q/A = 0.46 m/hour
Thickening: Q/A = 0.47 m/hour

By adopting the lowest value (Q/A = 0.46 m/hour), and since Q = 409 m^3/hour:

$$A = 409/0.46 = \textbf{889 m}^2$$

The A values obtained from the three methods are naturally the same (apart from a small difference in the result obtained from Table 4.5, as the table does not provide continuous solutions).

The example of Chapter 5 (Section 5.3, conventional activated sludge) adopts the settleability range between *fair* and *poor*. It is interesting to compare the results to have an idea of the influence of the settleability ranges on the final result.

Example 4.3

An activated sludge plant is working with a high sludge blanket level, and is facing problems concerning solids losses in the final effluent from the sedimentation tank. SVI tests indicated that the sludge settleability can be considered *fair*, according to the classification of Table 4.2. Analyse the loading conditions of the sedimentation tanks and propose control measures, using Table 4.5. Data are:

- average influent flow: Q = 250 m^3/hour
- return sludge flow: Q$_r$ = 150 m^3/hour

Example 4.3 (Continued)

- Mixed liquor suspended solids concentration: MLSS = 4,000 mg/L = 4.0 kg/m^3
- Surface area of the secondary sedimentation tanks: A = 200 m^2

Solution:

(a) Evaluation of the loading conditions

The return sludge ratio is $Q_r/Q = 150/250 = 0.6$

From Table 4.5, for MLSS = 4,000 mg/L, R = 0.6, and *fair* settleability, the maximum allowable values for Q/A are: 1.16 m^3/m^2·hour (clarification) and 1.01 m^3/m^2·hour (thickening). In this case, the thickening controls the process, since it is more restrictive than the clarification. The lowest value should be adopted (1.01 m^3/m^2·h). However, considering the present conditions of the wastewater treatment plant, the actual Q/A value is 250/200 = 1.25 m^3/m^2·hour. The sedimentation tanks are, therefore, overloaded due to the fact that the Q/A applied (1.25 m^3/m^2·hour) is higher than the maximum Q/A allowed (1.01 m^3/m^2·hour).

It is necessary to take operational control measures, which can involve one of the two following alternatives, or a combination between both: (a) reduce MLSS concentration and (b) increase R. Q cannot be altered, because it is independent of operational control. The surface area A also cannot be modified, because this is the existing available area. Thus, the Q/A applied remains the same (1.25 m^3/m^2·hour).

(b) Reduce the MLSS concentration

A reduction in MLSS implies a reduced applied solids load. The lowering of MLSS should be such that the maximum allowable Q/A value, extracted from Table 4.5, is higher than the applied Q/A value (1.25 m^3/m^2·hour). From Table 4.5, *fair* settleability, R = 0.6, the Q/A value for thickening immediately higher than 1.25 is 1.30 m^3/m^2·hour, which corresponds to MLSS of 3,600 mg/L. In these conditions, the maximum Q/A allowed for clarification is 1.42 m^3/m^2·hour (see Table 4.5), higher than the Q/A value for thickening. Therefore, the concentration of MLSS in the reactor should be decreased from 4,000 mg/L to 3,600 mg/L, by means of an increase in the removal of excess sludge, aiming at reducing the load of influent solids to the secondary sedimentation tank.

(c) Increase the return sludge ratio R

A reduction in R implies a higher solids absorption capacity by the sedimentation tank. A maximum allowable Q/A value higher than 1.25 m^3/m^2·hour should be obtained. From Table 4.5, *fair* settleability, MLSS = 4,000 mg/L,

---

Example 4.3 (Continued)

there is no Q/A value higher than 1.25 $m^3/m^2 \cdot$hour. The highest value, 1.22 $m^3/m^2 \cdot$hour, corresponding to R = 1.2, is slightly lower than the Q/A value applied. Hence, it would be necessary to increase R from 0.6 to more than 1.2, which may not be the best solution, in case there is no sufficient pumping capacity. Besides that, the Q/A value for clarification would still not be met, because the variation in R does not affect the clarification.

(d)   Reduce MLSS and simultaneously increase R

The joint action in MLSS and R allows different combinations, generating Q/A values higher than that of the applied Q/A (1.25 $m^3/m^2 \cdot$hour). A possible combination is MLSS = 3,800 mg/L and R = 0.8, which results in a maximum allowable Q/A equal to 1.29 $m^3/m^2 \cdot$hour (clarification) and 1.26 $m^3/m^2 \cdot$hour (thickening). In the two conditions (clarification and thickening), the maximum allowable Q/A values are higher than the applied Q/A value.

Example 4.4

An activated sludge plant is showing a weak performance in the BOD removal and in nitrification. The analysis of the process indicated that the concentration of MLSS is very low, and it needs to be increased. The plant has an oxygenation capacity sufficient to provide more oxygen, even with very high MLSS concentrations. Verify which concentration of MLSS can be maintained in the reactor without causing overloading problems to the secondary sedimentation tank. The sludge settleability in the plant is considered *fair*. The data of interest are:

- average influent flow: Q = 200 m^3/hour
- return sludge ratio: R = 1.0
- surface area of the secondary sedimentation tanks: 180 m^2

Solution:

The applied Q/A value is 200/180 = 1.11 $m^3/m^2 \cdot$hour. From Table 4.5, for *fair* settleability and R = 1.0, the maximum allowable Q/A value (thickening) immediately higher than the applied value of 1.11 $m^3/m^2 \cdot$hour is 1.17 $m^3/m^2 \cdot$hour, which is associated with a concentration of MLSS of 4,000 mg/L. For this concentration of MLSS, the maximum allowable value (clarification) is 1.16 $m^3/m^2 \cdot$hour, which is also satisfactory, as it is higher than the applied value. Therefore, the MLSS concentration in the reactor can be increased up to 4,000 mg/L, as long as the *fair* settleability conditions and the Q and R values are not altered.

Table 4.6. Minimum and recommended values for the
sidewater depth of secondary sedimentation tanks

Tank diameter (m)	Liquid sidewater depth (m)	
	Minimum	Recommended
<12	3.0	3.3
12–20	3.3	3.6
20–30	3.6	3.9
30–40	3.9	4.2
>40	4.2	4.5

Source: Adapted from WEF/ASCE (1992)

4.3 DESIGN DETAILS IN SECONDARY SEDIMENTATION TANKS

4.3.1 Sidewater depth

The liquid depth of a sedimentation tank is usually referred to as the sidewater
depth (SWD) of the cylindrical part (wall) in a circular sedimentation tank, and as
depth of the final end in a rectangular sedimentation tank.

The current tendency is to adopt high depths to ensure a better accommodation
of the sludge blanket in its occasional expansions, allowing a better quality of the
effluent. Table 4.6 presents values suggested by WEF/ASCE (1992), according to
the diameter of the tank.

It should be remembered that, in tropical countries, where nitrification occurs
almost systematically in the reactors, the occurrence of denitrification in secondary
sedimentation tanks is very likely to occur in case there is no intentional biological
nitrogen removal in the reactor. A long sludge detention time in the sedimentation
tank can allow denitrification, with the release of gaseous nitrogen bubbles, which
adhere in their upward movement to the settling sludge, thus carrying it to the
surface. Therefore, long **sludge** detention times should be avoided in the sedimentation tank, which means that high sludge blanket levels and low underflow rates
should be avoided.

In case the circular tank has a flat bottom, in view of the sludge removal
by suction, the design of the sedimentation tank should be more conservative.
WEF/ASCE (1992) suggest a 0.35 m/hour reduction in the hydraulic loading rate,
compared with the design of a conical-bottom sedimentation tank, which has a
higher absorption capacity of the variation of the sludge blanket level (due to the
additional volume provided by the conical section).

In sedimentation tanks with high bottom slopes and no mechanical sludge
removal (Dortmund-type tanks) the sidewater depth is lower, but should be higher
than 0.5 m, according to the Brazilian standards (ABNT, 1989).

4.3.2 Effluent weirs

The design of the effluent weirs is also an important item in the conception of secondary sedimentation tanks to minimise the transportation of solids with the final

Table 4.7. Maximum weir loading rate values

Sedimentation tank	Condition	Weir loading rate (m³/m·hour)	
		Average flow	Maximum flow
Small	–	5	10
Large	Outside the upturn zone of the current	–	15
	Inside the upturn zone of the current	–	10

Source: Metcalf and Eddy, 1991

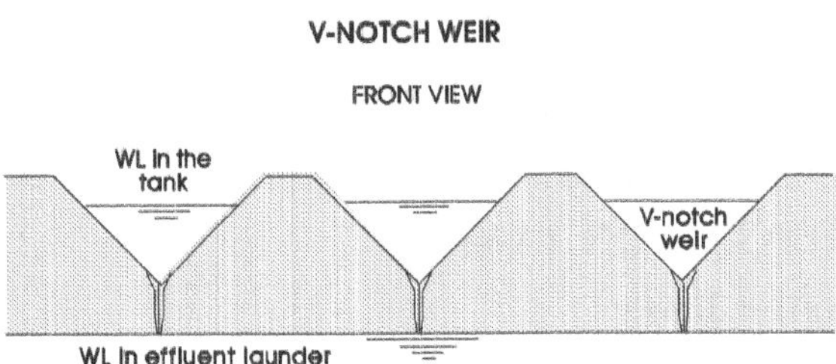

Figure 4.9.   Detail of a V-notch effluent weir

effluent. The weirs can be either continuous or, preferably, with V-notches (Figure 4.9). The latter one is more recommended as it is less influenced by occasional differences in the fixing level of the weirs.

The required length of the weirs is calculated based on the **weir loading rate** (m³/hour per metre of weir), which corresponds to the flow per unit length of the weir. The weir rate is associated with the liquid approaching velocity: high velocities could carry solids from the sludge blanket. Since the important factor is the approaching velocity, in V-notch weirs the flow that goes through the openings is not influential, and the velocity is dictated by the influent flow that approaches all the weir length.

Metcalf and Eddy (1991) suggest the weir loading rate values listed in Table 4.7. WEF/ASCE (1992) suggest 5 m³/m·hour for small plants and 8 m³/m·hour for larger plants.

In well-designed circular tanks, the perimeter of the tank usually meets the weir loading rate criteria. English experiences (Johnstone *et al.*, 1979) indicated that single-faced weir launders (even if with higher loading rates) are preferable to double-faced weir launders (internal to the tank). In double-faced weir launders, the face closer to the external wall receives the rising liquid current parallel to the wall, carrying a larger quantity of solids.

4.3.3 Other design details

The following comments can be made on circular and rectangular sedimentation tanks provided with mechanised sludge removal (Metcalf and Eddy, 1991;

WEF/ASCE, 1992):

Rectangular sedimentation tanks:

- The distribution of the influent flow should be homogeneous, to avoid excessive horizontal velocities and hydraulic short circuits.
- It is recommended that the length/depth ratio does not exceed the value of 10 to 15.
- If a tank is wider than 6 m, multiple sludge collectors can be adopted to allow a width up to 24 m.
- The sludge collecting mechanism should have a high capacity to avoid preferential routes of the liquid through the sludge. It should also be sturdy to remove and transport thicker sludges accumulated during interruptions in the operation.
- The most common sludge removal mechanisms are: (a) scraper with travelling bridge (see Figure 4.2), (b) scraper with submerged chains, and (c) suction removers. The scraping mechanisms usually transport the sludge to a hopper in the inlet end of the tank.

Circular sedimentation tank:

- The most usual diameters range from 10 m to 40 m.
- The diameter/sidewater depth ratio should not exceed the value of 10.
- The sludge can be removed either by rotating scrapers that direct the sludge to a hopper at the centre of the tank or by suction mechanisms supported by rotating bridges.
- The bottom of the tank should have a slope of approximately 1:12, in the case of sludge removal by scrapers, or be flat, in the case of removal by suction.

4.4 DESIGN OF PRIMARY SEDIMENTATION TANKS

Primary sedimentation tanks are used in conventional activated sludge plants. Their main function is to reduce the organic matter load to the biological treatment stage. The main design parameters are presented in Tables 4.8 and 4.9.

Weir loading rates are not important in primary sedimentation tanks followed by activated sludge systems.

The Brazilian standards for the design of wastewater treatment plants (ABNT, 1989) recommend the following points:

General considerations:

- A WWTP with a design maximum flow higher than 250 L/s should have more than one primary sedimentation tank.
- The sludge removal pipes should have a minimum diameter of 150 mm; the sludge gravity transport piping should have a minimum slope of 3%

Table 4.8. Typical design parameters for primary sedimentation tanks followed by activated sludge systems

Item	Primary settling followed by secondary treatment		Primary settling with excess activated sludge return	
	Range	Typical	Range	Typical
Hydraulic loading rate ($Q_{average}$) ($m^3/m^2 \cdot hour$)	1.4–2.1	–	1.0–1.4	–
Hydraulic loading rate (Q_{max}) ($m^3/m^2 \cdot hour$)	3.4–5.1	4.3	2.0–2.9	2.6
Detention time (hour)	1.5–2.5	2.0	1.5–2.5	2.0

Source: Metcalf and Eddy (1991)

Table 4.9. Typical design parameters for rectangular and circular primary sedimentation tanks

Item	Rectangular tank		Circular tank	
	Range	Typical	Range	Typical
Depth (m)	3.0–4.5	3.6	3.0–4.5	3.6
Length (m)	15–90	24–40	–	–
Width (m)	3–24	5–10	–	–
Diameter (m)	–	–	3–60	12–45
Bottom slope (%)	–	–	6–17	8

Source: Metcalf and Eddy (1991)

and the bottom sludge removal should be such that allows the observation and control of the sludge removed.

- The sludge accumulation hopper should have walls with slopes equal to or higher than 1.5 vertical to 1.0 horizontal, with a bottom base with a minimum dimension of 0.60 m.

Primary sedimentation tank with mechanised sludge removal:

- The removal device should have a velocity equal to or lower than 20 mm/s in the case of rectangular sedimentation tanks, and a peripheral velocity equal to or lower than 40 mm/s in the case of circular sedimentation tanks.
- The minimum sidewater depth should be equal to or higher than 2.0 m
- For rectangular sedimentation tanks, the length/sidewater depth ratio should be equal to or greater than 4:1; the width/sidewater depth ratio should be equal to or greater than 2:1 and the length/width ratio should be equal to or greater than 2:1.
- For rectangular sedimentation tanks, the horizontal flow velocity should be equal to or lower than 50 mm/s; when receiving excess activated sludge, the velocity should be equal to or lower than 20 mm/s.

Primary sedimentation tank without mechanised sludge removal (Dortmund-type tanks):

- The minimum sidewater depth should be equal to or higher than 0.5 m.
- The sedimentation tank can be either circular or square in plan, with a single conical or pyramidal sludge hopper, sludge discharge by gravity, wall slope equal to or greater than 1.5 vertical by 1.0 horizontal, and diameter or diagonal not exceeding 7.0 m.
- The sedimentation tank can be rectangular in plan, fed by the smaller side, provided that the tank consists totally of square-based pyramidal hoppers, with sides lower than 5.0 m and with individual sludge discharges.
- The minimum hydraulic load for sludge removal should be five times the head loss calculated for water, and not lower than 1.0 m.

5

Design example of an activated sludge system for organic matter removal

5.1 INTRODUCTION

Design the biological stage of an activated sludge system to treat the wastewater generated in the community. The treatment units should be designed for BOD removal, using the **conventional activated sludge** variant. The occurrence of nitrification should be taken into consideration, but the system should not be designed for biological nutrient (nitrogen and phosphorus) removal. The design should be made for the 20th year of operation. For that year, input data are as follows:

- Population equivalent: 67,000 inhabitants

- Influent flow:
  average: $9,820\ m^3/d$
  maximum: $19,212\ m^3/d$
  minimum: $4,003\ m^3/d$

- Influent loads:
  BOD: $3,350\ kg/d$
  SS: $3,720\ kg/d$ (per capita load of 60 gSS/ inhabitant·day)
  TKN: $496\ kg/d$ (per capita load of 8 gTKN/ inhabitant·day)

- Influent concentrations: BOD: 341mg/L
                              SS: 379mg/L
                              TKN: 51mg/L

- Temperature of the liquid: average of the coldest month: 20 °C
                              average of the warmest month: 25 °C

- Altitude: 800 m
- Desired characteristics for the effluent: BOD: 20mg/L
                                             SS: 30mg/L

5.2 MODEL PARAMETERS AND COEFFICIENTS

(a) Kinetic and stoichiometric parameters

According to Table 3.2, the following values were adopted:

$Y = 0.6$ gVSS/gBOD$_5$
$K_d = 0.08$ d^{-1} (20 °C)
$\theta = 1.07$ (temperature correction for K_d)
Ratio O$_2$/SS$_b$ = 1.42 gO$_2$ per g biodegradable VSS
Ratio BOD$_u$/BOD$_5$ = 1.46

(b) Relations between solids

According to Table 3.2, the following values were adopted:

Raw sewage:
SS$_b$/VSS = 0.60
VSS/SS = 0.80

Biological solids to be generated:
SS$_b$/SS = 0.80
VSS/SS = 0.90

After a time equal to the sludge age:
VSS/SS = 0.77 (conventional activated sludge; system with solids in the raw
   sewage and with primary sedimentation tank)
(initial estimate – see Table 2.6; exact value to be calculated later on):

(c) Aeration coefficients

The following values can be adopted:

$C_s = 9.02$ (clean water, 20 °C)
$\alpha = 0.85$

$\beta = 0.90$
$\theta = 1.024$ (for correction of $K_L a$ for the operating temperature)
Specific gravity of air $= 1.2$ kg/m^3 (20 °C, altitude $= 0$ m)
Fraction of O_2 in air (by weight) $= 0.23$ gO_2/g air

5.3 DESIGN OF THE CONVENTIONAL ACTIVATED SLUDGE SYSTEM

5.3.1 Design parameters

(a)   Removal efficiencies assumed for the primary sedimentation tank

$BOD_5 = 30\%$
$SS = 60\%$
$TKN = 20\%$

(b)   Reactor

Fully aerobic reactor (without anoxic or anaerobic zones)
$\theta_c = 6$ days
MLVSS $= 3,000$ mg/L
$R = 1.0$ (return sludge ratio)

(c)   Aeration system

Minimum DO (with Q_{max}): $C_L = 1.0$ mg/L
Average DO (with Q_{av}): $C_L = 2.0$ mg/L

Mechanical aeration (low speed):
OE (standard conditions) $= 1.8$ kgO_2/kWh

Diffused air (fine bubbles):
O_2 transfer efficiency $= 0.15$
Efficiency of the motor and the blower $\eta = 0.60$

(d)   Secondary sedimentation tank

Limiting flux (settleability ratios are between fair and poor) (see Table 4.2)

Taking the arithmetic mean for the values of v_o, K, m and n (Table 4.3) corresponding to *fair* and *poor* settleability, thus characterising settleability as *fair–poor*:

$v_o = 7.40$ m/hour
$K = 0.59$ m^3/kg
$m = 7.34$
$n = 0.71$

5.3.2 Effluent loads and concentrations from the primary sedimentation tank (influent to the reactor)

In the primary sedimentation tank, the main pollutants are removed according to the percentages given in Section 5.3.1.a. The effluent loads and concentrations are:

$$\text{Effluent} = \text{Influent} \cdot \frac{(100 - \text{Efficiency})}{100}$$

$$\text{Effluent BOD} = 3{,}350 \text{ kg/d} \cdot (100 - 30)/100 = 2{,}345 \text{ kg/d}$$
$$= 341 \text{ mg/L} \cdot (100 - 30)/100 = 239 \text{ mg/L}$$

$$\text{Effluent SS} = 3{,}720 \text{ kg/d} \cdot (100 - 60)/100 = 1{,}488 \text{ kg/d}$$
$$= 379 \text{ mg/L} \cdot (100 - 60)/100 = 152 \text{ mg/L}$$

$$\text{Effluent TKN} = 496 \text{ kg/d} \cdot (100 - 20)/100 = 397 \text{ kg/d}$$
$$= 51 \text{ mg/L} \cdot (100 - 20)/100 = 40 \text{ mg/L}$$

5.3.3 Soluble BOD of the final effluent

- Effluent SS concentration:

  $X_{effl} = 30$ mg/L (stated in the problem)

- Effluent VSS concentration:

  $Xv_{effl} = (\text{VSS/SS ratio with } \theta_c \text{ days}) \cdot X_{efl} = 0.77 \times 30 = 23 \text{ mgVSS/L}$

- Correction of K_d for the temperature of the coldest month (20 °C): no correction because the temperature of the coldest month coincides with the standard temperature of 20 °C: $K_d = 0.08 \text{ d}^{-1}$

- Coefficient f_b (SS_b/VSS ratio) (Equation 2.2):

  $$f_b = \frac{f_{b'}}{1 + (1 - f_{b'}) \cdot K_d \cdot \theta_c} = \frac{0.80}{1 + (1 - 0.80) \times 0.08 \times 6}$$
  $$= 0.73 \text{ mgSS}_b/\text{VSS}$$

- Concentration of biodegradable solids in the effluent:

  $$Xb_{effl} = f_b \cdot Xv_{effl} = 0.73 \times 23 = 17 \text{ mgSS}_b/\text{L}$$

- Particulate BOD_5 in the effluent (Section 2.6):

$$BOD_{5_{part}} = \frac{(BOD_u/X_b)\cdot Xb_{effl}}{(BOD_u/BOD_5)} = \frac{1.42 \times 17 \text{ mg/L}}{1.46} = 16 \text{ mg /L}$$

- Maximum soluble BOD to be obtained (Equation 2.10 rearranged):

$$BOD5_{sol} = BOD5_{tot} - BOD5_{part} = 20 - 16 = 4 \text{ mg/L}$$

5.3.4 Efficiency of the system in BOD removal

$$E = (BOD_{infl} - BOD_{effl}) \times 100/BOD_{infl}$$

In the system (primary sedimentation tank + biological stage): $E = (341 - 20) \times 100/341 = 94.1\%$

In the biological stage: $E = (239 - 20) \times 100/239 = 91.6\%$

5.3.5 BOD_5 load removed in the biological stage

$$S_r = Q_{average} \times (BODtot_{infl} - BODsol_{effl})$$
$$S_r = Q\cdot(S_0 - S)/1{,}000 = 9{,}820 \times (239 - 4)/1{,}000 = 2{,}308 \text{ kgBOD}_5\text{/d}$$

5.3.6 Distribution of the solids in the treatment

The purpose of this section is to estimate the total production of solids (raw sewage + biological solids) and the ratio VSS/SS. Since this is a laborious task, this section can be replaced (for predominantly domestic sewage) by simplified estimates of the production of solids and the VSS/SS ratio (Section 2.19 and Table 2.6). Thus, in a more straightforward and simplified version of the example, the present section can be omitted, and the calculations in the following paragraph may be adopted.

From Table 2.6, for a system with sedimentation tank, $\theta_c = 6$ d, $Y = 0.6$ and $K_d = 0.08$ d^{-1}:

- $VSS/SS = 0.76$
- $SS/S_r = 0.87$ kgSS/kgBOD$_5$ rem $\rightarrow$ $P_x = 0.87$ kgSS/kgBOD $\times$ 2,308 kgBOD/d = **2,008 kgSS/d**

For a detailed calculation, the following steps should be followed.

(a)   Influent solids to the reactor (effluent from the primary sedimentation tank)

- Total suspended solids:

$$P_x = 1,488 \text{ kgSS/d (Section 5.3.2)}$$

- Volatile suspended solids:

$$P_{xv} = (\text{VSS/SS ratio in the raw sewage}) \cdot P_x = 0.8 \times 1,488$$
$$= 1,190 \text{ kgVSS/d}$$

- Biodegradable volatile suspended solids (they are not added to the mass balance as they are already included in the influent BOD. They will be stabilised, causing the generation of biological solids – they are just used to compute P_{xnb}):

$$P_{xb} = (\text{SS}_b/\text{VSS in the raw sewage}) \cdot P_{xv} = 0.60 \times 1,190 = 714 \text{ kgSS}_b/\text{d}$$

- Non-biodegradable volatile suspended solids:

$$P_{xnb} = P_{xv} - P_{xb} = 1,190 - 714 = 476 \text{ kgSS}_{nb}/\text{d}$$

- Inorganic suspended solids (non-volatile):

$$P_{xi} = P_x - P_{xv} = 1,488 - 1,190 = 298 \text{ kgSS}_i/\text{d}$$

(b)   Biological solids generated in the reactor

- Volatile suspended solids produced:

$$\text{Produced } P_{xv} = Y \cdot S_r = 0.6 \times 2,308 \text{ kgBOD}_5/\text{d} = 1,385 \text{ kgVSS/d}$$

- Total suspended solids produced:

$$\text{Produced } P_x = P_{xv}/(\text{VSS/SS ratio in the generation of solids})$$
$$= 1,385/0.9 = 1,539 \text{ kgSS/d}$$

- Inorganic suspended solids produced:

$$\text{Produced } P_{xi} = \text{Produced } P_x - \text{Produced } P_{xv} = 1,539 - 1,385$$
$$= 154 \text{ kgSS}_i/\text{d}$$

- Biodegradable suspended solids produced:

$$\text{Produced } P_{xb} = f_b \cdot \text{produced } P_{xv} = 0.73 \times 1,385 = 1,011 \text{ kgSS}_b/\text{d}$$

- Non-biodegradable suspended solids produced:

$$\text{Produced } P_{xnb} = \text{Produced } P_{xv} - \text{Produced } P_{xb} = 1{,}385 - 1{,}011$$

$$= 374 \text{ kgSS}_{nb}/d$$

- Biodegradable suspended solids destroyed in the endogenous respiration:

$$\text{Destroyed } P_{xb} = \text{Produced } P_{xb} \cdot (K_d \cdot \theta_c)/(1 + f_b \cdot K_d \cdot \theta_c)$$

$$= 1{,}011 \times (0.08 \times 6)/(1 + 0.73 \times 0.08 \times 6)$$

$$= 359 \text{ kgSS}_b/d$$

- Remaining biodegradable suspended solids (net production):

$$\text{Net } P_{xb} = \text{Produced } P_{xb} - \text{Destroyed } P_{xb} = 1{,}011 - 359 = 652 \text{ kgSS}_b/d$$

- Remaining volatile suspended solids (net production):

$$\text{Net } P_{xv} = \text{Net } P_{xb} + \text{Produced } Px_{nb} = 652 + 374 = 1{,}026 \text{ kgVSS/d}$$

(c)   Summary of the reactor

Total production = Input from the influent sewage + Production of biological solids in the reactor

- Inorganic suspended solids:

$$P_{xi} = 298 + 154 = 452 \text{ kgSS}_i/d$$

- Non-biodegradable suspended solids:

$$P_{xnb} = 476 + 374 = 850 \text{ kgSS}_{nb}/d$$

- Biodegradable suspended solids:

$$P_{xb} = 0 + 652 = 652 \text{ kgSS}_b/d$$

- Volatile suspended solids:

$$P_{xv} = P_{xnb} + P_{xb} = 850 + 652 = 1{,}502 \text{ kgVSS/d}$$

- Total suspended solids:

$$P_x = P_{xv} + P_{xi} = 1{,}502 + 452 = \mathbf{1{,}954 \text{ kgSS/d}}$$

- Resultant VSS/SS ratio:
  VSS/SS $= 1{,}502/1{,}954 = \mathbf{0.77}$ (This value matches with the initially adopted value – see Section 5.2.b. If it had been substantially different, the initial value should be altered, and the calculations re-done)

- Ratio SS produced by BOD_5 removed:
  SS/$S_r = 1{,}954/2{,}308 = 0.85$ kgSS/kgBOD$_5$ rem (a value very close to the value of Section 2.19, Table 2.6)

The significant contribution represented by the solids of the raw sewage can be observed. To design the sludge treatment stage, the solids removed from the primary sedimentation tank (primary sludge) should be added to these values calculated for the solids produced in the reactor (secondary sludge).

5.3.7 Reactor volume

According to Equation 2.4 (S_r is the BOD load removed – see Section 2.4):

$$V = \frac{Y \cdot \theta_c \cdot S_r}{X_v \cdot (1 + f_b \cdot K_d \cdot \theta_c)} = \frac{0.6 \times 6 \times 2,308 \times 1,000}{3,000 \times (1 + 0.73 \times 0.08 \times 6)} = 2,051 \, m^3$$

Number of reactors to be used: **2**

Volume of each reactor: $V_1 = 2,051/2 = 1,026 \, m^3$
Depth: 4.0 m
Area required: $1,026/4.0 = 257 \, m^2$

Dimensions: length **L = 32.0 m**; width **B = 8.0 m**

The ratio $L/B = 32.0/8.0 = 4$ allows, in this example, the symmetrical allocation of four aerators. Reactors with a different number of aerators should have different ratios.

- Hydraulic detention time:

$$t = V/Q = 2,051/9,820 = 0.21 \, d = 5.0 \text{ hours}$$

- Substrate utilisation rate U:

$$U = \frac{S_r}{X_v \times V} = \frac{2,308 \times 1,000}{3,000 \times 2,051} = 0.38 \text{ kgBOD}_5/\text{kgMLVSS} \cdot d$$

- F/M ratio:

$$F/M = \frac{\text{Influent BOD load to the reactor}}{X_v \cdot V} = \frac{2,345 \times 1,000}{3,000 \times 2,051}$$

$$= 0.38 \text{ kgBOD}_5/\text{kgMLVSS} \cdot d$$

5.3.8 Excess sludge removal

Total SS produced (influent + produced in the reactor) (see Section 5.3.6.c) = 1,954 kgSS/d
SS leaving with the final effluent = $Q \cdot SS_{effluent}/1,000 = 9,820 \times 30/1,000 = 295$ kgSS/d
SS to be removed from the system = total SS − SS effluent = $1,954 − 295 = 1,659$ kgSS/d

(a)   Option: direct removal from the reactor

Concentration: MLSS = MLVSS/(VSS/SS) = 3,000/0.77 = 3,896 mg/L
Volume to be removed per day: $Q_{ex} = $ load/concentration = $1,659 \times 1,000/3,896 = $ **426 m^3/d**

(b) Option: removal from the sludge return line

Concentration: RASS = MLSS. $(1 + 1/R) = 3{,}896 \times (1 + 1/1) = 7{,}792$ mg/L
Volume to be removed per day: $Q_{ex} = $ load/concentration $ = 1{,}659 \times 1{,}000/$
   $7{,}792 = \mathbf{213\ m^3/d}$

Note that, with the return sludge ratio of $R = 1$, the excess sludge flow is double and the SS concentration is half, when the sludge is directly removed from the reactor, compared with the removal from the return sludge line. The solids load to be removed is, naturally, the same.

5.3.9 Oxygen requirements

(a) O_2 requirements in the field

See equations in Section 2.16.
$a' = (BOD_u/BOD_5) \quad - \quad (BOD_u/X_b) \cdot Y = 1.46 - 1.42 \times 0.6 = 0.608 \quad kgO_2/$
   $kgBOD_5$
$b' = (BOD_u/X_b) \cdot f_b \cdot K_d = 1.42 \times 0.73 \times 0.08 = 0.083\ kgO_2/kgVSS$

Demand for synthesis: $a' \cdot S_r = 0.608 \times 2{,}308 = 1{,}403\ kgO_2/d$
Demand for endogenous respiration: $b' \cdot X_v \cdot V = 0.083 \times 3{,}000 \times 2{,}051/1{,}000 =$
   $511\ kgO_2/d$
Demand for nitrification: $1{,}344\ kgO_2/d$ (see Item 'b' below – nitrification)
Saving with denitrification: $0\ kgO_2/d$ (there is no intentional denitrification in the reactor)

- OR average: total demand (for Q_{av}) $= 1{,}403 + 511 + 1{,}344 - 0 = \mathbf{3{,}258\ kgO_2/d}$
- Total demand (for Q_{max}): $OTR_{field} = (Q_{max}/Q_{av}) \cdot OR_{av} = (19{,}212/9{,}820) \times 3{,}258 = 1.96 \times 3{,}258 = \mathbf{6{,}374\ kgO_2/d}$

Demand to be satisfied in the field: total demand for Q_{max}

Average, O_2 required per $kgBOD_5$ removed: $(1{,}403 + 511)/2{,}308 = 0.83\ kgO_2/$
$kgBOD_5$ (very similar to the value shown in Table 2.6)

OTR/influent BOD to reactor ratio $= 6{,}374/2{,}345 = 2.72\ kgO_2/kgBOD_5$

(b) Nitrification

Assume 100% efficiency in the nitrification.
Ammonia fraction in the excess sludge: $0.1\ kgTKN/kgVSS$ (assumed)
Influent TKN load to the reactor (Section 5.3.2): $397\ kgTKN/d$
TKN load in the excess sludge: $0.1 \cdot P_{xv}$ net $= 0.1 \times 1{,}026 = 103\ kgTKN/d$ (see
   Section 5.3.6.b)
TKN load to be oxidised $=$ influent TKN load – excess sludge TKN load $=$
   $397 - 103 = 294\ kgTKN/d$
Stoichiometric O_2 demand ratio for nitrification: $4.57\ kgO_2/kgTKN$

O_2 demand for nitrification: $4.57 \times 294 = 1,344\ kgO_2/d$ (this value is included in Item 'a' above)

Chapters 6 and 7 provide a more detailed calculation for the estimation of the oxidised TKN load, which should be preferably adopted. The example in Chapter 7 shows the calculation for nitrification, according to this method.

(c)   Correction for standard conditions

DO saturation concentration as a function of the temperature:

$$C_s = 14.652 - 0.41022 \times T + 0.007991 \times T^2 - 0.000077774 \times T^3$$

C_s in the coldest month (20 °C): $C_s = 9.02\ mg/L$
C_s in the warmest month (25 °C): $C_s = 8.18\ mg/L$

- Standard oxygen transfer rate (SOTR or $OTR_{standard}$) required in the coldest month (see Section 5.2.c for parameters):

$$OTR_{standard} = \frac{OTR_{field}}{\frac{\beta \cdot f_H \cdot C_s - C_L}{C_s(20\ °C)} \cdot \alpha \cdot \theta^{T-20}} = \frac{6,374}{\frac{0.9 \times 0.92 \times 9.02 - 1.0}{9.02} \times 0.85 \times 1.024^{20-20}}$$

$$= \frac{6,374}{0.610} = 10,449\ kgO_2/d$$

$$
\begin{aligned}
f_H =\ & \text{correction factor of } C_s \text{ for the altitude} \\
& (= 1 - \text{altitude}/9{,}450) = 1 - 800/9{,}450 = 0.92 \\
C_L =\ & \text{oxygen concentration to be maintained in the reactor} = \\
& 1.0\ mg/L \\
C_s(20\ °C) =\ & \text{oxygen saturation concentration in clean water, under standard conditions: } 9.02\ mg/L \\
T =\ & \text{temperature of the liquid} = 20\ °C
\end{aligned}
$$

- Standard oxygen transfer rate (SOTR or $OTR_{standard}$) required in the warmest month:

$$OTR_{standard} = \frac{OTR_{field}}{\frac{\beta \cdot f_H \cdot C_s - C_L}{C_s(20\ °C)} \cdot \alpha \cdot \theta^{T-20}} = \frac{6.374}{\frac{0.9 \times 0.92 \times 8.18 - 1.0}{9.02} \times 0.85 \times 1.024^{25-20}}$$

$$= \frac{6,374}{0.613} = 10,398\ kgO2/d$$

From the values for the coldest month (10,449 kgO_2/d) and the warmest month (10,398 kgO_2/d), the larger should be chosen. Thus:

$$OTR_{standard} = \mathbf{10,449\ kgO_2/d = 435\ kgO_2/hour}$$

5.3.10　Alternative: mechanical aeration

(a)　Required power

Oxygenation efficiency under standard conditions (low speed, fixed vertical shaft mechanical aerators): OE = 1.8 kgO$_2$/kWh

Required power: OTR$_{standard}$/OE$_{standard}$ = 435/1.8 = 242 kW (323 HP)

Number of aerators for each reactor: **4**

Total number of aerators: 2 ×4 = **8**

Power required for each aerator: 323/8 = 40.4 HP. Use eight aerators of **50 HP** each

Total installed power: 8 ×50 = **400 HP** (294 kW)

Power level = Power (kW)·1,000/V = 294 × 1,000/2,051 = 143 W/m^3

(power level installed, but not necessarily used)

Resultant OTR$_{standard}$ = Power × OE = 294 × 1.8 = 529 kgO$_2$/hour (12,696 kgO$_2$/d)

(b)　Resultant DO concentration

By rearranging the OTR$_{standard}$ and making C$_L$ explicit, the DO concentration in the tank is obtained, for the values of OTR$_{field}$ and the resultant OTR$_{standard}$. This calculation is done because the supplied aeration capacity is slightly higher than that required, since more power was provided for the aerators (294 kW) compared with the value required (242 kW).

- Concentration of DO during Q$_{av}$ (average flow)

Warmest month:

$$C_L = \beta \cdot f_H \cdot C_s - \left(\frac{OTR_{field}}{OTR_{standard}} \cdot \frac{Cs_{20}}{\alpha \cdot \theta^{T-20}} \right)$$

$$= 0.9 \times 0.92 \times 8.18 - \left(\frac{3,258}{12,696} \cdot \frac{9.02}{0.85 \times 1.024^{(25-20)}} \right)$$

$$= 4.35 \text{ mgO}_2/\text{L}$$

Coldest month:

$$C_L = \beta \cdot f_H \cdot C_s - \left(\frac{OTR_{field}}{OTR_{standard}} \cdot \frac{Cs_{20}}{\alpha \cdot \theta^{T-20}} \right)$$

$$= 0.9 \times 0.92 \times 9.02 - \left(\frac{3,258}{12,696} \cdot \frac{9.02}{0.85 \times 1.024^{(20-20)}} \right)$$

$$= 4.75 \text{ mgO}_2/\text{L}$$

To save energy, lower DO concentrations than this can be reached, if the aeration capacity is reduced by turning off the aerators or lowering the aerator submergence or speed. Reducing the oxygenation capacity can maintain DO in the desirable range of around 2.0 mg/L (see Section 5.3.1.c).

- Concentration of DO during Q_{max} (maximum flow)

Warmest month:

$$C_L = \beta \cdot f_H \cdot C_s - \left(\frac{OTR_{field}}{OTR_{standard}} \cdot \frac{Cs_{20}}{\alpha \cdot \theta^{T-20}} \right)$$

$$= 0.9 \times 0.92 \times 8.18 - \left(\frac{6{,}374}{12{,}696} \cdot \frac{9.02}{0.85 \times 1.024^{(25-20)}} \right)$$

$$= 2.04 \text{ mgO}_2/\text{L}$$

Coldest month:

$$C_L = \beta \cdot f_H \cdot C_s - \left(\frac{OTR_{field}}{OTR_{standard}} \cdot \frac{Cs_{20}}{\alpha \cdot \theta^{T-20}} \right)$$

$$= 0.9 \times 0.92 \times 9.02 - \left(\frac{6{,}374}{12{,}696} \cdot \frac{9.02}{0.85 \times 1.024^{(20-20)}} \right)$$

$$= 2.14 \text{ mgO}_2/\text{L}$$

These DO values for Q_{max} are higher than the minimum allowable design value of 1.0 mg/L (Section 5.3.1.c), since the installed power is higher than that required. If the aerators had been adopted with a power identical to that required, the preceding calculations would have led to a DO concentration in the warmest month (in the present case, the most critical month) equal to 1.0 mg/L.

5.3.11 Alternative: aeration by diffused air

- Theoretical amount of air required in the field:

$$R_{air} \text{ theoretical}$$

$$= \frac{OTR_{standard}}{\text{specific gravity air (20 °C, altit.0 m)} \times \text{fraction O}_2 \text{ air (by weight)}}$$

$$= \frac{10{,}449 \text{ kgO}_2/\text{d}}{1.2 \text{ kg/m}^3 \times 0.23 \text{ gO}_2/\text{g air}} = 37.859 \text{ m}^3\text{air/d}$$

- Actual amount of air required (including O_2 transfer efficiency):
For an efficiency of 15% (see Section 5.3.1.c):

$$R_{air} \text{ actual} = R_{air} \text{ theoretical/efficiency} = 37{,}859/0.15$$

$$= 252{,}393 \text{ m}^3 \text{ air/d}$$

- Quantity of air to be used (with safety factor):
  Apply a safety factor. Metcalf and Eddy (1991) suggest a value of 2 for sizing the blowers. Since the current calculation has already been made computing the oxygen demand for maximum flow, a lower value of the safety factor could be adopted (say, 1.5).

  Adopted R_{air} = actual R_{air} × safety factor = 252,393 × 1.5 = 378,590 m^3 air/d (= 265 m^3 air/min) (= 4.4 m^3/s)
- Energy requirements:
  Assume that the head loss in the air piping (ΔH) is 0.4 m. In a real design, the head loss ΔH should be calculated along the air distribution system.

$$P = \frac{Q_g \cdot \rho \cdot g \cdot (d_i + \Delta H)}{\eta} = \frac{4.4 \times 1,000 \times 9.81 \times (4.0 + 0.4)}{0.60}$$

$$= 316,536 \ W = 317 \ kW = 431 \ HP$$

- Resultant oxygenation efficiency:

$$EO = \frac{OTR_{standard} \times \text{Safety factor}}{P} = \frac{435 \ kgO_2/h \times 1.5}{317 \ kW}$$

$$= 2.06 \ kgO_2/kW \cdot hour$$

- Resultant DO concentrations:

  Follow the methodology used in 5.3.10.b.

Note: In the diffused air alternative, a larger depth can be adopted for the reactor (5 to 6 m), thus optimising the transfer of oxygen and reducing the area required.

5.3.12 Area required for the secondary sedimentation tank

Use equations presented in Chapter 4.

(a)   Input data

Q = 9,820 m^3/d = 409 m^3/hour
Q_{max} = 19,212 m^3/d = 801 m^3/hour
Q_r = 9,820 m^3/d = 409 m^3/hour
MLSS = 3896 mg/L = 3.896 kg/m^3

(b)   Surface area required based on the simplified limiting flux theory

The relevant coefficients, for *fair–poor* settleability, are (see Section 5.3.2·d):

$$v_o = 7.40 \ m/hour; \ K = 0.59 \ m^3/kg; \ m = 7.34; \ n = 0.71$$

Q/A for the *clarification* criteria:

$$Q/A = v_0 \cdot e^{-K \cdot C} = 7.40 \cdot e^{-0.59 \times 3.896} = 0.743 \text{ m/hour}$$

Q/A for the *thickening* criteria:

$$Q/A = \frac{m \cdot \left[R \cdot v_0 \cdot e^{(-K \cdot C_o)} \right]^n}{(R+1) \cdot C_o} = \frac{7.34 \cdot \left[1.0 \times 7.40 \cdot e^{(-0.59 \times 3.896)} \right]^{0.71}}{(1.0+1) \times 3.896}$$

$$= 0.763 \text{ m/hour}$$

Using the smallest of the Q/A values (0.743 m/hour for clarification and 0.763 m/hour for thickening) and knowing that $Q = 409$ m^3/hour:

$$A = 409/0.743 = \textbf{550 m}^2$$

5.3.13 Alternative: circular secondary sedimentation tanks

(a) Diameter

Number of sedimentation tanks to be used: **2**
Area required for each sedimentation tank: $550/2 = 275$ m^2
Diameter:

$$D = \sqrt{\frac{4A}{\pi}} = \sqrt{\frac{4 \times 275}{3.14}} = 18.7 \text{ m} \qquad \text{Adopt \textbf{19.0 m}.}$$

Resultant area of each sedimentation tank: $A = \pi \cdot D^2/4 = 3.14 \times 19.0^2/4 = 283$ m^2
Total resultant area: $2 \times 283 = 566$ m^2

(b) Resultant loading rates

- Resultant hydraulic loading rate with Q_{av}: HLR $= Q/A = 409/566 = 0.72$ m^3/m$^2 \cdot$hour
- Resultant hydraulic loading rate with Q_{max}: HLR $= Q_{max}/A = 801/566 = 1.42$ m^3/m$^2 \cdot$hour
- Resultant solids loading rate with Q_{av}: SLR $= (Q+Q_r) \cdot X/A = (409 + 409) \times 3.896/566 = 5.6$ kgSS/m$^2 \cdot$hour
- Resultant solids loading rate with Q_{max}: SLR $= (Q_{max}+Q_r) \cdot X/A = (801 + 409) \times 3.896/566 = 8.3$ kgSS/m$^2 \cdot$hour

All the loading rates are within typical ranges reported by Metcalf and Eddy (1991) and WEF/ASCE (1992) (see Table 4.1).

(c) Other dimensions

Sidewater depth (cylindrical part of the tank): **H = 3.5 m** (adopted)

Bottom slope: 8% (= 1/12 vertical/horizontal)

Depth of the conical part of the tank: $H_{cone} = (D/2) \cdot (slope/100) = (19.0/2) \times (8/100) = 0.76$ m

Volume of each sedimentation tank:

$$V = A \cdot (H + H_{cone}/3) = 283 \times (3.5 + 0.76/3) = 1,064 \text{ m}^2$$

Total volume of the sedimentation tanks: $2 \times 1,064 = 2,128$ m^3

(d) Hydraulic detention time

For average flow + recirculation: $t = V/(Q + Q_r) = 2,128/(409 + 409) = 2.6$ hours

For maximum flow + recirculation: $t = V/(Q_{max} + Q_r) = 2,128/(801 + 409) = 1.7$ hours

(e) Effluent weir

Available weir length (for each sedimentation tank; assume that the crest of the weir is 0.5 m from the side wall, into the sedimentation tank):

$$L_{weir} = \pi \cdot (D - 2 \cdot distance) = 3.14 \times (19.0 - 2 \times 0.5) = 56.5 \text{ m}$$

Resultant weir loading rate in each of the 2 sedimentation tanks:

For influent Q_{av}: Weir rate = $(Q/2)/L_{weir} = (409/2)/56.5 = 3.6$ m^3/m·hour

For influent Q_{max}: Weir rate = $(Q_{max}/2)/L_{weir} = (801/2)/56.5 = 7.1$ m^3/m·hour

These rates are within recommended values (Table 4.7)

5.3.14 Alternative: rectangular secondary sedimentation tanks

Number of sedimentation tanks: **4**

Area required for each sedimentation tank: $550/4 = 138$ m^2

Dimensions:

Depth: **H = 4.0 m**

Length: **L = 20.0 m**

Width: **B = 6.9 m**

Other calculations: similar approach to the circular sedimentation tanks

5.3.15 Primary sedimentation tanks

The primary sedimentation tanks can be designed based on the loading rates and criteria presented in Section 4.4. The sizing of the tanks is similar to that presented in the current example (Sections 5.3.13 and 5.3.14).

5.4 SUMMARY OF THE DESIGN

Characteristics of the influent

Influent	Characteristic	Item	Value
Raw sewage	Flow (m^3/d)	Average	9,820
		Maximum	19,212
		Minimum	4,003
	Flow (l/s)	Average	114
		Maximum	222
		Minimum	46
	Average concentration (mg/L)	BOD_5	341
		TKN	51
		SS	379
	Average load (kg/d)	BOD_5	3,350
		TKN	496
		SS	3,720
Settled sewage (effluent from primary sedimentation tanks and influent to biological stage)	Average concentration (mg/L)	BOD_5	239
		TKN	40
		SS	152
	Average load (kg/d)	BOD_5	2,345
		TKN	397
		SS	1,488

Biological reactors (alternative: mechanical aeration)

Characteristic	Value
Sludge age (d)	6
Adopted MLVSS (mg/L)	3,000
Resultant MLSS (mg/L)	3,896
Dimensions	
• Total volume of the reactors (m^3)	2,051
• Number of reactors (−)	2
• Length (m)	32.0
• Width (m)	8.0
• Depth (m)	4.0
Average detention time (hours)	5.0
Return sludge ratio (−)	1.0
Aeration	
• O_2 requirements (field) (for Q_{av}) (kgO_2/d)	3,258
• O_2 requirements (field) (for Q_{max}) (kgO_2/d)	6,374
• O_2 requirements (standard) (kgO_2/d)	10,449
• Oxygenation efficiency (kgO_2/kWh)	1.8
• Required power (HP)	323
• Number of aerators (−)	8
• Power of each aerator (HP)	50
• Total installed power (HP)	400

(*Continued*)

Characteristic	Value
Excess sludge	
Removal from the reactor	
• Flow (m³/d)	426
• Load (kgTSS/d)	1,659
• Concentration (mgTSS/L)	3,896
Removal from the return sludge line	
• Flow (m³/d)	213
• Load (kgTSS/d)	1,659
• Concentration (mgTSS/L)	7,792
Estimated concentrations in the final effluent (mg/L)	
BOD$_5$	20
SS	30

Secondary sedimentation tanks (alternative: circular sedimentation tanks)

Characteristic	Value
Dimensions	
• Number of sedimentation tanks	2
• Diameter of each sedimentation tank (m)	19.0
• Sidewater depth (m)	3.5
• Depth of the conical part (m)	0.76
• Bottom slope (vertical/horizontal) (%)	8
• Total resultant volume of the sedimentation tanks (m³)	2.128
Detention times (hour)	
• For $Q_{av} + Q_r$	2.6
• For $Q_{max} + Q_r$	1.7
Resultant average loading rates	
• Hydraulic loading rate (m³/m²·hour)	0.7
• Solids loading rate (kg/m²·hour)	5.6
• Weir loading rate (m³/m·hour)	3.6

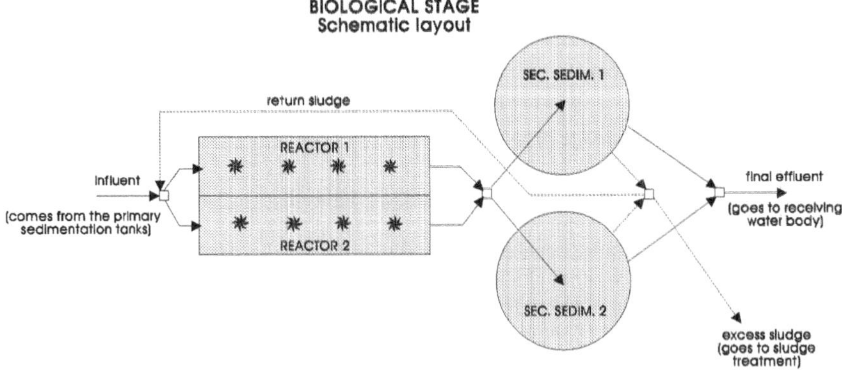

6

Principles of biological nutrient removal

6.1 INTRODUCTION

Biological nutrient removal (BNR) is a topic that is proving increasingly important in the design of activated sludge systems. The nutrients of interest, in this case, are *nitrogen* and *phosphorus*. In many regions, BNR is being used in a systematic way in new projects, and existing treatment plants are being converted to enable the occurrence of BNR.

Naturally, the need or desirability to have nitrogen and phosphorus removal depends on a broader view of the treatment objectives and the final effluent and receiving-body water quality. In *sensitive bodies*, such as lakes, reservoirs and estuaries subject to eutrophication problems, BNR assumes a great importance. The discharge and water-body standards can influence the decision on whether nutrient removal is needed and to what degree it should be performed.

The European Community's guidelines (CEC, 1991) for discharge into sensitive water bodies, that is, subject to eutrophication, establish the following limits:

Total phosphorus:

- populations between 10,000 and 100,000 inhabitants: concentration of less than 2 mg/L or minimum removal of 80%

- populations above 100,000 inhabitants: concentration of less than 1 mg/L or minimum removal of 80%

Total nitrogen:

- populations between 10,000 and 100,000 inhabitants: concentration of less than 15 mg/L or minimum removal of 70–80%
- population above 100,000 inhabitants: concentration of less than 10 mg/L or minimum removal of 70–80%

When analysing the desirability of incorporating BNR, following a trend observed in more developed countries, a scale of priorities should always be kept in mind. Many of the developed countries have already solved most of the problems of carbonaceous matter (BOD and COD) in their effluents and now need to move to a second stage of priorities, which concerns BNR. In developing countries, there is still the need to solve the basic problems of carbonaceous matter and pathogenic organisms, obviously without losing the perspective of applying, whenever necessary, nutrient removal.

Besides the aspects of the receiving body, the inclusion of intentional nutrient removal can lead to an improvement in the operation of the WWTP. In the case of nitrogen removal, there are savings on oxygen and alkalinity, besides the reduction of the possibility of having rising sludge in the secondary sedimentation tanks.

This chapter focuses on the *basic principles* of the following topics associated with biological nutrient removal:

- *nitrification* (oxidation of ammonia to nitrite and then to nitrate)
- *denitrification* (conversion of nitrate into gaseous nitrogen)
- *phosphorus removal* (biological phosphorus removal)

It should be stressed that nitrification does not result in the removal of nitrogen, but only in a conversion in its form from ammonia to nitrate. Thus, nitrification should be understood as removal of ammonia, but not of nitrogen. Nitrification takes place almost systematically in activated sludge plants operating in warm-climate conditions. Thus, the design should take its occurrence into consideration, mainly in the estimation of the oxygen requirements. The design example presented in Chapter 5 was based on this assumption. In the main conversion route of N, for denitrification to occur, it is necessary that nitrification occurs first (there are other routes not covered in this book).

The accumulated knowledge and operational experience in this area is already high, and the designs can be made with a satisfactory degree of reliability. Presently, research efforts are made mainly to understand the interaction among the various microorganisms involved and how they affect plant operation (e.g., sedimentation), as well as to produce reliable mathematical models for the process, mainly in the case of phosphorus.

The *design* of *BNR* systems is dealt with in Chapter 7. It should be noted that the aim of this chapter is only to introduce the main aspects of BNR, and not to discuss it thoroughly, because of the wide amplitude of the theme. The books by Barnes and Bliss (1983), WRC (1984), Sedlak (1991), Randall *et al.* (1992), EPA (1987b, 1993) and Orhon and Artan (1994) and the reports by the International Water Association (IAWPRC, 1987; IAWQ, 1995, IWA, 2000) are excellent and specific literature on biological nutrient removal.

6.2 NITROGEN IN RAW SEWAGE AND MAIN TRANSFORMATIONS IN THE TREATMENT PROCESS

The nitrogen present in raw sewage, as well as the processes that occur in interaction with the biomass, can be characterised as illustrated in Figure 6.1.

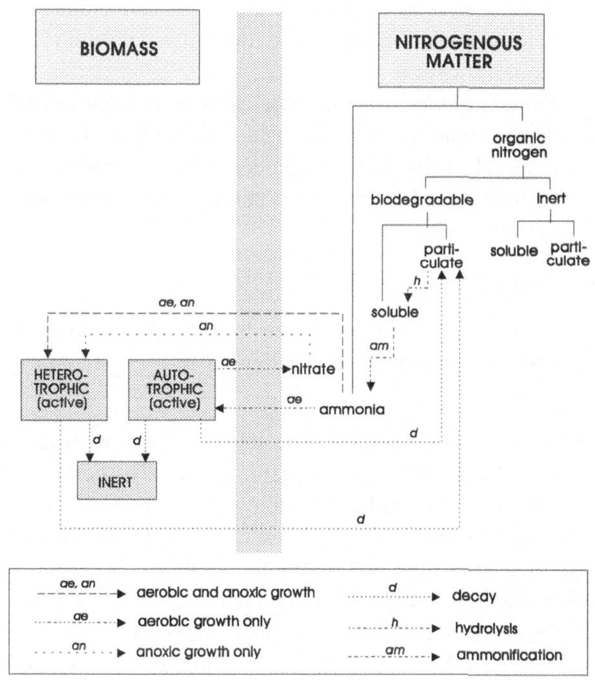

Figure 6.1.   Subdivisions and transformations of the nitrogenous matter in the activated sludge process

(a)   Characterisation of the nitrogenous matter

- The **inorganic nitrogenous matter** is represented by *ammonia*, both in a free (NH_3) and in an ionised form (NH_4^+). In reality, ammonia is present in the influent wastewater due to the fact that hydrolysis and ammonification reactions begin in the collection and interception sewerage system, as described below. Ammonia is used by the heterotrophic and autotrophic bacteria.
- The **organic nitrogenous matter** is also divided, in a similar way to the carbonaceous matter, into two fractions in terms of its biodegradability: (a) *inert* and (b) *biodegradable*.
    - *Inert*. The inert fraction is divided into two fractions, based on its physical state:
        - *Soluble*. This fraction is usually negligible and can be disregarded.
        - *Particulate*. This fraction is associated with the non-biodegradable carbonaceous organic matter, involved in the biomass and removed with the excess sludge.
    - *Biodegradable*. The biodegradable fraction can be subdivided into the following components:
        - *Rapidly biodegradable*. The quickly biodegradable nitrogenous organic matter is found in a *soluble* form, and is converted by heterotrophic bacteria into ammonia, through the process of ammonification.
        - *Slowly biodegradable*. The slowly biodegradable nitrogenous organic matter is found in a *particulate* form, being converted into a soluble form (quickly biodegradable) through *hydrolysis*. This hydrolysis takes place in parallel with the hydrolysis of the carbonaceous matter.

(b)   Characterisation of the biomass

The **active biomass** is responsible for the biological degradation. In terms of the carbon source, the biomass can be divided into (a) *heterotrophic* and (b) *autotrophic*:

- *Heterotrophic active biomass*. The source of carbon of the heterotrophic organisms is the carbonaceous organic matter. The heterotrophic biomass uses the rapidly biodegradable soluble carbonaceous matter. Part of the energy associated with the molecules is incorporated into the biomass, while the rest is used to supply the energy for synthesis. In aerobic treatment, the growth of the heterotrophic biomass occurs in aerobic (use of oxygen as electron acceptors) or anoxic (absence of oxygen, with the use of nitrate as electron acceptors) conditions. This growth is very low in anaerobic conditions (absence of oxygen and nitrate). Heterotrophic bacteria use the nitrogen in the form of ammonia for synthesis (in aerobic and anoxic conditions) and the nitrogen in the form of nitrate as an electron acceptor (in anoxic conditions). The decay of the heterotrophic biomass also generates, besides the inert residue, carbonaceous and nitrogenous matter of slow degradation. This material needs to subsequently undergo a hydrolysis process to become a rapidly biodegradable matter, which can be used again by the heterotrophic and autotrophic biomass.

- *Autotrophic active biomass.* The source of carbon for the autotrophic organisms is carbon dioxide. The autotrophic biomass uses ammonia as an energy source (they are chemoautotrophic organisms, that is, that use *inorganic* material as an energy source). Under aerobic conditions, these bacteria use ammonia in the nitrification process, in which ammonia is converted into nitrite and subsequently into nitrate. Similar to that for the heterotrophic organisms, the decay of the autotrophic biomass also generates, besides the inert residue, carbonaceous and nitrogenous matter of slow degradation. This material needs to subsequently undergo a hydrolysis process to become a rapidly biodegradable material, which can be used again by the heterotrophic and autotrophic biomass.

The **inert residue** is formed by the decay of the biomass involved in the wastewater treatment. Biomass decay can occur by the action of several mechanisms, which include endogenous metabolism, death, predation and others. As a result, products of slow degradation are generated, as well as particulate products, which are inert to biological attack.

As mentioned, the microorganisms involved in the **nitrification** process are *chemoautotrophs*, for which carbon dioxide is the main source of carbon, and energy is obtained through the oxidation of an inorganic substrate, such as ammonia, into mineralised forms.

The transformation of ammonia into nitrites is accomplished by bacteria, such as those from the genus *Nitrosomonas*, according to the following reaction:

$$2NH_4^+\text{-}N + 3O_2 \xrightarrow{\text{\textit{Nitrosomonas}}} 2NO_2^-\text{-}N + 4H^+ + 2H_2O + \text{Energy}$$

$$(6.1)$$

The oxidation of nitrites into nitrates occurs by the action of bacteria, such as those from the genus *Nitrobacter*, expressed by:

$$2NO_2^-\text{-}N + O_2 \xrightarrow{\text{\textit{Nitrobacter}}} 2NO_3^-\text{-}N + \text{Energy} \qquad (6.2)$$

The **overall nitrification reaction** is the sum of Equations 6.1 and 6.2:

$$NH_4^+\text{-}N + 2O_2 \longrightarrow NO_3^-\text{-}N + 2H^+ + H_2O + \text{Energy} \qquad (6.3)$$

In reactions 6.1 and 6.2 (as well as in the overall reaction 6.3), the following points should be noted:

- *consumption of oxygen*. This consumption is generally referred to as *nitrogenous demand*
- *release of H^+*, consuming the alkalinity of the medium and possibly reducing the pH

The energy liberated in these reactions is used by the nitrifying microorganisms in the synthesis of organic compounds from inorganic carbon sources, such as carbon dioxide, bicarbonate and carbonate. Therefore, nitrification is intimately associated with the growth of nitrifying bacteria (Barnes and Bliss, 1983).

The growth rate of the nitrifying microorganisms, mainly *Nitrosomonas*, is very slow, and much lower than that of the microorganisms responsible for the conversion of the carbonaceous matter. Thus, in a biological treatment system where nitrification is desired, the mean cell residence time, or sludge age, should be such that it enables the development of the nitrifying bacteria, before they are washed out from the system. The system is controlled, therefore, by the organism with the slowest growth rate, in this case, *Nitrosomonas*. The bacteria of the *Nitrobacter* genus have a faster growth rate and, for this reason, there is practically no accumulation of nitrites in the system.

In anoxic conditions (absence of oxygen but presence of nitrates), the nitrates are used by heterotrophic microorganisms as electron acceptors, as a replacement for oxygen. In this process named **denitrification**, the *nitrate is* reduced to *gaseous nitrogen*, according to the following reaction:

$$2NO_3^- \text{-N} + 2H^+ \longrightarrow N_2 + 2.5O_2 + H_2O \qquad (6.4)$$

In the denitrification reaction the following should be noted:

- *economy of oxygen* (the organic matter can be stabilised in the absence of oxygen)
- *consumption of H^+*, implying savings in alkalinity and an increase in the buffer capacity of the medium

When representing stoichiometric relations, the difference between, for example, NH_4^+ and NH_4^+-N, should be clearly distinguished. The first form expresses the concentration of the ammonium ion, while the second represents the nitrogen in the form of the ammonium ion. The molecular weights vary, as shown below:

NH_4^+ : molecular weight $= 18$ g/mol
NH_4^+-N : molecular weight $= 14$ g/mol ($=$ molecular weight of N)

The second form is more convenient because it allows comparisons among relations, based always on nitrogen, irrespective of whether it is in the organic, ammonia, nitrite or nitrate forms. In this book, including the equations and chemical reactions, the use of the second concept is implied, that is, the nitrogen forms are expressed, in terms of mass, as nitrogen.

6.3 PRINCIPLES OF NITRIFICATION

6.3.1 Kinetics of nitrification

The growth rate of the nitrifying bacteria can be expressed in terms of Monod's relation as follows:

$$\mu = \mu_{max} \cdot \left[\frac{NH_4^+}{K_N + NH_4^+} \right] \tag{6.5}$$

where:

μ = specific growth rate of the nitrifying bacteria (d^{-1})

μ_{max} = maximum specific growth rate of the nitrifying bacteria (d^{-1})

NH_4^+ = ammonia concentration, expressed in terms of nitrogen (mg/L)

K_N = half-saturation constant (mg/L)

For further details concerning Monod's kinetics, see Chapter 8. For simplification purposes in the model structure, nitrification is assumed to take place in a single stage (ammonia-nitrate), instead of two stages (ammonia-nitrite, nitrite-nitrate). Typical values of the kinetic and stoichiometric coefficients for the unified nitrifying biomass are shown in Table 6.1.

The value of K_O (oxygen) in Table 6.1 can be explained by the fact that oxygen is also a limiting factor in the growth of nitrifying bacteria, and could also be expressed by Monod's relation (see Section 6.3.2.c). θ is the temperature coefficient for the correction of the growth rate of the nitrifying bacteria (see Section 6.3.2.a). Y_N is the yield coefficient, which indicates the mass of nitrifying bacteria that is produced per unit mass of ammonia used (see Section 6.3.4). K_d is the bacterial decay coefficient and is frequently ignored in modelling, because of its low value and the fact that most of the growth rates reported in the literature have been calculated without taking K_d into account (Randall *et al.*, 1992; EPA, 1993).

Frequently, ammonia is replaced in stoichiometric relations by TKN (total Kjeldahl nitrogen), assuming that the organic nitrogen will be transformed into

Table 6.1. Typical values of the kinetic and stoichiometric coefficients for nitrification (unified nitrifying biomass)

Coefficient	Unit	Wider range	Typical range or value
μ_{max}(20 °C)	d^{-1}	0.3–2.2	0.3–0.7
K_N (ammonia)	$mgNH_4^+/L$	0.1–5.6	0.5–1.0
K_O (oxygen)	mgO_2/L	0.3–2.0	0.4–1.0
θ	–	1.08–1.13	1.10
Y_N	mg cells/$mgNH_4^+$ oxidised	0.03–0.13	0.05–0.10
K_d	d^{-1}	0.04–0.16	≈ 0

Source: Arceivala (1981), Barnes and Bliss (1983), Sedlak (1991), Randall *et al.* (1992), EPA (1993) and Orhon and Artan (1994)

ammonia in the treatment line and that, for this reason, the influent TKN will be a good representation of the ammonia available for the nitrifying bacteria. The conversion of organic nitrogen into ammonia is nearly total, even with reduced sludge ages. This adaptation is used mainly when calculating oxygen (Section 6.3.5) and alkalinity (Section 6.3.6) requirements, leading to safer estimates.

Example 6.1

Calculate the growth rate of the nitrifying bacteria in a complete-mix reactor based on the following data:

- Desired effluent TKN $= 2.0$ mg/L (this concentration will also be prevalent in the whole reactor, since there are complete-mix conditions)
- $\mu_{max} = 0.5$ d^{-1} (adopted – Table 6.1)
- $K_N = 0.7$ mg/L (adopted – Table 6.1)

Solution:

According to Equation 6.5, the growth rate of the nitrifying bacteria, under ideal conditions and at a temperature of 20 °C, is:

$$\mu = \mu_{max} \cdot \left[\frac{NH_4{}^+}{K_N + NH_4{}^+} \right] = 0.5 \cdot \left[\frac{2.0}{0.7 + 2.0} \right] = 0.37 \text{ d}^{-1}$$

Thus, the specific growth rate with the TKN concentration in the reactor of 2.0 mg/L is 0.37 d^{-1}. If the TKN in the reactor were still lower, for example 1.0 mg/L, μ would be still more reduced, and reach 0.29 d^{-1}. The lower the μ, the greater the sludge age should be, so that the nitrifying organisms would have conditions to develop without being washed out from the system. In this example, an arbitrary concentration of effluent TKN was selected, without taking into consideration the influent TKN. The removal of TKN according to the operational conditions in the reactor is discussed in Section 6.3.4.

6.3.2 Environmental factors of influence on nitrification

The following environmental factors influence the growth rate of the nitrifying organisms and, as a consequence, the oxidation rate of ammonia:

- temperature
- pH
- dissolved oxygen
- toxic or inhibiting substances

(a) Temperature

Temperature significantly affects the maximum growth rate (μ_{max}) of the nitrifying organisms. According to Downing (1978), the effect of temperature can be described as follows:

$$\mu_{max(T)} = \mu_{max(20\,^\circ C)} \cdot \theta^{(T-20)} \tag{6.6}$$

where:

$\mu_{max(T)}$ = maximum growth rate at a temperature T (d^{-1})
θ = temperature coefficient
T = temperature $(^\circ C)$

The temperature coefficient θ is reported in a range from 1.08 to 1.13, and the value of 1.10, supported by a large number of data, seems reasonable (Barnes and Bliss, 1983). Thus, for each increment of approximately 7 °C in the temperature, the growth rate doubles and, conversely, each drop of 7 °C implies a reduction in the growth rate by half.

The half-saturation coefficients K_N and K_O also increase with an increase in temperature, although the data available in the literature are not conclusive. EPA (1993) and Orhon and Artan (1994) suggest adopting a constant value for the half-saturation coefficients, irrespective of the temperature.

The occurrence of nitrification was observed in a range from 5 to 50 °C, but the optimal temperature is in the order of 25 to 36 °C (Arceivala, 1981; Barnes and Bliss, 1983).

(b) pH

According to Downing (1978), the nitrification rate is at its optimal and approximately constant in the pH range from 7.2 to 8.0. Below 7.2, μ_{max} decreases with pH according to the following relation:

$$\mu_{max(pH)} = \mu_{max}[1 - 0.83(7.2 - pH)] \tag{6.7}$$

where:

$\mu_{max(pH)}$ = maximum growth rate of the nitrifying bacteria for a given pH (d^{-1})
μ_{max} = maximum growth rate of the nitrifying bacteria at a pH of 7.2 (d^{-1})

Equation 6.7 has a validity range of pH from 6.0 to 7.2. For a stable performance, it is advisable to maintain the pH in the range from 6.5 to 8.0 (EPA, 1993).

It is important to know that nitrification is responsible for the decrease of pH and generates H^+ as a final product (see Equation 6.3 and Section 6.3.6). The decrease of the pH is a function of the buffer capacity of the medium or, in other words, of its alkalinity. This aspect can be of great importance for the adequate nitrification performance in an activated sludge system.

(c) Dissolved oxygen

Dissolved oxygen in the reactor is an indispensable pre-requisite for the occurrence of nitrification. It seems that the critical DO concentration, below which no nitrification is expected to occur, is around 0.2 mg/L (Barnes and Bliss, 1983). However, higher values should be maintained in the aeration tank to ensure that, in points where oxygen access is more difficult, such as inside the activated sludge flocs, a higher than critical concentration is maintained. Downing (1978) recommends that the DO concentration in the reactor should not be reduced to less than 0.5 mg/L. However, EPA (1993) recommends that a minimum DO of 2.0 mg/L is specified to avoid problems with the influent ammonia peaks.

The effect of the DO concentration on the specific growth rate can also be represented by Monod's kinetics, as follows:

$$\mu = \mu_{max} \cdot \left[\frac{DO}{K_O + DO} \right] \tag{6.8}$$

where:

DO = dissolved oxygen concentration in the reactor (mg/L)
K_O = half-saturation constant for oxygen (mg/L) (see Table 6.1)

The presence of oxygen is more important to nitrification than it is to the removal of carbonaceous matter. In the removal of the carbonaceous matter, the absorption phase, which precedes metabolism, can store energy in some way until oxygen becomes available again. In contrast, nitrification ceases the moment oxygen is reduced below the critical level. On the other hand, nitrification resumes very fast as soon as DO rises.

(d)   Toxic or inhibiting substances

Toxic substances can seriously inhibit the growth of nitrifying bacteria, mainly *Nitrosomonas*, which are more sensitive. A large list of inhibiting substances and products, expressed in terms of the percentage inhibition that they cause, is known. The references Sedlak (1991), Randall *et al.* (1992) and EPA (1993) provide lists including several of these compounds.

One of the aspects to be analysed in the planning of a WWTP receiving industrial effluent is the possible influence of these on nitrification. A pre-treatment in the industry may be often necessary.

Example 6.2

Calculate the specific growth rate of the nitrifying bacteria, according to the data from Example 6.1 ($\mu_{max} = 0.5$ d^{-1}), and under the following environmental conditions:

- temperature: T = 20 °C
- pH = 6.9
- DO = 2.0 mg/L
- absence of toxic or inhibiting substances

Example 6.2 (*Continued*)

Solution:

(a) Effect of ammonia concentration

$$\mu_{max} = 0.50\ d^{-1}$$

$$\mu = 0.37\ d^{-1}\ \text{(calculated in Example 6.1)}$$

μ_{max} correction factor $= 0.37/0.50 = 0.74$ (reduction of 26%)

(b) Temperature

According to Equation 6.6 and using $\theta = 1.10$:

$$\mu_{max(T)} = \mu_{max(20\ °C)}\cdot\theta^{(20-20)} = 0.50 \times 1.10^{(20-20)} = 0.50\ d^{-1}$$

μ_{max} correction factor $= 0.50/0.50 = 1.00$ (reduction of 0%) (unchanged, because the temperature is the same as the standard temperature)

(c) pH

According to Equation 6.7:

$$\mu_{max\,(pH)} = \mu_{max}[1 - 0.83\,(7.2 - pH)] = 0.5 \times [1 - 0.83 \times (7.2 - 6.9)]$$
$$= 0.38$$

μ_{max} correction factor $= 0.38/0.50 = 0.76$ (reduction of 24%)

(d) Dissolved oxygen

According to Equation 6.8 and Table 6.1:

$$\mu = \mu_{max}\cdot\left[\frac{DO}{K_O + DO}\right] = 0.5\cdot\left[\frac{2.0}{0.6 + 2.0}\right] = 0.38$$

μ_{max} correction factor $= 0.38/0.50 = 0.76$ (reduction of 24%)

(e) Combined effect of the environmental conditions

Multiple correction factor:

$$0.74 \times 1.00 \times 0.76 \times 0.76 = 0.43$$

The specific growth rate of the nitrifying bacteria under these environmental conditions is 43% of the maximum rate ($\mu = 0.43\mu_{max}$). Under these environmental conditions, μ is:

$$\mu = 0.43 \times \mu_{max} = 0.43 \times 0.50 = 0.22\ d^{-1}$$

Table 6.2. Minimum sludge age required for nitrification

Temperature of the liquid in the reactor (°C)	Minimum θ_c for complete nitrification (days)
5	12
10	9.5
15	6.5
20	3.5

Source: Arceivala (1981)

6.3.3 Sludge age required for nitrification

As mentioned, the reproduction rate of the nitrifying organisms is much smaller than that of the heterotrophic organisms responsible for the stabilisation of the carbonaceous matter. This suggests that the concept of sludge age is extremely important for nitrification to be achieved in the activated sludge process.

Nitrification will happen if the sludge age is such that it will allow the development of the nitrifying bacteria before they are washed out of the system. The sludge age is the reciprocal of the specific growth rate in an activated sludge system in equilibrium ($\theta_c = 1/\mu$). As the growth rate of the nitrifying bacteria is lower than that of the heterotrophic bacteria, the sludge age should be equal to or higher than the reciprocal of their growth rate to allow the nitrifying bacteria to develop, that is:

$$\theta_c \geq \frac{1}{\mu_N} \tag{6.9}$$

Thus, if the specific growth rate of the nitrifying bacteria is known, a minimum sludge age can be established to ensure proper nitrification.

Arceivala (1981) proposes that, for sewage without any specific inhibiting factors, the minimum sludge age values presented in Table 6.2 should be considered.

The required sludge age can also be calculated, if data are available, based on the value of μ determined according to the prevalent environmental conditions in the reactor, as described in the previous section and illustrated in Example 6.3.

Some authors still recommend including a safety factor in the order of 1.5 to 2.5 to cover the peaks in influent ammonia load and other unexpected environmental variations.

Example 6.3

Calculate the minimum sludge age required for nitrification to occur in the system described in Example 6.2. Data: $\mu = 0.22$ d^{-1} (as calculated in Example 6.2).

Solution:

According to Equation 6.9:

$$\theta_c \geq \frac{1}{\mu_N} = \frac{1}{0.22} = 4.5\,\text{d}$$

Example 6.3 (*Continued*)

Thus, a minimum sludge age of 4.5 days is required to ensure full nitrification. With a design safety factor of 1.5, the recommended sludge age will be $4.5 \times 1.5 = 6.8$ days.

For comparison purposes, if the temperature of the liquid were 10 °C (common in temperate-climate countries), the correction factor for the temperature would decrease from 1.00 (see Example 6.2) to 0.39. The overall correction factor would be 0.17 and the specific growth rate $0.5 \times 0.17 = 0.09$ d^{-1}. In these environmental conditions, the minimum sludge age required would be $1/0.09 = 11.1$ days, which, with a design safety factor of 1.5, would rise to 16.7 days. The great influence of a non-controllable variable, such as the temperature, on nitrification is observed, thus requiring larger sludge ages in cold climates. In tropical countries, the high temperatures greatly facilitate nitrification, which takes place almost systematically, even in conventional activated sludge systems, with a reduced sludge age.

6.3.4 Nitrification rate

Once the growth of the nitrifying bacteria is ensured by using a satisfactory sludge age based on the specific growth rate, it becomes necessary to calculate the nitrification rate, that is, the rate at which ammonia is converted into nitrate. The nitrification rate is a function of the mass of nitrifying organisms present in the aerated zones of the reactor and can be expressed as follows:

$\Delta TKN/\Delta t$ = (unitary nitrification rate) × (concentration of nitrifying bacteria)

$$\frac{\Delta TKN}{\Delta t} = \left(\frac{\mu_N}{Y_N}\right) \cdot X_N \tag{6.10}$$

where:

$\Delta TKN/\Delta t$ = nitrification rate (oxidised $gTKN/m^3 \cdot d$)

μ_N = specific growth rate of the nitrifying bacteria, determined based on μ_{max} and the environmental conditions (d^{-1})

Y_N = yield coefficient of the nitrifying bacteria ($gX_N/gTKN$)

X_N = concentration of the nitrifying bacteria in the aerated zones of the reactor (g/m^3)

Usually, it is preferable to express the concentration of the nitrifying bacteria in terms of the volatile suspended solids in the reactor. Therefore, it is necessary to determine which fraction of VSS is represented by the nitrifying bacteria. The fraction of nitrifying bacteria in the VSS (f_N) can be estimated through the relation between the growth rates (Barnes and Bliss, 1983):

$$f_N = \frac{\text{growth rate of nitrifying bacteria } (gX_N/m^3 \cdot d)}{\text{growth rate of the total biomass (heterotrophs and nitrifiers) } (gVSS/m^3 \cdot d)}$$

$$f_N = \frac{\Delta X_N / \Delta t}{\Delta X_V / \Delta t} \qquad (6.11)$$

The denominator of Equation 6.11 is the VSS production (P_{Xv}) and can be calculated by:

$$\frac{\Delta X_V}{\Delta t} = \frac{X_V}{\theta_c} \qquad (6.12)$$

The numerator of Equation 6.11, relative to the production of the mass of nitrifying bacteria, can be expressed as:

$$\frac{\Delta X_N}{\Delta t} = Y_N \cdot \left[TKN_{removed} - TKN_{incorporated\ in\ excess\ sludge} \right] = Y_N \cdot TKN_{oxidised}$$

$$(6.13)$$

The yield coefficient (Y_N) can be obtained from Table 6.1.

The fraction of TKN incorporated into the excess sludge is 12% of the VSS mass produced per day (N is 12% in mass of the composition of the bacterial cell, represented by $C_5H_7NO_2$; molecular weight of $N = 14\ g/mol$; molecular weight of $C_5H_7NO_2 = 5 \times 12 + 7 \times 1 + 1 \times 14 + 2 \times 16 = 113\ g/mol$; $14/113 = 0.12 = 12\%$). In Section 2.17 (*Nutrient Requirements*), a more advanced formula is presented for the estimation of the TKN fraction in the excess sludge, and the value of 10% was used in the example of Chapter 5. The three approaches lead to similar results. For the purposes of this section, the value of 12% is used.

The TKN to be removed corresponds to the product of the flowrate multiplied by the difference between the influent and effluent TKN. Thus, Equation 6.11 can be finally presented as follows:

$$\boxed{f_N = \frac{Y_N \cdot [Q \cdot (TKN_o - TKN_e) - 0.12 \cdot V \cdot (\Delta X_V / \Delta t)]}{V \cdot (\Delta X_V / \Delta t)}} \qquad (6.14)$$

where:

Q = influent flow (m^3/d)
TKN_o = influent TKN to the reactor (g/m^3)
TKN_e = effluent TKN from the reactor (g/m^3)
V = total volume of the reactor (m^3)

Once the fraction f_N is known, the mass of nitrifying bacteria can be expressed in terms of the total biomass (X_v).

The **nitrification rate** can then be expressed as follows:

$$\boxed{\frac{\Delta TKN}{\Delta t} = f_N \cdot \frac{X_V \cdot \mu_N}{Y_N}} \qquad (g/m^3 \cdot d) \qquad (6.15)$$

The **TKN load oxidised per day** is:

$$\boxed{L_{TKN} = \frac{V_{aer}}{10^3} \cdot \frac{\Delta TKN}{\Delta t}} \qquad (kg/d) \qquad (6.16)$$

where:
L_{TKN} = load of oxidised TKN (kg/d)
V_{aer} = volume of the aerated zone of the reactor (m^3)

Example 6.4

Calculate the nitrification rate based on the conventional activated sludge system data provided in the example of Chapter 5 and on the environmental conditions of Examples 6.1 and 6.2, that is:

- $Q = 9{,}820$ m^3/d
- $V = 2{,}051$ m^3
- $V_{aer} = 2{,}051$ m^3/d (the reactor is totally aerobic without anoxic zones)
- $\theta_c = 6$ d
- $X_V = 3{,}000$ g/m^3
- Influent TKN to the reactor $= 40$ g/m^3 (after the primary sedimentation tank, where a removal of 20% was assumed)
- Effluent TKN $= 2$ g/m^3 (desired)
- $T = 20\,^\circ$C
- $\mu = 0.22$d^{-1} (calculated in Example 6.2)
- $Y_N = 0.08$ gX$_N$/gX$_V$ (Table 6.1)

Solution:

(a)   Analysis of the sludge age

Considering the environmental conditions of Examples 6.1 and 6.2, the minimum sludge age required for nitrification is 4.5 d (as calculated in Example 6.3). In this example, θ_c is equal to 6 d, which ensures the development of nitrifying bacteria. The safety factor for the sludge age is 6.0/4.5 = 1.33.

(b)   Production of solids

The production of VSS in the reactor ($\Delta X_V/\Delta t$), even though it has already been determined in the example of Chapter 5, can be calculated using Equation 6.13:

$$\frac{\Delta X_V}{\Delta t} = \frac{X_V}{\theta_c} = \frac{3{,}000}{6} = 500 \text{ gVSS/m}^3 \cdot d$$

Example 6.4 (*Continued*)

The load of VSS produced is:

$$P_{Xv} = (500 \text{ gVSS/m}^3 \cdot \text{d}) \times (2051 \text{ m}^3) = 1,025,500 \text{ gVSS/d}$$

(c)   Fraction of nitrifying bacteria f_N

TKN to be removed $= Q \cdot (TKN_o - TKN_e) = 9820 \times (40 - 2) = 373,160 \text{ g/d}$

TKN incorporated into the excess sludge $= 0.12 \cdot V \cdot (\Delta X_V / \Delta t)$
$$= 0.12 \times 2051 \times 500 = 123,060 \text{ g/d}$$

TKN to be oxidised $= 373,160 - 123,060 = 250,100 \text{ g/d}$

According to Equation 6.13, the production of nitrifying bacteria is:

$$\Delta X_N / \Delta t = Y_N . TKN_{oxidised} = 0.08 \times 250,100 = 20,008 \text{gX}_N/\text{d}$$

The fraction f_N can then be calculated as the quotient between the production of X_N and the production of X_V (Equation 6.11):

$$f_N = \frac{20,008}{1,025,500} = 0.020 \text{ gX}_N/\text{gX}_V$$

The fraction f_N can also be calculated directly using Equation 6.14:

$$f_N = \frac{Y_N \cdot [Q \cdot (TKN_o - TKN_e) - 0.12 \cdot V \cdot (\Delta X_V / \Delta t)]}{V \cdot (\Delta X_V / \Delta t)}$$

$$= \frac{0.08 \times [9,820 \times (40 - 2) - 0.12 \times 2,051 \times 500]}{2,051 \times 500}$$

$$= \frac{0.08 \times (373,160 - 123,060)}{1,025,500} = \frac{20,008}{1,025,500}$$

$$= 0.020 \text{ gX}_N/\text{gX}_V$$

In this case, the nitrifying bacteria represent 2.0% of the total biomass (expressed as volatile suspended solids).

(d)   Nitrification rate

According to Equation 6.15, the nitrification rate is given by:

$$\frac{\Delta TKN}{\Delta t} = f_N \cdot \frac{X_V \cdot \mu_N}{Y_N} = 0.020 \times \frac{3,000 \times 0.22}{0.08} = 165 \text{ gTKN/m}^3 \cdot \text{d}$$

Example 6.4 (Continued)

The TKN load capable of being oxidised is (Equation 6.16):

$$L_{TKN} = \frac{V_{aer}}{10^3} \cdot \frac{\Delta TKN}{\Delta t} = \frac{2,051}{1,000} \times 165 = 338\,kgTKN/d$$

(e) Comments

In the conditions assumed, the TKN load capable of being oxidised in the system (nitrification capacity) is 338 kg/d, much higher than the load available to be oxidised, which is 250 kg/d (see item (c)). Thus, nitrification will be complete, all the available load will be oxidised and the TKN effluent concentration is expected to be less than 2 g/m³, which was initially assumed. Given the degree of uncertainty in several design input data and considering that 2 g/m³ is already sufficiently low and close to zero, there is no need to redo the calculations, for a new lower effluent TKN concentration.

In summary, the mass balance is:

- TKN load to be removed: 373 kg/d
- TKN load incorporated into the excess sludge: 123 kg/d
- TKN load oxidised (nitrified): 250 kg/d

If the nitrification capacity was smaller than the load to be oxidised, the non-removed load should be calculated and, accordingly, the effluent concentration. For example, if the load capable of being oxidised (L_{TKN}) were 150 kg/d, the non-oxidised load would be: $250 - 150 = 100$ kg/d. For a flow rate of 9,820 m³/d, the effluent concentration would be $100/9,820 = 0.010$ kgTKN/m³ $= 10\,gTKN/m^3$. Since this value is much higher than the 2 g/m³ initially assumed, the μ growth rate calculations in Examples 6.1, 6.2, 6.3 and 6.4 should be redone until a satisfactory convergence is obtained.

In extended aeration systems, due to the larger sludge age, the sludge production is lower and the withdrawal route of TKN in the excess sludge is also smaller. On the other hand, the nitrification capacity can be higher due to the larger sludge age.

6.3.5 Oxygen requirements for nitrification

From the overall nitrification reaction (Equation 6.3), it can be seen that 1 mol of ammonia-N requires 2 moles of oxygen for its oxidation, that is, 4.57 kgO₂ are required for 1 kg of N (MW of N = 14 g/mol; MW of O₂ = 64 g/mol; 64/14 = 4.57). In summary:

$$\boxed{\text{oxidation of } 1\ mgNH_4^+\text{-N/L consumes } 4.57\ mgO_2/L}$$

The O_2 required for the nitrification in an activated sludge system is therefore:

$$\boxed{O_{2\,required}\,(kgO_2/d) = 4.57\,(kgO_2/kgTKN) \times TKN_{oxidised}\,(kgTKN/d)}$$

(6.17)

The determination of the load of oxidised TKN was discussed in Section 6.3.4.

In the design example of Chapter 5 it was assumed for simplicity that the load of influent TKN would be totally oxidised. This was done due to the fact that the concepts of nitrification had still not been introduced in that chapter. However, the approach described in this section is preferable and should be adopted.

In terms of demand, the O_2 consumption for nitrification corresponds to a significant fraction of the overall oxygen requirement, which includes the oxidation of the carbonaceous and nitrogenous material.

Example 6.5

Calculate the O_2 requirements for nitrification based on the data included in Example 6.4. The relevant data are: oxidised TKN = 250 kg/d.

Solution:

According to Equation 6.17:

$$O_{2\,required}\,(kgO_2/d) = 4.57\,(kgO_2/kgTKN) \times TKN_{oxidised}\,(kgTKN/d)$$

$$= 4.57 \times 250 = 1,143\ kgO_2/d$$

For comparison purposes, the value calculated in the example in Chapter 5, for conventional activated sludge, was 1,344 kgO_2/d, 18% higher than the value calculated in this example. The preferable value to be adopted is that in this example (1,143 kgO_2/d), since it has been calculated using a method that takes into consideration a larger number of interacting factors. In the examples in question, the difference in the **overall** O_2 requirements according to the two approaches is, however, small (7%).

6.3.6 Alkalinity requirements for nitrification

When analysing the overall nitrification reaction (Equation 6.3), it is observed that:

oxidation of 1 mol of NH_4^+-N produces 2 moles of H^+

It is known that in sewage, due to the presence of alkalinity, H^+ will not generate acidity directly, and the buffering bicarbonate – carbon dioxide system will be activated:

$$H^+ + HCO_3^- \leftrightarrow H_2O + CO_2 \tag{6.18}$$

Thus, each mol of H^+ consumes 1 mol of HCO_3^- (bicarbonate). Therefore, the 2 moles of H^+ generated in nitrification will consume 2 moles of HCO_3^-, that is, in the end, the oxidation of 1 mol of NH_4^+ implies the consumption of 2 moles of HCO_3^-. In terms of concentration, one has:

1 mol NH_4^+-N $\rightarrow$ 2 moles HCO_3^- or
14 mgNH_4^+-N/L $\rightarrow$ 122 mgHCO_3^-/L or
1 mgNH_4^+-N/L $\rightarrow$ 8.7 mgHCO_3^-/L

Alkalinity is given by (Schippers, 1981):

$$\text{alkalinity} = 100 \cdot \left\{ \left[CO_3^{2-}\right] + \frac{1}{2}\left[HCO_3^-\right] + \frac{1}{2}\left[OH^-\right] \right\} \qquad (6.19)$$

(concentrations in millimoles)

In the usual pH range, the terms corresponding to OH^- and CO_3^{2-} may be ignored. Hence, the alkalinity, after conversion to mg/L, is simply given by:

$$\text{alkalinity} = \frac{HCO_3^-}{1.2} \qquad (6.20)$$

where:
HCO_3^- = bicarbonate concentration (mg/L)

Consequently, 8.7 mgHCO_3^-/L corresponds to $8.7/1.2 = 7.1$ mg/L of alkalinity. In other words:

> oxidation of 1 mgNH_4^+-N/L consumes 7.1 mg/L of alkalinity

The decrease in alkalinity and, as a result, the decrease in the buffer capacity of the mixed liquor favour subsequent pH reductions. The consequence of this, which justifies this whole analysis, is that the nitrification rate will be reduced, as it is dependent on pH (see Section 6.3.2.b). Depending on the alkalinity of the raw sewage, it may be necessary to add some alkaline agent (100 mg$CaCO_3$/L of alkalinity are equivalent to 74 mg/L of $Ca(OH)_2$ – hydrated lime). The alkalinity usually available in raw sewage is in the order of 100 to 250 mg$CaCO_3$/L.

Example 6.6

Calculate the alkalinity requirements based on data from Example 6.4. Assume that the alkalinity of the raw sewage is 150 mg/L. Other relevant data are:

Oxidised TKN = 250 kg/d
Average inflow rate: $Q = 9,820$ m^3/d

Example 6.6 (Continued)

Solution:

(a)   Alkalinity requirements

Knowing that 1 mgTKN/L implies a consumption of 7.1 mg/L of alkalinity, the alkalinity load required is:

$$\text{alkalinity load required} = 7.1\frac{\text{kg alkalinity}}{\text{kgTKN}} \times 250\frac{\text{kgTKN}}{\text{d}} = 1{,}775 \text{ kgCaCO}_3/\text{d}$$

(b)   Available alkalinity in the influent

The available alkalinity load in the influent is:

$$\text{available alkalinity load} = 9{,}820\frac{\text{m}^3}{\text{d}} \times 150\frac{\text{g}}{\text{m}^3} \times \frac{1}{10^3}\frac{\text{kg}}{\text{g}} = 1{,}473 \text{ kgCaCO}_3/\text{d}$$

(c)   Comments

The available alkalinity load is lower than that required, and there is a deficit of 1,775 − 1,473 = 302 kgCaCO$_3$/day. This will lead to a reduction in the nitrification rate, due to the resulting decline in the pH. For this reason, nitrification may not be complete, which will in its turn result in a decrease in the required alkalinity load, with a point of balance being reached.

    If nitrification is to be achieved according to the conditions specified in the previous examples, there are two possible solutions: (a) to stimulate denitrification to take place in the system to reduce alkalinity requirements (see Section 6.4.2) or (b) to add an alkaline agent, lime for instance.

    If lime is added, the consumption will be (knowing that 100 mgCaCO$_3$/L of alkalinity is equivalent to 74 mg/L of Ca(OH)$_2$ – hydrated lime):

$$\text{lime consumption} = \frac{74\,\text{kgCa(OH)}_2}{100\,\text{kgCaCO}_3} \times 302\frac{\text{kgCaCO}_3}{\text{d}}$$

$$= 223 \text{ kgCa(OH)}_2 \text{ per day}$$

6.4 PRINCIPLES OF BIOLOGICAL DENITRIFICATION

6.4.1 Preliminaries

As seen in other chapters in this book, under aerobic conditions the microorganisms use the oxygen as "electron acceptors" in the respiration processes. In these conditions, there is a process of oxidation of the organic matter, in which the following reactions take place (simplified analysis):

$$H_2 \rightarrow 2H^+ + 2e^- \text{ (hydrogen is oxidised, that is, gives out electrons)} \quad (6.21)$$

$$O_2 + 4e^- \rightarrow 2O_2^- \text{ (oxygen is reduced, that is, gains electrons)} \quad (6.22)$$

Therefore, the oxygen is the electron acceptor in the processes of aerobic respiration. However, in the absence of oxygen, there is a predominance of organisms that have the capacity to use other inorganic anions as electron acceptors, such as the nitrates, sulfates and carbonates. The first to be used will be that which is available in the medium and whose reaction releases the largest amount of energy. In sewage treatment, both of these requirements can be satisfied by the nitrates, which are generated by the nitrification process. Thus, in conditions of total depletion of dissolved oxygen, the microorganisms start to use the nitrates in their respiration. Such conditions are not properly anaerobic, but are named *anoxic*. A simple distinction among the three conditions is:

- *aerobic conditions*: presence of oxygen
- *anoxic conditions*: absence of oxygen, presence of nitrate
- *anaerobic conditions*: absence of oxygen and nitrates, presence of sulphates or carbonates

Denitrification corresponds to the reduction of nitrates to gaseous nitrogen. The main route for biological denitrification starts with the nitrates, and this is the reason why in sewage treatment denitrification should be preceded by nitrification. The microorganisms involved in denitrification are *facultative heterotrophic* and are usually abundant in domestic sewage; examples are *Pseudomonas*, *Micrococcus* and others (Arceivala, 1981).

For denitrification to occur, the heterotrophic microorganisms require a source of organic carbon (electron donor), such as methanol, that can be added artificially or be available internally in the domestic sewage. For the organic carbon in the sewage, the denitrification reaction is (Arceivala, 1981):

$$C_5H_7NO_2 + 4NO_3^- \rightarrow 5CO_2 + 2N_2 + NH_3 + 4OH^- \quad (6.23)$$

In the reaction above, $C_5H_7NO_2$ corresponds to the typical composition of the bacterial cell. Including assimilation, the consumption is approximately 3 $mgC_5H_7NO_2/mgNO_3^--N$, or approximately 4.5 $mgBOD_5/mgNO_3^--N$. As most of the domestic sewage has a BOD_5:N ratio that is larger than that mentioned, the use of internally available carbon becomes an attractive and economic method of achieving denitrification (Arceivala, 1981). However, it should be remembered that, depending on the location in the treatment line, most of the BOD will have already been removed, thus reducing the availability of organic carbon for denitrification.

6.4.2 Reasons for and advantages of intentionally induced denitrification in the treatment system

In activated sludge systems where nitrification occurs, it is interesting to include a denitrification stage to be intentionally accomplished in the reactor. This intentional denitrification is made possible through the incorporation of anoxic zones in the reactor, as detailed in Chapter 7. The reasons are usually associated with some of the following aspects:

- economy of oxygen (savings on energy)
- reduced alkalinity requirements (preservation of the buffer capacity of the mixed liquor)
- operation of the secondary sedimentation tank (to avoid rising sludge)
- control of nutrients (eutrophication)

(a)   Economy of oxygen

A great advantage of intentional denitrification taking place in the activated sludge system is that the oxygen released by nitrate reduction can become immediately available for the biological oxidation of the organic matter in the mixed liquor. The release of oxygen through the reduction of nitrates occurs according to the denitrification reaction (Equation 6.4, described in Section 6.2):

$$2NO_3^--N + 2H^+ \longrightarrow N_2 + 2.5O_2 + H_2O \qquad (6.4)$$

Thus, each 2 moles of nitrate release 2.5 moles of oxygen, that is:

the reduction of 1 mg/L of nitrogen in the form of nitrate releases 2.86 mgO$_2$/L

As seen in Section 6.3.5, the oxidation of 1 mg of nitrogen in the form of ammonia implies the consumption of 4.57 gO$_2$. As a result, if total denitrification is achieved, a **theoretical saving of 62.5%** can be obtained (2.5/4.0 or 2.86/4.57) in the consumption of the oxygen used in the nitrification.

In the design of the treatment plants this economy can be taken into consideration, if a reduction in the required power for the aerators is desired. In the operation of the plant, the denitrification will make it possible to reduce the consumption of energy, provided that the aeration level is controlled to maintain the desired DO concentration in the reactor.

(b)   Economy of alkalinity

As seen in Section 6.3.2, the maintenance of a satisfactory level of alkalinity in the mixed liquor is of great importance to keep the pH within the adequate range for nitrification. From the denitrification reaction (Equation 6.4), it can be seen that the reduction of 1 mol of nitrate occurs along with the consumption of 1 mol of

H^+. During nitrification, the formation of 1 mol of nitrate implies the production of 2 moles of H^+ (see Equation 6.3).

Thus, if denitrification is incorporated into the treatment system, a theoretical reduction of 50% in the release of H^+ can be obtained, that is to say, an **economy of 50% in alkalinity consumption**. Thus, if 7.1 mg/L of alkalinity are consumed for the nitrification of 1 g NH_4^+-N/L (see Section 6.3.6), only 3.5 mg/L of alkalinity will be consumed if denitrification is included in the system. Some authors (Barnes and Bliss, 1983; Eckenfelder Jr and Argaman, 1978) indicate a lower practical economy, in the order of 3 mg/L of alkalinity (alkalinity consumption of approximately 4.1 mg/L for nitrification combined with denitrification). Example 6.6 can be analysed from this new perspective and, in this case, the available alkalinity in the raw sewage will be sufficient.

(c) Operation of the secondary sedimentation tank

In secondary sedimentation tanks, the sludge has a certain detention time. For nitrified mixed liquors, under certain conditions, such as high temperatures, the situation becomes favourable for the occurrence of denitrification in the sedimentation tank. As a result, the nitrates formed in the reactor are reduced to gaseous nitrogen in the secondary sedimentation tank (see Equation 6.4). This implies the production of small bubbles of N_2 that adhere to the sludge, thus preventing it from settling, and carrying it to the surface. This is the so-called **rising sludge**. This sludge will leave with the final effluent, deteriorating its quality in terms of SS and BOD. This effect is particularly common in warm-climate regions, where high temperatures favour nitrification and denitrification.

Therefore, it is an appropriate strategy to prevent denitrification from taking place in the secondary sedimentation tank, while allowing it to occur in controlled locations, where the additional advantages of oxygen and alkalinity economy can be achieved.

(d) Nutrient control

Usually, when dealing with denitrification, the first point to come to mind is the control of eutrophication of water bodies through the removal of nutrients in wastewater treatment. This aspect, even though of great importance in some situations, is not always the decisive factor, for two reasons. The first is that not all the effluents from wastewater treatment plants go to sensitive water bodies, such as lakes, reservoirs or estuaries. For disposal to rivers, the control of nutrients is usually not necessary. The second reason, also very important, is that cyanobacteria, which are usually associated with the more developed stages of eutrophication, in which they proliferate in great numbers, have the capacity to absorb the atmospheric nitrogen and convert it into a form that can be assimilated. Thus, the nitrogen in the liquid medium is not the limiting factor for these organisms and the reduction in the amount conveyed by the wastewater will have a lower influence. In these conditions, the truly limiting nutrient is phosphorus.

If phosphorus is really the limiting factor for algal growth, all the efforts in the wastewater treatment should be concentrated on its removal. However, the potential advantage of nitrogen removal should not be disregarded for the control of the trophic status of water bodies that still have a certain species diversity, with different requirements in terms of N and P.

6.4.3 Kinetics of denitrification

The denitrification rate can be obtained from the growth rate of the denitrifying microorganisms, similar to the calculations of the nitrification rate (Section 6.3.1). The growth rate can be expressed in terms of Monod's kinetics, according to the electron acceptor (nitrate) and donor (organic matter) concentration, as follows:

$$\mu = \mu_{max} \cdot \left[\frac{S}{K_S + S} \right] \cdot \left[\frac{NO_3^-}{K_{NO_3^-} + NO_3^-} \right] \tag{6.24}$$

where:

S = concentration of carbonaceous matter (mgBOD/L)
K_S = half-saturation coefficient for the carbonaceous matter (mgBOD/L)
NO_3^- = concentration of nitrogen in the form of nitrate (mgN/L)
K_{NO_3} = half-saturation coefficient for the nitrogen in the form of nitrate (mgN/L)

Usually $NO_3^- \gg K_{NO_3}$ (EPA, 1993), which makes the term in the second bracket in Equation 6.24 negligible, that is, it can be considered that the growth rate of the denitrifying bacteria does not depend on the nitrate concentration in the medium (zero-order reaction with relation to the nitrate).

However, the value of K_S for the carbonaceous matter depends fundamentally on the type of organic carbon, which is a function of the denitrification system adopted and the characteristics of the process, such as the sludge age. Depending on the value of K_S, the growth rate can be of order 0 or 1 for the organic carbon. With this range of variations, and aiming at keeping a simple model structure, it is not very practical to design the activated sludge system by expressing the denitrification rate in terms of the growth rate of the denitrifying organisms, according to Monod's kinetics.

A simplified way to express the denitrification rate is through the relation with the volatile suspended solids in the reactor (denitrification rate = μ_{denit}/Y_{denit}). Typical values of the denitrification rate are given in Table 6.3.

The denitrification rate in the anoxic zone upstream of the reactor is higher than in the anoxic zone downstream of the reactor. This is because in the first anoxic zone the raw sewage contains high levels of organic carbon, which are necessary for the denitrifying bacteria. On the other hand, in the second anoxic zone most of the organic carbon has been already removed in the reactor, leading to a predominance of the endogenous metabolism, with low denitrification rates.

Table 6.3. Typical ranges of the specific denitrification rate

Position of the anoxic zone	Source of organic carbon	Specific denitrification rate $(mgNO_3{}^- \text{-N} /mgVSS\cdot d)$
Anoxic zone upstream of the aerated zone	Raw sewage	0.03–0.11
Anoxic zone downstream of the aerated zone	Endogenous metabolism	0.015–0.045

Source: Eckenfelder and Argaman (1978); Arceivala (1981); Metcalf and Eddy (1991); EPA (1993)

The denitrification rate decreases with the increase in the sludge age (or the reduction in the F/M ratio). In Table 6.3, within each range, the smallest values correspond to the highest sludge ages. EPA (1993) includes two equations that correlate the denitrification rate with F/M and θ_c:

- Anoxic zone upstream of the aerated zone:

$$SDR = 0.03 \times (F/M_{anox}) + 0.029 \qquad (6.25)$$

- Anoxic zone downstream of the aerated zone

$$SDR = 0.12 \times \theta_c{}^{-0.706} \qquad (6.26)$$

where:

SDR = specific denitrification rate $(mgNO_3{}^- \text{-N}/mgVSS\cdot d)$

F/M_{anox} = food/microorganism ratio in the first anoxic zone (not in the reactor as a whole) (kgBOD/kgMLSS in the first anoxic zone per day)

θ_c = sludge age (d)

The processes for achieving denitrification in the activated sludge system are discussed in Chapter 7, where an analysis is made of different flowsheets, the position of the anoxic and aerated zones, the recirculations and the differences between the use of raw sewage and the carbon from the endogenous respiration. Relevant examples are provided in this chapter.

6.4.4 Environmental factors of influence on denitrification

Compared with the nitrifying bacteria, the denitrifying bacteria are much less sensitive to environmental conditions. However, the following environmental factors influence the denitrification rate:

- dissolved oxygen
- temperature
- pH
- toxic or inhibiting substances

(a)   Dissolved oxygen

The absence of oxygen is obviously a fundamental pre-requisite for the occurrence of denitrification. Anoxic conditions are needed in the floc, that is, in the immediate vicinity of the denitrifying bacteria. Hence, it is possible that there is dissolved oxygen at low concentrations in the liquid medium and, even so, denitrification takes place, because of the fact that the denitrifying bacteria are in an anoxic micro-environment within the floc.

Metcalf and Eddy (1991) present the following equations for correcting the denitrification rate for the presence of DO. It should be noted that the rate decreases linearly with the increase of DO and reaches zero when DO is equal to 1.0 mg/L.

$$SDR_{DO} = SDR_{20°C} \times (1.0 - DO) \tag{6.27}$$

where:

SDR = specific denitrification rate, as determined in Section 6.4.3 $(mgNO_3^--N/mgVSS\cdot d)$

$SDR_{20°C}$ = specific denitrification rate with inhibition due to the presence of DO $(mgNO_3^--N/mgVSS\cdot d)$

DO = dissolved oxygen (mg/L)

The specific growth rate of the denitrifying bacteria and, in other words, the denitrification rate, can also be modelled according to Monod's kinetics, with the inhibition term for DO included (IAWPRC, 1987; EPA, 1993). Equation 6.28 corresponds to Equation 6.24, with the DO inhibition term. Note that the term for DO, since it is related to inhibition, is given with inverted numerators and denominators compared to the nutrient terms (S and N).

$$\mu = \mu_{max} \cdot \left[\frac{S}{K_S + S}\right] \cdot \left[\frac{NO_3^-}{K_{NO_3^-} + NO_3^-}\right] \cdot \left[\frac{K_O}{K_O + DO}\right] \tag{6.28}$$

where:

K_O = half-saturation coefficient for oxygen (mg/L). A value of K_O equal to 1.0 is suggested by the IAWPRC (1987) model.

Naturally, in a properly designed and operated anoxic zone, the DO should be equal or very close to zero, since there is no aeration in this zone. Denitrification can still happen in the reactor in a predictable way, such as in the anoxic zones in an oxidation ditch. It can also occur in a manner that was not predicted in the design, such as in poorly aerated zones in the reactor (bottom and corners).

(b)   Temperature

Temperature has an effect on the growth rate of denitrifying bacteria and, as a consequence, on the denitrification rate. The denitrification reaction takes place in a wide temperature range, from 0 °C to 50 °C, reaching its optimal level in the range of 35 °C to 50 °C (Barnes and Bliss, 1983).

The influence of temperature can be expressed in the conventional Arrhenius form, that is:

$$SDR_T = SDR_{20°C}\theta^{(T-20)} \tag{6.29}$$

where:

SDR = specific denitrification rate at a temperature T ($mgNO_3^--N/mgVSS\cdot d$)

$SDR_{20°C}$ = specific denitrification rate at the temperature of 20 °C ($mgNO_3^--N/mgVSS\cdot d$)

T = temperature of the liquid (°C)

θ = temperature coefficient

Very broad ranges are given in the literature for the temperature coefficient θ. Arceivala (1981) mentions values between 1.15 and 1.20. EPA (1993) lists values ranging from 1.03 to 1.20, with the predominance of values close to 1.08. Metcalf and Eddy (1991) use the value of 1.09.

(c) pH

There is a certain variation in the literature regarding the ideal pH for denitrification. Arceivala (1981) indicates values in the range of 7.5 to 9.2, while Barnes and Bliss (1983) suggest a range from 6.5 to 7.5, with 70% decline in the denitrification rate for a pH of 6 or 8. EPA (1993) presents four curves for the variation of the denitrification rate with pH. The general tendency in these curves is that the maximum rate occurs at a pH between 7.0 and 7.5 and decreases approximately linearly with both the reduction and the increase in pH. For a pH of 6.0, the denitrification rates vary between 40% and 80% of the maximum value. For a pH of 8.0, the denitrification rates vary between approximately 70 and 90% of the maximum rate.

In spite of the variation of the information, it can be concluded that the pH should be close to neutrality and values below 6.0 and above 8.0 should be avoided.

(d) Toxic or inhibiting substances

The major route for the occurrence of denitrification is after nitrification. As already discussed, the nitrifying bacteria are much more sensitive to toxic or inhibiting substances than the heterotrophic bacteria responsible for denitrification. In addition, the denitrifying bacteria are present in a larger diversity of species, which reduces the impact of some specific inhibiting agent. Thus, if toxic or inhibiting substances are present, it is very likely that denitrification will be very reduced (or eliminated) for the simple reason that nitrification is inhibited.

6.5 PRINCIPLES OF BIOLOGICAL PHOSPHORUS REMOVAL

6.5.1 Mechanisms of biological phosphorus removal

For biological phosphorus removal, it is *essential to have anaerobic and aerobic zones* in the treatment line. The most convenient arrangements of both zones are discussed in Chapter 7.

The early explanations for the mechanism of biological phosphorus removal referred to the anaerobic zone as causing a condition of bacterial stress that would result in phosphorus being released in this zone. After that, high assimilation of the phosphorus available in the liquid medium would occur in the aerobic zone at a higher level than the normal metabolic requirements of the bacteria. When removing the excess biological sludge, bacteria with high phosphorus levels are also removed.

As more information has become available through intense research in this area in the past years, a mechanistic model has been developed, which includes fundamental biochemical aspects. It should be noted that, in spite of the great progresses made in this area, some knowledge gaps still need to be filled in. Several of the organisms involved are taxonomically unknown. The current mathematical models for biological phosphorus removal, such as the IWA models (IAWQ, 1995, and subsequent versions), are extremely complex and are still being tested in full-scale activated sludge plants. However, the merit exists as the increasing knowledge in the area has allowed better designs and operational control strategies to be developed.

This book deals with biological phosphorus removal in a simplified way. More recent and deeper information should be obtained from specific publications, since the theme has developed significantly.

Biological phosphorus removal is based on the following fundamental points (Sedlak, 1991, IAWQ, 1995):

- Certain bacteria are capable of accumulating excess amounts of phosphorus in the form of polyphosphates. These microorganisms are named phosphorus accumulating organisms (PAOs). The bacteria most frequently mentioned as an important PAO is *Acinetobacter*.
- These bacteria are capable of removing simple fermentation substrates produced in the anaerobic zone and then assimilate them as products stored inside their cells.
- In the aerobic zone, energy is produced by the oxidation of these stored products. The storage of polyphosphates in the cell increases.

The anaerobic zone is considered a *biological selector* for the phosphorus accumulating microorganisms. This zone has an advantage in competition terms for the phosphorus accumulating organisms, since they can assimilate the substrate in this zone before other microorganisms, which are not phosphorus accumulating organisms. Thus, this anaerobic zone allows the development or selection of a large

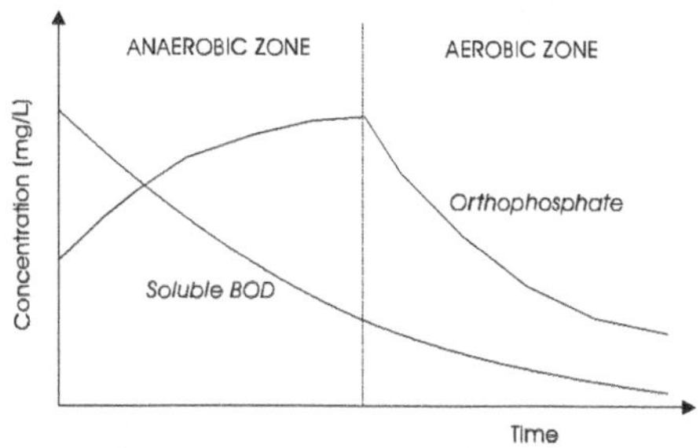

Figure 6.2. Variation of the soluble BOD and orthophosphate concentrations in the anaerobic and aerobic zones in an activated sludge system designed for biological phosphorus removal (adapted from EPA, 1987)

population of phosphorus accumulating organisms (PAOs) in the system, which absorb substantial amounts of phosphorus in the liquid medium. Phosphorus is then removed from the system with the excess sludge (Sedlak, 1991).

Figure 6.2 presents typical profiles of soluble BOD and orthophosphates in anaerobic and aerobic zones in an activated sludge system designed for phosphorus removal. The concentration of soluble BOD decreases in the anaerobic zone, even if there are no aerobic or anoxic electron acceptors. In the anaerobic zone, while the soluble BOD concentration decreases, the soluble phosphorus concentration increases. Subsequently, in the aerobic zone, the phosphorus concentration decreases, while the soluble BOD concentration continues in its decline.

The biological phosphorus removal mechanism is summarised in Figure 6.3 and is described in the following paragraphs (EPA, 1987b; Sedlak, 1991; Henze, 1996).

Alternation between anaerobic and aerobic conditions

- *Alternation of conditions*. The PAO require the alternation between anaerobic and aerobic conditions, to build their internal energy, organic molecules and polyphosphate storage components.

Anaerobic conditions

- *Production of volatile fatty acids by facultative bacteria*. Part of the biodegradable organic matter (soluble BOD) is converted, through fermentation processes in the raw sewage or in the anaerobic zone, into simple organic molecules of low molecular weight, such as volatile fatty acids. This conversion is usually made by facultative organisms that normally occur in the sewage and in the anaerobic zone. The volatile fatty acids become available in the liquid medium. There is not enough time for hydrolysis and the conversion of the particulate influent organic matter.
- *Accumulation of the volatile fatty acids by the PAOs*. The phosphate accumulating organisms give preference to these volatile fatty acids, which are

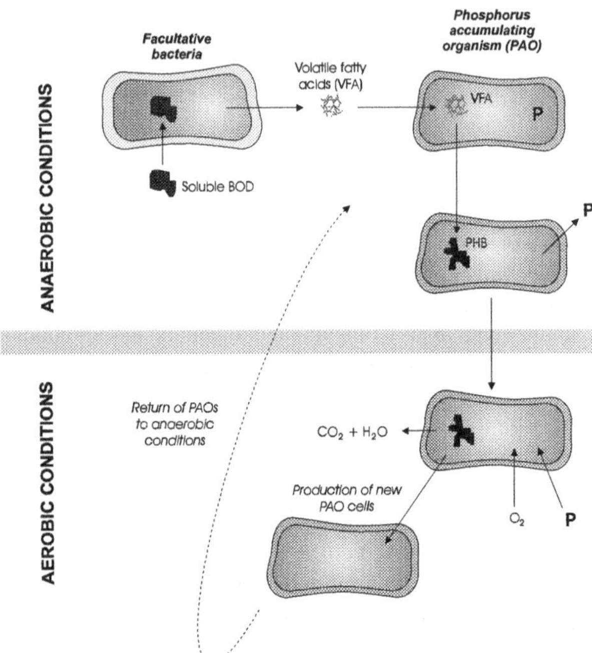

Figure 6.3.   Schematic mechanism of biological phosphorus removal

quickly assimilated and accumulated inside the cells. PAOs assimilate these fermentation products better than the other organisms usually occurring in the activated sludge process. As a consequence, there is a selection of the population of these phosphorus accumulating organisms in the anaerobic zone.

- *Phosphate release.* The release of phosphate that was previously accumulated by the organisms (in the aerobic stage) supplies energy for the transport of the substrate and for the formation and storage of organic metabolic products, such as PHB (polyhydroxybutyrate).

Aerobic conditions

- *Consumption of the stored substrate and assimilation of phosphate.* PHB is oxidised into carbon dioxide and water. The soluble phosphate is removed from the solution by the PAOs and is stored in their cells for generation of energy in the anaerobic phase.
- *Production of new cells.* Due to the use of substrate, the PAO population increases.

Phosphorus removal

- *Phosphorus removal by the excess sludge*. The phosphorus is incorporated in large amounts into the PAOs cells and is removed from the system through the removal of the biological excess sludge, which discards a fraction of the mixed liquor containing all the organisms in the activated sludge, including PAOs.

6.5.2 Factors of influence on biological phosphorus removal

The following factors influence the performance of biological phosphorus removal (EPA, 1987b; Sedlak, 1991):

- *environmental factors*
  - DO
  - temperature
  - pH
  - nitrate in the anaerobic zone
- *design parameters*
  - sludge age
  - detention time and configuration of the anaerobic zone
  - detention time in the aerobic zone
  - excess sludge treatment methods
- *characteristics of the influent sewage*
- *suspended solids in the effluent*

(a) Dissolved oxygen

Biological phosphorus removal depends on the alternation between anaerobic and aerobic conditions. Naturally, there will be no dissolved oxygen available in the anaerobic zone. The presence of DO in anaerobic zones has been reported to decrease phosphorus removal and cause the growth of filamentous bacteria. DO can come from the raw sewage through infiltration, screw pumps, turbulence and cascading in the inlet structures, aeration in grit chambers and vortices created by stirrers in the anaerobic zone.

For the aerobic zone, there are no generally accepted studies that describe the effects of the DO concentration on the phosphorus removal efficiency. The mechanism of biological phosphorus removal suggests that the DO concentration can affect the phosphorus removal rate in the aerobic zone, but not the possible degree of removal, provided there is enough aerobic time.

However, there is evidence that in treatment plants the DO concentration in the aerobic zone should be kept between 1.5 and 3.0 mg/L. If the DO is very low, the phosphorus removal can reduce and the nitrification will be limited, possibly leading to the development of sludge with poor settleability. If the DO is very high, the denitrification efficiency can be reduced due to DO entering the first anoxic zone. As a consequence, an increased concentration of nitrates can occur, which affects the release of phosphorus in the anaerobic zone.

The control of DO in the aerobic zone is very important and usually plants with biological phosphorus removal are provided with automated control of the aeration capacity and the DO concentration.

(b) Temperature

Biological phosphorus removal has been successfully applied in a wide range of temperatures, and it seems that the phosphorus removal capacity is not affected by low temperatures. However, there are indications that the phosphorus release rate is lower for low temperatures, and longer detention times are needed in the anaerobic zone for fermentation to be completed and/or the substrate consumed.

(c) pH

Studies on the influence of pH on phosphorus removal suggest the following points:

- phosphorus removal is more efficient at a pH between 7.5 and 8.0
- phosphorus removal is reduced significantly at pH values lower than 6.5, and all activity is lost at a pH close to 5.0

(d) Nitrate in the anaerobic zone

The entrance of nitrate into the anaerobic zone reduces the phosphorus removal efficiency. This is because the nitrate reduction in the anaerobic zone uses substrate that, otherwise, would be available for assimilation by phosphorus accumulating organisms. As a consequence, the nitrate has the effect of reducing the BOD/P ratio in the system. The influence depends on the influent BOD and the phosphorus concentration, as well as on the sludge age. Item (i) below includes other considerations about this topic.

The various processes available for biological phosphorus removal have different internal recirculation methods and, therefore, the potential for nitrates to return or not to the anaerobic zone will differ among them. Care should also be taken in respect to the return of nitrates through the return sludge from the secondary sedimentation tanks.

(e) Sludge age

Systems operating with higher sludge ages produce less excess sludge. The main phosphorus removal route in the system is through the excess sludge, since phosphorus is accumulated in high concentrations in the bacterial cells. Thus, the larger the sludge age, the lower the sludge production, the lesser the wastage of excess sludge, and the smaller the phosphorus removal from the system. Therefore, extended aeration systems are less efficient in phosphorus removal than conventional activated sludge systems.

Systems with a high sludge age require higher BOD/P ratios in the influent to reach concentrations of soluble phosphorus in the effluent below 1.0 mg/L.

To maximise phosphorus removal, the systems should not operate with sludge ages above those required for the overall treatment requirements.

(f)   Detention time and configuration of the anaerobic zone

Detention times in the anaerobic zone have been traditionally established between 1 and 2 hours. This period of time is needed for fermentation to produce the volatile fatty acids consumed by the phosphorus accumulating organisms. After 2 hours, most of the applied BOD is already removed from the solution.

Extended periods of time should be avoided in the anaerobic zone, because they can cause the release of phosphorus without the consumption of volatile fatty acids. When this happens, there are not enough carbon storage products inside the cells to produce the necessary energy for the total absorption of the phosphorus released.

The configuration of the anaerobic zone also affects phosphorus removal. The division of the anaerobic zone into two or more compartments in series improves phosphorus removal. Naturally, the costs of the dividing walls and increased mixing equipment requirements need to be considered.

(g)   Detention time in the aerobic zone

The aerobic zone plays an important role, creating conditions for the absorption of phosphorus after its release in the anaerobic zone. As the aerobic stage is designed to allow enough time for BOD removal and nitrification, it is expected that there will be enough time for phosphorus absorption. This aspect becomes critical if the aerobic zone is not always entirely oxygenated. There are still no conclusive findings about the aerobic detention time required, but there are some indications that 1 to 2 hours are enough.

(h)   Excess sludge treatment methods

Special care should be taken in the sludge treatment stage to avoid anaerobic conditions that favour the release into the liquid of the phosphorus stored in the biomass. In this respect, the following points should be noted:

- adoption of thickening by dissolved air flotation is preferable to gravity thickening
- aerobic digestion is preferable to anaerobic digestion
- dewatering of the sludge by fast and continuous processes is preferable to the dewatering by equipment with intermittent operation or with time-consuming methods

(i)   Characteristics of the influent sewage

For biological phosphorus removal, organic fermentation products need to be available for the phosphorus accumulating organisms. The more they are available in the anaerobic zones, the larger the phosphorus removal. It is important that the organic matter is available in the soluble form (soluble BOD) to make fermentation possible, since the short hydraulic detention times in the anaerobic zone hinder the assimilation of the slowly-biodegradable organic matter, such as the particulate BOD.

Sedlak (1991) mentions an advisable minimum ratio of soluble BOD: P in the influent of 15:1, to obtain low concentrations of soluble phosphorus in the effluent from systems with relatively low sludge ages.

The Water Research Commission (1984) makes the following comments. The mentioned treatment processes are described in Chapter 7:

- If the rapidly biodegradable COD concentration (approximately equivalent to the soluble COD) in the influent is less than 60 mg/L, irrespective of the total COD concentration, it is not very likely that a significant phosphorus removal will be achieved with any process.
- If the rapidly biodegradable COD concentration is higher than 60 mg/L, phosphorus removal can be achieved, provided that the nitrate is excluded from the anaerobic zone. The removal of P increases quickly with the increase in the biodegradable COD concentration.
- The ability to prevent nitrates from going into the anaerobic zone will depend on the TKN/COD ratio in the influent and the process adopted for phosphorus removal. Some limits are indicated below for typical domestic sewage (from South Africa):
  - **COD/TKN > 13** mgCOD/mgN. Complete removal of nitrate is possible. The Phoredox process is recommended.
  - **COD/TKN: 9 to 13** mgCOD/mgN. Complete removal of nitrates is no longer possible, but the nitrates can be excluded from the anaerobic zone by using the modified UCT process.
  - **COD/TKN: 7 to 9** mgCOD/mgN. The modified UCT process cannot exclude the nitrate from the anaerobic compartment. The UCT process is recommended, provided that the internal recirculation from the aerobic to the anoxic zone is carefully controlled.
  - **COD/TKN < 7** mgCOD/mgN. Biological phosphorus removal in systems with nitrification is unlikely to occur.

If BOD is adopted instead of COD, and a COD/BOD$_5$ ratio in the influent of around 2 is assumed, the values of the above relations are approximately half of those stated (e.g., a COD/TKN ratio of 10 corresponds to approximately BOD$_5$/TKN = 5).

Primary settling is unfavourable when trying to reach high efficiencies in N and P removal, because it increases the TKN/COD and P/COD ratios substantially, by reducing the COD concentration in the influent to the biological stage (although the concentration of soluble COD is little affected) (WRC, 1984).

(j) Suspended solids in the effluent

Since biological phosphorus removal is based on the incorporation of phosphorus in excessive amounts into the bacterial biomass, the loss of suspended solids in the effluent results in the increase of the phosphorus concentrations in this effluent. The phosphorus levels in the MLSS of biological P-removal processes range between

2 and 7% (and, under very favourable conditions, even more). Thus, if the effluent has a SS concentration equal to 20 mg/L and a proportion of P equal to 4%, this would imply that the P concentration discharged with the effluent SS is $20 \times 0.04 =$ 0.8 mg/L. This value is high when considering that total P concentration usually desired for the final effluent in systems with BNR is around 1.0 mg/L. In these conditions, the soluble P concentration in the effluent should be no more than 0.2 mg/L $(= 1.0 - 0.8)$, which is a very reduced value.

Thus, in situations where very low levels of P in the effluent are desired, it is very common to adopt polishing stages for the removal of suspended solids, such as filtration or flotation.

6.5.3 Modelling of biological phosphorus removal

The mechanistic models available for biological phosphorus removal have been developed substantially in the last years, as a result of intensive investigations in several parts of Europe, North America and South Africa. However, their degree of complexity is very high in view of the great number of variables and parameters involved, some of which are not directly measurable. The IWA models are an example of widely accepted models for the activated sludge process, including BNR. However, their degree of complexity is outside the scope of this book.

For this reason, the following simplified approach is presented for the estimation of the effluent phosphorus concentration, based mainly on the research by Professor Marais and co-workers, in South Africa (WRC, 1984).

(a) Determination of the fraction of P in the suspended solids

The main phosphorus removal route from the system is through its incorporation, in excessive amounts, into the biological excess sludge. With the removal of the excess sludge from the system, phosphorus removal is also achieved. Therefore, it is important to quantify the phosphorus fraction in the excess sludge solids (mgP/mgSS). Usually, this fraction is from 2% to 7% in systems with biological phosphorus removal. However, this value can be estimated using the methodology described below.

The *propensity factor of excess phosphorus removal* (P_f) is a parameter that reflects the system's ability to remove phosphorus. The value of P_f can be estimated using the following equation (WRC, 1984):

$$P_f = (f_{rb} \times COD - 25) \cdot f_{an} \qquad (6.30)$$

where:

f_{rb} = fraction of rapidly biodegradable COD in the influent
COD = total COD of the influent wastewater (mg/L)
f_{an} = mass fraction of the anaerobic sludge

The rapidly biodegradable fraction f_{rb} usually represents 15 to 30% of the total COD of the raw sewage, and 20 to 35% of the total COD of the sewage after primary settling (Orhon and Artan, 1994).

Influent BOD is converted into COD by simply multiplying it by a factor (COD/BOD$_5$ ratio) between 1.7 and 2.4.

With respect to the anaerobic sludge fraction f_{an}, if the concentration of solids is the same in all zones of the reactor, f_{an} can be considered equal to the ratio between the volume of the anaerobic zone and the total volume of the reactor (V_{anaer}/V_{tot}). Values of this anaerobic fraction vary between 0.10 and 0.25 (V_{anaer} varies between 10% and 25% of the total volume of the reactor).

The *phosphorus fraction in the active biomass* (mgP/mgX$_a$) can be expressed using the following relation (WRC, 1984):

$$P/X_a = 0.35 - 0.29 \cdot e^{-0.242 \cdot P_f} \qquad (6.31)$$

The active fraction of the mixed liquor volatile suspended solids (X_a/X_v) is given by:

$$f_a = \frac{1}{1 + 0.2 \cdot K_d \cdot \theta_c} \qquad (6.32)$$

where:
   K_d = coefficient of endogenous respiration (0.08 to 0.09 d^{-1})
   θ_c = total sludge age (d)

The ratio between the volatile suspended solids and the total suspended solids in the reactor (X_v/X) can be calculated, as shown in the example in Chapter 5, or be obtained from Table 2.8. Typical values are: (a) conventional activated sludge: 0.70 to 0.85, (b) extended aeration: 0.60 to 0.75. A quick way of calculating the ratio for the treatment of domestic sewage is to use the regression equations with the sludge age contained in Table 2.9, namely:

- system with primary sedimentation:

$$X_v/X = 0.817 \cdot \theta_c^{-0.043} \qquad (6.33)$$

- system without primary sedimentation:

$$X_v/X = 0.774 \cdot \theta_c^{-0.038} \qquad (6.34)$$

Thus, the phosphorus fraction in the suspended solids can be calculated through the following equations, whose terms can be obtained from Equations 6.31 to 6.34:

- Fraction of P in the *volatile* suspended solids in the excess sludge (mgP/mgVSS):

$$P/X_v = f_a \cdot (P/X_a) \qquad (6.35)$$

- Fraction of P in the *total* suspended solids in the excess sludge (mgP/mgSS):

$$P/X = \left(\frac{VSS}{SS}\right) \cdot f_a \cdot (P/X_a)$$

(6.36)

Depending on the values of the influent COD and the rates and coefficients adopted, it is possible to obtain P/X values much higher than the value of 7% mentioned by EPA (1987b) and Orhon and Artan (1994). For safety reasons, it is suggested that, for design purposes, a maximum value of 7% is assigned for this relation.

(b) Removal of P with the excess sludge

The ratio of the phosphorus removed per unit of BOD removed (mgP/mgBOD) can be expressed as follows (EPA, 1987b):

$$P/BOD = Y_{obs} \cdot (P/X_v)$$

(6.37)

or

$$P/BOD = \frac{Y}{1 + f_b \cdot K_d \cdot \theta_c} \cdot (P/X_v)$$

(6.38)

where:
P/X_v = fraction of P in VSS (calculated from Equation 6.35) (mgP/mgVSS)
Y = yield coefficient (0.4 to 0.8 mgVSS/mgBOD)
f_b = biodegradable fraction of the VSS (mgSS$_b$/mgVSS)

The f_b value can be calculated as follows:

$$f_b = \frac{0.8}{1 + 0.2 \cdot K_d \cdot \theta_c}$$

(6.39)

Typical values of f_b are: (a) conventional activated sludge: 0.55 to 0.70 and (b) extended aeration: 0.40 to 0.65.

The amount of phosphorus removed in the excess sludge, taking into consideration the amount of BOD removed, can be determined by multiplying the result of Equation 6.38 by the removed BOD concentration ($S_o - S$):

$$P_{rem} = \frac{Y}{1 + f_b \cdot K_d \cdot \theta_c} \cdot (P/X_v) \cdot (S_o - S)$$

(6.40)

where:
P_{rem} = concentration of P removed in the excess sludge (mg/L)
S_o = total influent BOD concentration to the biological stage (mg/L)
S = soluble effluent BOD concentration from the biological stage (mg/L)

(c)   Effluent P concentration

The concentration of the effluent *soluble phosphorus* is given by the difference between the total effluent concentration of P and the removed concentration of P (given by Equation 6.40):

$$P_{sol\,eff} = P_{tot\,inf} - P_{rem} \qquad (6.41)$$

The concentration of the effluent *particulate phosphorus* (present in the effluent SS) is determined by multiplying the SS concentration in the effluent from the system by the fraction of P in the suspended solids (P/X). P/X is given in Equation 6.36.

$$P_{part\,eff} = SS \cdot (P/X) \qquad (6.42)$$

The total effluent phosphorus concentration is the sum of the concentrations of soluble P and particulate P in the effluent:

$$P_{tot\,eff} = P_{sol\,eff} + P_{part\,eff} \qquad (6.43)$$

The example in Section 7.2 illustrates this calculation method for biological phosphorus removal.

7

Design of continuous-flow systems for biological nutrient removal

7.1 BIOLOGICAL NITROGEN REMOVAL

7.1.1 Processes most frequently used

The main flowsheets for nitrification and denitrification combined in a single reactor are as follows (see Figure 7.1):

- pre-denitrification (removal of nitrogen with carbon from the raw sewage)
- post-denitrification (removal of nitrogen with carbon from endogenous respiration)
- four-stage Bardenpho process
- oxidation ditch
- intermittent operation reactor (sequencing batch reactor)

There are still other processes, with nitrification and denitrification in separate lines from carbon removal, as well as other processes that use an external carbon source (usually methanol) for denitrification. However, these systems are more complex, which makes the single reactor systems without external carbon source more frequently used. Each of the main variants presented in Figure 7.1 are described below. There are still other interesting processes in which N removal follow other routes (e.g. Sharon-Anammox process), but these are outside the scope of this book.

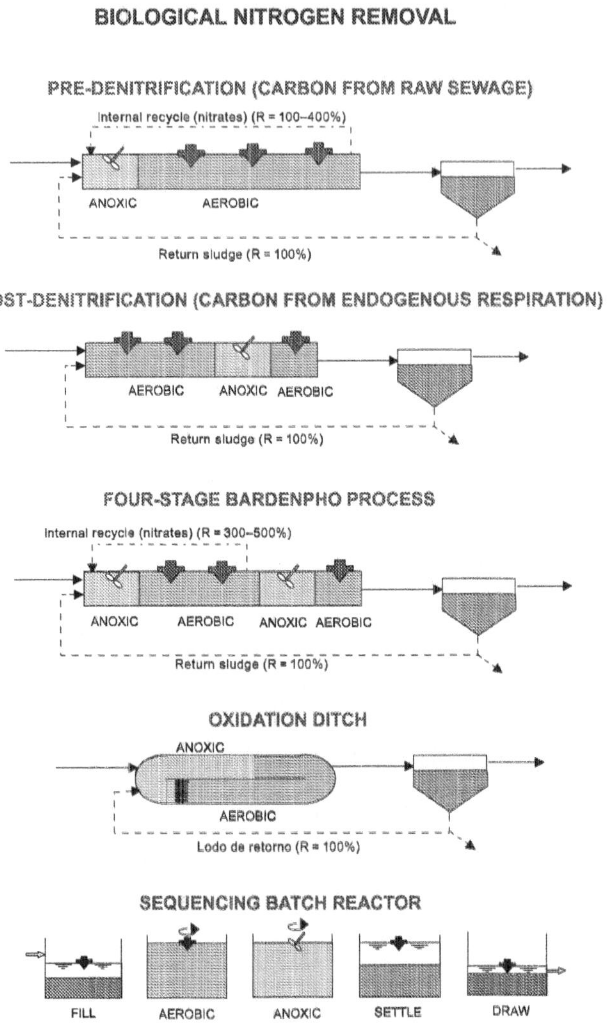

BIOLOGICAL NITROGEN REMOVAL

PRE-DENITRIFICATION (CARBON FROM RAW SEWAGE)

POST-DENITRIFICATION (CARBON FROM ENDOGENOUS RESPIRATION)

FOUR-STAGE BARDENPHO PROCESS

OXIDATION DITCH

SEQUENCING BATCH REACTOR

Figure 7.1.   Main processes for biological nitrogen removal

(a)   Pre-denitrification (removal of nitrogen with carbon from raw sewage)

The reactor has an anoxic zone followed by the aerobic zone. Nitrification occurs in the aerobic zone, leading to the formation of nitrates. The nitrates are directed to the anoxic zone by means of an internal recirculation. In the anoxic zone, the nitrates are converted into gaseous nitrogen, which escapes to the atmosphere. Should there be no internal recirculation, the only form of return of the nitrates would be through the return sludge, with the possible operational risks of denitrification in the secondary sedimentation tank (formation of N_2 bubbles, causing rising sludge). This process is also named modified Ludzack-Ettinger.

The internal recirculation is done with high recycle ratios, ranging from 100 to 400% of the influent flow. The efficiency of denitrification is highly associated with the quantity of nitrate that returns to the anoxic zone. For example, if 80% of the nitrates are returned to the anoxic zone, their potential removal is 80%. The other 20% leave with the final effluent. The formula that determines the amount of nitrate to be returned to the anoxic zone is:

$$F_{NO3\ rec} = \frac{R_{int} + R_{sludge}}{R_{int} + R_{sludge} + 1} \tag{7.1}$$

where:

$F_{NO3\ rec}$ = fraction of the nitrates formed that are recirculated to the anoxic zone (corresponds to the maximum theoretical NO_3^- removal efficiency)

R_{int} = internal recirculation ratio

R_{sludge} = sludge recirculation ratio (return sludge ratio)

For example, if the internal recirculation ratio were 0% ($R_{int} = 0$) and the sludge recirculation ratio were 100% ($R_{sludge} = 1.0$), only 50% ($F_{NO3\ rec} = 0.5$) of the nitrates would return to the anoxic zone, and the remaining 50% would leave with the final effluent. With an internal recirculation ratio of 300% ($R_{int} = 3.0$) and a sludge recirculation ratio of 100% ($R_{sludge} = 1.0$), 80% of the formed nitrates would return to the anoxic zone ($F_{NO3\ rec} = 0.8$), where they would have the chance to be converted into gaseous nitrogen. In this latter case, the maximum theoretical nitrate removal efficiency would be of 80%.

Figure 7.2 presents the maximum theoretical nitrate removal efficiency values ($F_{NO3\ rec}$) as a function of the total recirculation ratio ($R_{int} + R_{sludge}$).

In the anoxic zones of pre-denitrification systems, the denitrification rate is high (0.03 to 0.11 $mgNO_3^--N/mgVSS\cdot d$), due to the high concentration of organic carbon in the anoxic zone, brought by the raw sewage. Primary sedimentation can be omitted to allow the input of a higher load of organic carbon in the anoxic zone.

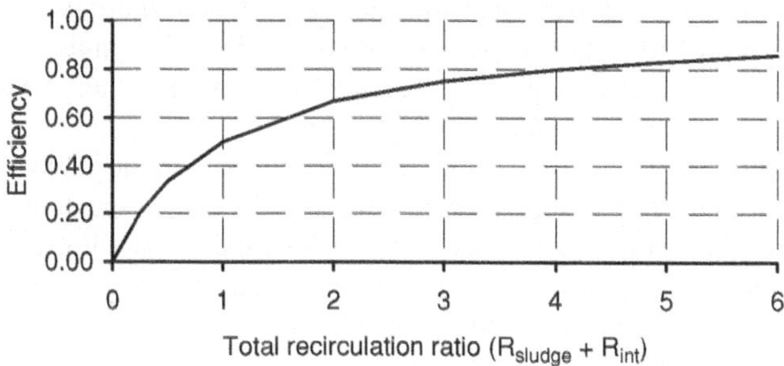

Figure 7.2.   Maximum theoretical nitrate removal efficiency values in systems with pre-denitrification as a function of the total recirculation ratio ($R_{int} + R_{sludge}$)

The advantages of the pre-denitrification systems are:

- low detention time in the anoxic zone, compared to the post-denitrification systems
- reduction in the oxygen consumption in view of the stabilisation of the organic matter using nitrate as electron acceptor in the anoxic zone
- possibility of the reduction of the volume of the aerobic zone, as a result of the stabilisation of part of the BOD in the anoxic zone (the reduction in the volume should be such as not to affect nitrification)
- there is no need for a separate reaeration tank, like in the post-denitrification arrangement

The disadvantage is that, to reach high denitrification efficiencies, very high internal recirculation ratios are needed, which is not always economically advisable. For this reason, the internal recirculation ratios are limited to 400% or 500%. The internal recirculation pumping stations are designed to work under low heads (the water level in the anoxic and aerobic zones is practically the same) and high flows.

(b)   Post-denitrification (removal of nitrogen with carbon from endogenous respiration)

The reactor comprises an aerobic zone followed by an anoxic zone and, optionally, a final aerobic zone. The removal of carbon and the production of nitrates occur in the aerobic zone. The nitrates formed enter the anoxic zone, where they are reduced to gaseous nitrogen. Thus, there is no need of internal recirculations, as in the pre-denitrification system. This process, without the final aerobic zone, is named Wuhrmann process.

The disadvantage is that denitrification is carried out under endogenous conditions, since most of the organic carbon to be used by the denitrifying bacteria has been removed in the aerobic zone. Therefore, the denitrification rate is slower (0.015 to 0.045 mgNO$_3^-$-N/mgVSS·d), which implies longer detention times in the anoxic zone, compared with the pre-denitrification alternative.

A possibility to increase the denitrification rate is by the addition of an external carbon source, such as methanol. Although this practice leads to high denitrification rates, it is less frequently applied in developing countries, since it requires the continuous addition of a chemical product.

Another possibility to increase the denitrification rate in the anoxic zone is by directing part of the raw sewage straight to the anoxic zone, by-passing the aerobic zone. Even if a considerable fraction of BOD from the by-pass line could still be removed in the anoxic zone, the introduction of a non-nitrified nitrogen (ammonia) into the anoxic zone could be a problem, as it could deteriorate the effluent quality.

The final zone is for reaeration, with a short detention time (approximately 30 minutes). The main purposes are the release of gaseous nitrogen bubbles and the addition of dissolved oxygen prior to sedimentation.

(c)   Four-stage Bardenpho process

The Bardenpho process corresponds to a combination of the two previous arrangements, comprising pre-denitrification and post-denitrification, besides the final reaeration zone. The nitrogen removal efficiency is high, of at least 90%, since the nitrates not removed in the first anoxic zone have a second opportunity to be removed, in the second anoxic zone. The disadvantage is that it requires reactors with a larger total volume. However, when high nitrogen removal efficiencies are required, this aspect should not be considered a disadvantage, but a requirement of the process.

(d)   Oxidation ditch

The liquid circulates in the oxidation ditch, passing many times (70 to 100 times a day) through the zones with and without aeration. Aerobic conditions prevail in the aerated zones and a certain distance downstream them. However, as the liquid becomes more distant from the aerator, the oxygen concentration decreases, being liable to reach anoxic conditions at a certain distance. This anoxic zone is limited by the next aerator, where the aerobic conditions restart.

This alternation between aerobic and anoxic conditions allows the occurrence of BOD removal and nitrification in the reactor, besides denitrification itself. The nitrifying and denitrifying bacteria are not harmed by these alternating environmental conditions, so that where there is dissolved oxygen available, nitrates will be formed, and where it lacks, nitrates will be reduced.

The oxidation ditches may have more than one aerator, a condition in which there may be more than one aerobic zone and more than one anoxic zone. Naturally, for the occurrence of denitrification, there should be no overlapping of aerobic zones, leading to a suppression of the anoxic zones, in view of an excessive number of aerators in the reactor.

Conventional ditches (Pasveer ditches) have horizontal-shaft aerators (rotors), while the Carrousel-type ditches have vertical-shaft aerators.

The behaviour of the ditches regarding nitrogen removal takes place according to dynamics different from the other systems, due to the DO gradient and the fast alternation between aerobic and anoxic conditions. Figure 7.3 shows the close relation between DO concentration and nitrification in two ditches in England (von Sperling, 1993b). During the total sampling period, there were successive reductions and increases in the nitrification capacity. The increased DO concentration implies a reduced concentration of ammonia in the ditch, and the decreased DO causes an increase in the concentration of ammonia. The observation of the time series of ammonia and DO presented in Figure 7.3 indicates a fast recovery of the nitrification, after the increase in the DO. Within a certain range, increases in the DO concentration, even if small, imply an almost immediate decrease in the concentration of ammonia. The fast recovery of the nitrification cannot be explained by Monod's conventional kinetics (von Sperling, 1990). It is probable that the frequent alternation between high and low DO concentration zones along the course of the liquid in the ditch creates satisfactory conditions for a

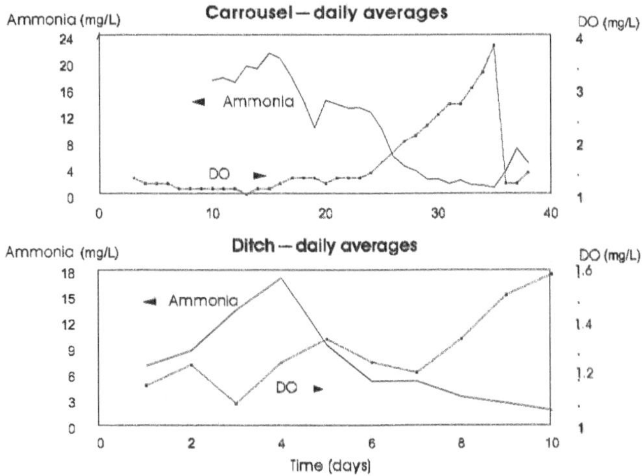

Figure 7.3. Relation between DO and nitrification in two oxidation ditches (von Sperling, 1993b)

fast increase in the growth rate of the nitrifying organisms, as soon as the average DO concentration in the tank (or the size of the higher DO concentration zones) increases. This same behaviour was noticed by the author in other ditches in England.

To obtain a higher denitrification efficiency in the ditches, there should be an automated control of the dissolved oxygen, altering the oxygen transfer rate by means of (a) turning on/off the aerators, (b) varying the aerator submergence (acting on the outlet weir level or on the vertical shaft of the aerators), or (c) varying the rotation speed of the aerators. This is due to the fact that, with a variable influent load over the day, the size of the aerobic zone would vary if the oxygen transfer rate were constant. As the aerators are usually designed for conditions of peak organic load, there could be a good balance between oxygen production and consumption in those moments, thus allowing the coexistence of aerobic and anoxic zones. However, in periods of lower load, such as during the night, the oxygen production would become larger than the consumption, making the anoxic zone decrease or contingently disappear, thus reducing substantially the overall nitrogen removal efficiency. For this reason, it is important that the aeration capacity is variable, allowing the oxygen production rate to follow the consumption rate, generating relatively stable DO concentrations and anoxic zone sizes. However, the selection of the DO set point is not simple: sufficient aerobic zones are needed for nitrification, but, at the same time, sufficient anoxic zones are needed for denitrification. In other words, enough oxygen should be provided for nitrification, but not excessively to inhibit the denitrification.

(e)   Intermittent operation reactors (sequencing batch reactors)

The sequencing batch systems have a cyclic operation. Each cycle consists of a sequence of fill, reaction, settle, draw and, if necessary, idle stages. Depending on the load generation profile over the day, the system may have just one tank or more than one (two, three or more) in parallel, each one in a different stage of the cycle. Further details on sequencing batch reactors are presented in Chapter 8.

During the fill period, some nitrates remaining from the previous cycle may be removed, if the aerators are turned off. Therefore, pre-denitrification with organic carbon from the raw sewage occurs. An anoxic stage follows the aerobic reaction stage, in which post-denitrification occurs under endogenous conditions.

The advantage of the system is its conceptual simplicity, which does not require separate recirculations and sedimentation tanks.

7.1.2 Comparison between the performances of the biological nitrogen removal systems

Table 7.1 presents a comparison between the capacities of the systems described to meet different discharge objectives. If the aerobic sludge age is greater than approximately 5 days (or even greater, if the temperature, the DO, and the pH in the reactor are low), all the processes are capable to nitrify and meet an effluent ammonia level of 5 mg/L. In terms of total nitrogen, all the variants presented can meet targets ranging from 8 to 12 mg/L, but only the four-stage Bardenpho system can produce an effluent between 3 and 6 mg/L, or even less.

7.1.3 Design criteria for biological nitrogen removal

The main criteria, coefficients and rates for the design of systems with pre-denitrification, post-denitrification and four-stage Bardenpho are presented in Tables 7.2 and 7.3. The values of Table 7.3 refer to the N removal mathematical

Table 7.1. Capacity of several processes to meet different discharge targets for ammonia and total nitrogen

Process	Ammonia <5 mg/L[a]	Total nitrogen		
		8–12 mg/L	6–8 mg/L	3–6 mg/L
Reactor fully aerobic	X	–	–	–
Reactor with pre-denitrification	X	X	X[b]	–
Reactor with post-denitrification	X	X	–	–
Four-stage Bardenpho	X	X	X	X
Oxidation ditch	X	X	X[c]	–
Sequencing batch reactor	X	X	–	–

[a] nitrification will occur consistently provided that aerobic θ_c is higher than approximately 5 d
[b] with high internal recirculation ratios (R_{int} between 200 and 400%)
[c] with efficient automatic control of dissolved oxygen
Source: Table prepared based on information from EPA (1993)

Table 7.2. Design criteria for biological nitrogen removal

Parameter	System with pre-denitrification	System with post-denitrification	Four-stage Bardenpho
MLVSS (mg/L)	1500–3500	1500–3500	1500–4000
Total θ_c(d)	6–10	6–10	10–30
Aerobic θ_c(d)	$\geq$5	$\geq$5	$\geq$8
HDT – 1st anoxic zone (hour)	0.5–2.5	–	1.0–3.0
HDT – aerobic zone (hour)	4.0–10.0	5.0–10.0	5.0–10.0
HDT – 2nd anoxic zone (hour)	–	2.0–5.0	2.0–5.0
HDT – final aerobic zone (hour)	–	–	0.5–1.0
BOD removal ratio – anoxic zone/ aerobic zone	0.7	0.7	0.7
Sludge recirculation ratio R_{sludge} (Q_r/Q) (%)	60–100	100	100
Internal recirculation ratio R_{int} (Q_{int}/Q) (%)	100–400	–	300–500
Power level in the anoxic zone (W/m^3)	5–10	5–10	5–10
Average DO in the aerobic zone	2.0	2.0	2.0

Source: Adapted from IAWPRC (1987), Metcalf and Eddy (1991), Randall *et al.* (1992), EPA (1987, 1993)

Table 7.3. Typical values of the rates and kinetic and stoichiometric coefficients for the modelling of nitrification and denitrification

Stage	Coefficient or rate	Unit	Typical values or range
Nitrification	Spec. nitrifiers growth rate μ_{max} (20 °C)	d^{-1}	0.3–0.7
	Half-saturation coefficient K_N (ammonia)	mgNH$_4^+$/L	0.5–1.0
	Half-saturation coefficient K_O (oxygen)	mgO$_2$/L	0.4–1.0
	Temperature coefficient θ for μ_{max}	–	1.10
	Yield coefficient for nitrifiers Y_N	mg cells/mgNH$_4^+$ oxidation	0.05–0.10
	O$_2$ consumption	mg O$_2$/mgNH$_4^+$ oxidation	4.57
	Alkalinity consumption	mg CaCO$_3$/mgNH$_4^+$ oxidation	7.1
Denitrification	Denitrification rate SDR – 1st anoxic zone	mgNO$_3^-$/mgVSS·d	0.03–0.11
	Denitrification rate SDR – 2nd anoxic zone	mgNO$_3^-$/mgVSS·d	0.015–0.045
	Fraction of ammonia in the excess sludge	mgNH$_4^+$/mg VSS	0.12
	Temperature coefficient θ for denitrif. rate	–	1.08–1.09
	O$_2$ economy	mgO$_2$/mgNO$_3^-$	2.86
	Alkalinity economy	mgCaCO$_3$/mgNO$_3^-$	3,5

See Sections 6.3 and 6.4 for interpretation of the values
Source: Eckenfelder and Argaman (1978), Arceivala (1981), Barnes and Bliss (1983), Sedlak (1991), Metcalf and Eddy (1991), Randall *et al.* (1992), EPA (1993) and Orhon and Artan (1994)

modelling, discussed in Sections 6.3 and 6.4. The design criteria for sequencing batch reactors are presented in Chapter 8.

7.1.4 Design considerations

Specific design aspects for activated sludge plants with biological nitrogen removal are presented next. The information was extracted from Randall *et al.* (1992) and EPA (1993).

(a)   Primary sedimentation

Primary sedimentation offers the usual advantages related to systems without biological nutrient removal, such as reduced volume of the reactor and reduced aeration capacity needs, besides reduced floating materials and solids in the supernatant and drained liquids from the sludge processing units. However, the primary sedimentation reduces the BOD:TKN ratio, which may reduce the denitrification rate to be achieved. This may not be a problem if a large part of the influent BOD is soluble. A BOD_5:TKN ratio >5 favours denitrification. In case primary sedimentation is included, the detention time should be reduced, and conditions should be provided so that part of the raw sewage can be directly by-passed to the reactor to increase the organic carbon necessary for denitrification.

(b)   Aeration systems

Mechanical and diffused air aeration systems can be used. The aeration capacity estimated in the design should comprise the carbonaceous and the nitrogenous demand under peak conditions. Plug-flow reactors should provide a larger aeration capacity in the inlet end of the tank. Point aerators, such as mechanical aerators, allow the occurrence of denitrification in the reactor itself, due to the possible presence of anoxic zones in the reactor, on the bottom and at corners of the reactor. Automatic control of the dissolved oxygen is advisable and, in most of the cases, necessary.

(c)   Stirrers

In the anoxic zones, the stirrers should maintain the solids in suspension, but should avoid the aeration of the liquid mass. The most used types of stirrers are low speed devices, with either vertical shaft or submerged horizontal shaft. Submersible stirrers are more flexible, as they allow the adjustment of the level and direction of the mixing, although some models have not shown a good performance, making the vertical shaft stirrers to be more frequently used (Randall *et al.*, 1992). Stirrers are not essential in systems with intermittent aeration if the time with the aerators turned off is short. The power level of the stirrers varies from 5 to 10 W/m^3, but the range of lower values does not guarantee good mixing between the influent and the recirculated liquids. It is advisable to mix them when they enter the anoxic zone. The location of the stirrers is crucial for the operation, and manufacturers should be consulted about that.

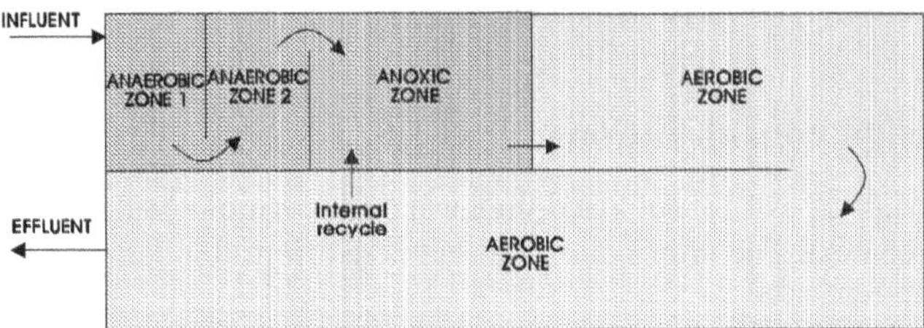

Figure 7.4.   Configuration of a U-shaped reactor, with internal recirculation through the dividing wall between the anoxic and the aerobic zones

(d)   Internal recirculation

Nitrate pumping from the aerobic zone to the anoxic zone is a characteristic of systems with pre-denitrification. This frequently requires pumping from the final end to the initial end of the reactor. The pumping line can be above the tank wall or even through it. Pumping through the wall can occur in U-shaped reactors, in which the inlet is located close to the outlet (see Figure 7.4).

The water level in the aerobic and anoxic zones is frequently approximately the same, which implies a very low pumping head. Centrifugal pumps can be used in smaller plants, but it is more advantageous to use low-speed, axial-flow pumps in larger plants, thus reducing the energy required and the introduction of oxygen into the anoxic zone. It is usually preferable to adopt a larger number of small pumps to allow a variable recycle flow.

(e)   Reactor

The design of the anoxic and aerobic zones should allow flexibility in the entrance of the influent and recirculation lines. The anoxic zone can be divided into compartments by submerged walls. The U-reactor facilitates the internal recirculation, which can be achieved through the dividing wall between the anoxic and aerobic zones (see Figure 7.4).

(f)   Secondary sedimentation tanks

Activated sludge plants with biological nutrient removal are susceptible to the same operational problems as those of the typical activated sludge system, besides other problems associated with the existence of the anoxic zone. Bulking sludge can occur, associated with several possible causes, including low DO concentrations and excessive detention times in the anoxic zone. The presence of scum is also possible, and plants with biological nutrient removal should be designed assuming the presence of scum, thus providing conditions for its removal in the secondary sedimentation tanks.

7.1.5 Design example of a reactor with nitrification and pre-denitrification

Design the reactor for biological nitrogen removal (nitrification and denitrification), in a conventional activated sludge system with *pre-denitrification* (anoxic zone followed by aerobic zone). The input data are the same as those of the example in Chapter 5. The data of interest are:

Raw sewage (see Section 5.1):

- Average influent flow: $Q = 9{,}820$ m^3/d
- Influent TKN load $= 496$ kg/d
- Influent TKN concentration $= 51$ mg/L

Final effluent:

- Effluent TKN $= 2$ mg/L (desired)

Primary sedimentation tank (see Section 5.3.2):

- TKN removal efficiency in the primary sedimentation tank $= 20\%$

Reactor (see Section 5.3.3):

- Sludge age $= 6$ d
- MLVSS $= 3{,}000$ mg/L
- DO in the reactor: OD $= 2$ mg/L
- pH in the reactor: pH $= 6.8$
- Temperature of the liquid (average in the coldest month): $T = 20\,°C$

Nitrification coefficients (adopted in this example – see Table 7.3):

- Maximum specific growth rate (μ_{max}) $(20\,°C) = 0.5$ d^{-1}
- Ammonia half-saturation coefficient $(K_N) = 0.70$ gNH$_4^+$/m^3
- Oxygen half-saturation coefficient $(K_O) = 0.80$ gO$_2$/m^3
- Yield coefficient for nitrifiers $(Y_N) = 0.08$ gNitrif/gNH$_4^+$ oxidised
- Temperature coefficient for $\mu_{max}(\theta) = 1.1$
- O$_2$ demand for nitrification $= 4.57$ gO$_2$/gNH$_4^+$ oxidised

Denitrification coefficients (adopted in this example – see Table 7.3):

- Denitrification rate in the pre-anoxic zone $(20\,°C) = 0.08$ kgNO$_3^-$/kgVSS·d
- Temperature coefficient for the denitrification rate $(\theta) = 1.09$
- O$_2$ production in denitrification $= 2.85$ g O$_2$/gNO$_3^-$ reduced
- Fraction of ammonia in the excess sludge $= 0.12$ kgNH$_4^+$/kgVSS

Reactor (values adopted in this example – see Table 7.2):

- Fraction of the reactor as pre-anoxic zone: 0.25 (25% of the volume of the reactor is a pre-anoxic zone)
- Fraction of the reactor as aerobic zone: 0.75 (75% of the volume of the reactor is an aerobic zone)

- Ratio between the BOD removal rate under anoxic and aerobic conditions: 0.7 (the BOD removal rate under anoxic conditions is 70% of the rate under aerobic conditions)
- Sludge recirculation ratio: 100%
- Internal recirculation ratio (aerobic zone to anoxic zone): 300%

All the TKN, NH_4^+ and NO_3^- concentrations are expressed in terms of nitrogen. The example uses indistinctively TKN and NH_4^+ to represent the ammonia at any point of the process.

Solution:

(a)   TKN removal in the primary sedimentation

TKN removal efficiency in the primary sedimentation tank = 20% (input data)

$$TKN_{eff.\ primary} = TKN_{inf.\ prim.}(100 - E)/100 = 496\,kg/d(100 - 20)/100$$
$$= 397\,kg/d = 51\,mg/L.(100 - 20)/100 = 40\,mg/L$$

The considerations made in Section 7.1.4.a, regarding the desirability of not having primary sedimentation in systems with biological nutrient removal, are also applicable here. However, for compatibility with the design already made in Chapter 5, the primary sedimentation tank is maintained here in the flowsheet of the plant.

(b)   Volume of the reactor

According to the conception of the reactor, 25% of the total volume is represented by the anoxic zone and 75% is represented by the aerobic zone (see input data of the problem).
   The sludge age can be divided as follows:

- *Total* sludge age = 6 d (input data to the problem)
- *Aerobic* sludge age = 6 × 0.75 = 4.5 d

Volume required for the reactor (calculated in Section 5.3.7): V = 2,051 m³
   According to Table 7.2 and to what is stated in the example, the BOD removal rate in the anoxic zone is slower, being 70% of the removal rate in the aerobic zone. As 25% of the volume of this reactor consists of an anoxic zone, the total volume required should be multiplied by a correction factor:

$$V_{tot} = V \cdot \frac{(F_{anox} + F_{aer})}{(0.7 \times F_{anox} + F_{aer})} = 2{,}051 \times \frac{(0.25 + 0.75)}{(0.7 \times 0.25 + 0.75)}$$
$$= 2{,}051 \times 1.08 = 2{,}215\,m^3$$

Therefore, the total volume of the reactor should be multiplied by the correction factor 1.08, resulting in **2,215 m³**, instead of 2,051 m³.

The volumes of the anoxic and aerobic zones are:

- $V_{anox} = 0.25 \times 2{,}215 = 554\ m^3$
- $V_{aer} = 0.75 \times 2{,}215 = 1{,}661\ m^3$

The total hydraulic detention time is: $t = V/Q = 2{,}215/9{,}820 = 0.226\ d = 5.4$ hours
The detention times in the anoxic and aerobic zones are:

- $t_{anox} = 0.25 \times 5.4 = 1.35$ hours
- $t_{aer} = 0.75 \times 5.4 = 4.05$ hours

The detention time in the pre-anoxic zone is within the range presented in Table 7.2. The resultant sludge ages should also be multiplied by the correction factor 1.08:

- *Total* sludge age $= 6.0 \times 1.08 = $ **6.5 d**
- *Aerobic* sludge age $= 4.5 \times 1.08 = $ **4.9 d**

(c)   Calculation of the growth rate of the nitrifying bacteria (μ_{max}) according to the environmental conditions in the reactor

The calculations below follow the methodology presented in Example 6.2 (although with data different from those of the referred to example).

Maximum specific growth rate: $\mu_{max} = 0.5\ d^{-1}$ (statement of the problem)

Influencing factors on μ_{max} (see statement of the problem):

- Ammonia concentration in the reactor: $NH_4^+ = 2$ mg/L (desired concentration for the effluent)
- DO concentration in the reactor: $DO = 2$ mg/L
- pH in the reactor: $pH = 6.8$
- Temperature: $T = 20\ °C$
- Effect of the ammonia concentration:

$$\mu = \mu_{max} \cdot \left[\frac{NH_4^+}{K_N + NH_4^+} \right] = 0.5 \cdot \left[\frac{2.0}{0.7 + 2.0} \right] = 0.37\ d^{-1}$$

(μ_{max} correction factor $= 0.37/0.50 = 0.74$)

- Effect of the DO concentration in the reactor:
  According to Equation 6.8 and Table 6.5:

$$\mu = \mu_{max} \cdot \left[\frac{DO}{K_O + DO} \right] = 0.5 \cdot \left[\frac{2.0}{0.6 + 2.0} \right] = 0.36$$

(μ_{max} correction factor $= 0.36/0.50 = 0.72$)

- Effect of the pH in the reactor:

  According to Equation 6.7:

  $$\mu_{max(pH)} = \mu_{max}[1 - 0.83(7.2 - pH)]$$
  $$= 0.5 \times [1 - 0.83 \times (7.2 - 6.8)] = 0.33$$

  (μ_{max} correction factor $= 0.33/0.50 = 0.66$)

- Effect of the temperature:

  According to Equation 6.6, and adopting $\theta = 1.10$:

  $$\mu_{max(T)} = \mu_{max(20\,°C)} \cdot \theta^{(20-20)} = 0.50 \times 1.10^{(20-20)} = 0.50 \text{ d}^{-1}$$

  (μ_{max} correction factor $= 0.50/0.50 = 1.00$) (without alteration, because the temperature is equal to the standard temperature)

- Integrated effect of the environmental conditions (multiple correction factor):

  $$0.74 \times 0.72 \times 0.66 \times 1.00 = 0.35$$

  The specific growth rate of the nitrifying bacteria under these environmental conditions is 35% of the maximum rate ($\mu_N = 0.35 \cdot \mu_{max}$). Therefore, μ_N is:

  $$\mu_N = 0.35 \times \mu_{max} = 0.35 \times 0.50 = 0.18 \text{ d}^{-1}$$

(d) Minimum aerobic sludge age required for total nitrification

According to Equation 6.9, the minimum aerobic sludge age required for total nitrification is:

$$\theta_c = \frac{1}{\mu_N} = \frac{1}{0.18} = \textbf{5.6 d}$$

The aerobic sludge age obtained in the design is 4.9 days, therefore being lower than the minimum required value of 5.6 days to ensure full nitrification under the specified environmental conditions. The aerobic sludge age can be increased by increasing the volume of the aerobic zone, by increasing the MLVSS concentration, or by increasing the aerobic fraction of the reactor, until the minimum value required is reached. However, no changes are made in this example, and it is only verified whether the effluent ammonia concentration is still acceptable under these conditions.

(e) Calculation of the fraction of nitrifiers in the mixed liquor volatile suspended solids

The calculations below follow the methodology presented in Example 6.4 (although with data different from those of the referred to example).

- Net production of biological solids in the reactor:

Net P_{xv} = 1.026 kgVSS/d (calculated in Section 5.3.6.b; this value is not affected by the increased volume of the reactor, because the BOD load removed remained the same)

- Ammonia load to be oxidised:

Influent TKN load = $Q \cdot TKN_o$ = 9820 × 40/1000 = 393 kg/d
Effluent TKN load = $Q \cdot TKN_e$ = 9820 × 2/1000 = 20 kg/d
TKN load in the excess sludge = (ammonia fraction in the excess sludge) × P_{xv} = 0.12 × P_{xv} = 0.12 × 1,026 = 123 kg/d
TKN load to be oxidised = influent TKN – effluent TKN – TKN in excess sludge = 393 – 20 – 123 = 250 kg/d

- Production of nitrifying bacteria:

According to Equation 6.13, the production of nitrifying bacteria is:

$$P_{xN} = \Delta X_N/\Delta t = Y_N \cdot TKN_{oxidised} = 0.08 \times 250 = 20 kgX_{N/d}$$

- f_N ratio
The f_N ratio, which corresponds to the fraction of nitrifying bacteria in the volatile suspended solids (X_N/X_v), can then be calculated by the quotient between the production of X_N and the production of X_V (Equation 6.11):

$$f_N = \frac{P_{xN}}{P_{xv}} = \frac{20}{1,026} = \mathbf{0.019\, gX_N/gX_V}$$

In this case, the nitrifying bacteria represent 1.9% of the total biomass (expressed as MLVSS).

(f)   Calculation of the nitrification rate

According to Equation 6.15, the nitrification rate is given by:

$$\frac{\Delta TKN}{\Delta t} = f_N \cdot \frac{X_V \cdot \mu_N}{Y_N} = 0.019 \cdot \frac{3,000 \times 0.18}{0.08} = 128\, gTKN/m^3 \cdot d$$

The TKN load susceptible to being oxidised is (Equation 6.16):

$$L_{TKN} = \frac{V_{aer}}{10^3} \cdot \frac{\Delta TKN}{\Delta t} = \frac{1,661}{1,000} \times 128 = 213\, kgTKN/d$$

This value of 213 kgTKN/d is lower than the expected value to be oxidised (250 kgTKN/d, calculated in Item f above). Therefore, the concentration of effluent TKN will be higher than the concentration initially assumed (2 mg/L). If this value were higher than 250 kgTKN/d, the load liable to be oxidised would naturally be 250 kgTKN/d.

(g)   Calculation of the concentration of effluent ammonia

- Calculation of the TKN loads:

  Influent TKN load = 393 kg/d (calculated in Item e)
  TKN load in the excess sludge = 123 kg/d (calculated in Item e)
  TKN load liable to be oxidised = 213 kg/d (calculated in Item f)

  Effluent TKN load = influent TKN – TKN in the excess sludge – TKN liable to be oxidised = 393 – 123 – 213 = 57 kg/d

- Concentration of effluent TKN:

$$TKN_e = \frac{\text{effluent load}}{\text{flow}} = \frac{57 \times 1,000}{9,820} = \mathbf{6\ mgTKN/L}$$

The concentration of effluent TKN (or ammonia) in the system is, therefore, 6 mg/L. The value initially assumed had been 2 mg/L. Since this value influences the calculation of μ_N, the calculations of Item d can be redone by using the 6 mg/L concentration, and so forth, until a convergence is obtained, with an ammonia value between 2 and 6 mg/L. However, the difference obtained in this first iteration is not great, which justifies the fact that the iterative calculations are not made in this example.

(h)   Ammonia removal efficiencies

The efficiency of the system in the removal of TKN is:

$$E = (TKN_o - TKN_e)/TKN_o = (51 - 6)/51 = 0.88 = \mathbf{88\%}$$

(i)   Mass of VSS in the pre-anoxic zone

Volume of the pre-anoxic zone: $V_{anox} = 554\ m^3$ (calculated in Item b)

Mass of VSS in the pre-anoxic zone = $V_{anox} \cdot X_v/1000 = 554 \times 3,000/1,000 = 1,662\ kgVSS$

(j)   Recirculation of nitrates to the anoxic zone

According to the statement of the problem:

- Sludge recirculation ratio: $R_{sludge} = 1.0$ (100%)
- Internal recirculation ratio (from the aerobic zone to the anoxic zone): $R_{int} = 3.0$ (300%)
- Total recirculation ratio: $R_{tot} = 1.0 + 3.0 = 4.0$

(l)   Specific denitrification rate

$$SDR = 0.08\ kgNO_3^-/kgVSS \cdot d\ (20\,^\circ C)$$

Correction for temperature (Equation 6.29):

$$SDR_T = SDR_{20\,°C}\cdot\theta^{(T-20)} = 0.08 \times 1.09^{(20-20)} = 0.08 \ kgNO_3^-/kgVSS\cdot d$$

No correction was necessary due to the fact that the average temperature of the liquid in the coldest month is 20 °C. Accordingly, there is no need for correction due to the presence of DO (Equation 6.27), since it is assumed that the DO in the anoxic zone is equal to zero.

(m)   Nitrate loads

- Load of NO_3^- produced in the aerobic zone = load of oxidised TKN = 213 kg/d (calculated in Item g)
- Load of NO_3^- recirculated to the anoxic zone by the return of sludge = $213 \times R_{sludge}/(R_{tot} + 1) = 213 \times 1.0/(4.0 + 1) = 43$ kg/d
- Load of NO_3^- recirculated to the anoxic zone by the internal recirculation = $213 \times R_{int}/(R_{tot} + 1) = 213 \times 3.0/(4.0 + 1) = 128$ kg/d
- Load of total NO_3^- recirculated = $43 + 128 = 171$ kg/d
- Load of NO_3^- liable to reduction in the pre-anoxic zone = SDR $\times$ VSS mass = $0.08 \times 1662 = 133$ kg/d

As this value of 133 kg/d is lower than the total load recirculated (171 kg/d), the nitrate load to be really reduced will be **133 kg/d**. If the value of the load susceptible to reduction were higher than 171 kg/d, the load to be really reduced would naturally be 171 kg/d.

A means to increase the load susceptible to reduction would be to increase the MLVSS concentration or the volume of the anoxic zone. In this example, such changes are not made, and the concentration of effluent nitrate is calculated taking into account the conditions initially assumed.

- Load of effluent NO_3^- = Load of NO_3^- produced – Load of NO_3^- to denitrify = $213 - 133 = 80$ kg/d

(n)   Concentration of effluent nitrate

$$NO_{3eff} = \frac{\text{effluent load}}{\text{flow}} = \frac{80 \times 1,000}{9,820} = \textbf{8 mgNO}_3^-\textbf{/L}$$

Removal efficiency of the nitrate formed:

$$E = (\text{load produced} - \text{effluent load})/\text{load produced}$$

$$= (213 - 80)/213 = 0.62 = \textbf{62\%}$$

(o)   Summary of the nitrogen concentrations

- Influent (raw sewage):

Total nitrogen = **51 mg/L** (assuming that the total nitrogen in the influent is the same as TKN)

- Final effluent:

  Ammonia = **6 mg/L** (calculated in Item g)
  Nitrate = **8 mg/L** (calculated in Item n)
  Total nitrogen = 6 + 8 = **14 mg/L**

(p) Summary of the removal efficiencies

- Ammonia removal efficiency: E = **88%** (calculated in Item h)
- Nitrate removal efficiency: E = **62%** (calculated in Item n)
- Total nitrogen removal efficiency: E = (51 − 14)/51 = **73%**

These values meet the European Community's Directive (CEC, 1991) for discharge of urban wastewater in sensitive water bodies, for the population range between 10,000 inhabitants and 100,000 inhabitants, which means a total nitrogen concentration lower than 15 mg/L or a minimum removal efficiency between 70 and 80%. If the population were larger than 100,000 inhabitants, the total nitrogen removal should be optimised to allow the compliance with the stricter standard, which is, in this case, 10 mg/L of total nitrogen.

(q) Oxygen consumption

O_2 consumption for nitrification = 4.57 × load of TKN oxidised (Equation 6.17)
$$= 4.57 \times 213 \text{ kg/d} = \textbf{973 kgO}_2\textbf{/d}$$

O_2 economy with denitrification = 2.86 × load of reduced NO_3^- (Section 6.4.2.a)
$$= 2.86 \times 133 \text{ kg/d} = \textbf{380 kgO}_2\textbf{/d}$$

7.2 BIOLOGICAL REMOVAL OF NITROGEN AND PHOSPHORUS

7.2.1 Processes most frequently used

This section presents a description of the main processes used for the combined removal of nitrogen and phosphorus. The processes employed for the removal of phosphorus alone are not discussed here, due to the difficulties they face in the presence of nitrates in the anaerobic zone. In warm-climate regions, nitrification occurs almost systematically in activated sludge plants. Thus, if an efficient denitrification is not provided in the reactor, a considerable amount of nitrates will be returned to the anaerobic zone through the recirculation lines, hindering the maintenance of strictly anaerobic conditions. For this reason, the removal of nitrogen is encouraged, even if, under some conditions, the removal of only phosphorus would be necessary in terms of the receiving body requirements.

The main processes used for the combined removal of N and P are (see Figure 7.5):

- A^2O process (3-stage Phoredox)
- 5-stage Bardenpho process (Phoredox)

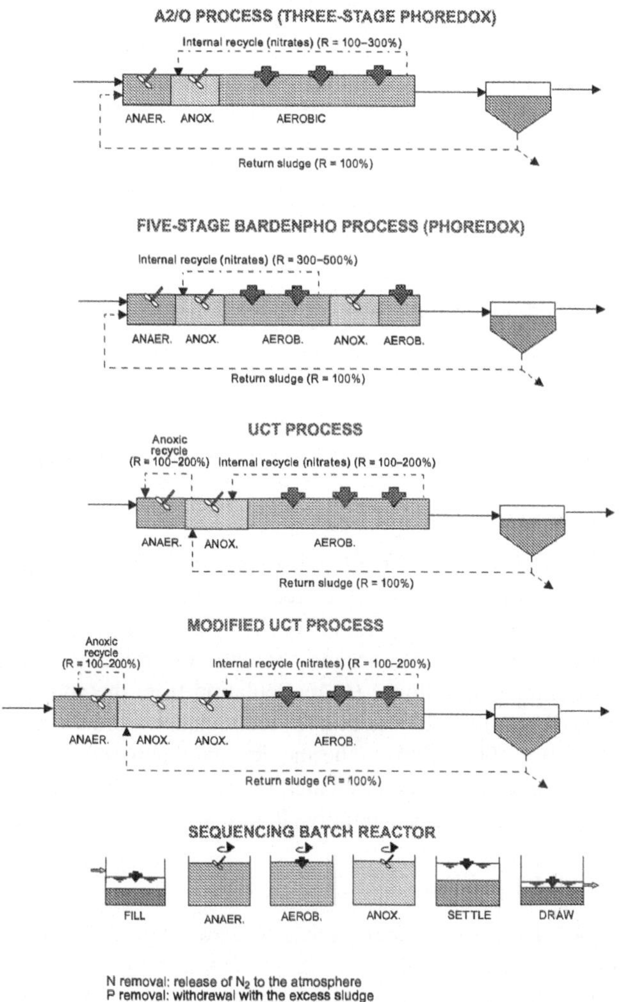

Figure 7.5.   Main processes for the combined removal of nitrogen and phosphorus

- UCT Process
- Modified UCT Process
- Intermittent operation reactors (sequencing batch reactors)

The literature presents divergent nomenclature with relation to some processes, in view of variations between commercial and technical names. A brief description of the main variants is presented below (WRC, 1984; Sedlak, 1991).

(a) A^2O Process (3-stage Phoredox)

The name A^2O originates from 'anaerobic/anoxic/aerobic', which describes the basic flow line of the process. In other locations, this variant has been named Phoredox. Nitrogen removal results mainly from the internal recirculation from the aerobic zone to the anoxic zone. The alternation between anaerobic and aerobic conditions, necessary for phosphorus removal, is reached by means of the sludge recirculation, which is directed to the anaerobic zone. A high denitrification efficiency is required because the return of nitrates to the anaerobic zone can harm phosphorus removal.

(b)   Five-stage Bardenpho process (Phoredox)

This process is similar to the four-stage Bardenpho (Section 7.1.1.c), with the inclusion of one anaerobic zone upstream. The returned sludge is directed to the anaerobic zone. The sludge age is usually higher than in other processes, ranging from 10 to 30 days.

(c)   UCT Process (University of Cape Town)

The UCT process consists of three zones (anaerobic, anoxic and aerobic). The main aspect that distinguishes the UCT process from the others is that it prevents nitrates from returning to the anaerobic zone. In view of that, the recirculation of sludge is directed to the anoxic zone, and not to the anaerobic zone. There are two internal recirculations, as follows: (a) recirculation from the anoxic zone to the anaerobic zone (R = 100 to 200%), and (b) recirculation from the aerobic zone to the anoxic zone (R = 100 to 200%). The anoxic zone should provide denitrification capacity to the load of nitrates recirculated to avoid their return to the anaerobic zone. The VIP process (Virginia Initiative Plant) is similar to the UCT process.

(d)   Modified UCT process

The modified UCT process separates the anoxic zone into two. The first zone receives the return sludge, and allows the recirculation from the anoxic zone to the anaerobic zone. This first zone is used to reduce only the nitrates from the sludge recirculation line. The second anoxic zone receives the internal recirculation from the aerobic zone, being the zone where most of the denitrification occurs. In separating this second anoxic zone from the first one, which recirculates to the anaerobic zone, the nitrate in excess can be recirculated without harming the process. To allow flexibility to operate as UCT or modified UCT process, the recirculation system to the anaerobic zone should be installed from both anoxic zones.

(e)   Sequencing batch reactors

The sequencing batch process is similar to that described in Section 7.1.1.e., and includes, at the beginning of the operational cycle, an anaerobic stage. The sequence comprises the aerobic reaction, anoxic reaction, sedimentation and

Table 7.4. Capacity of the processes to meet different phosphorus discharge standards

Process	Effluent: 0.5 mgP/L				Effluent: 1.0 mgP/L				Effluent: 2.0 mgP/L			
	Biol	Biol + C	Biol + F	Biol + C + F	Biol	Biol + C	Biol + F	Biol + C + F	Biol	Biol + C	Biol + F	Biol + C + F
A^2O/3-stage Phoredox.	N	N	N	Y	V	Y*	V	Y	Y	Y	Y	Y
5-stage Bardenpho/ Phoredox	N	N	N	Y	V	Y*	V	Y	Y	Y	Y	Y
UCT/VIP/ Modif. UCT	N	N	N	Y	V	Y*	V	Y	Y	Y	Y	Y
Sequencing batch reactor	N	N	N	Y	V	Y*	V	Y	Y	Y	Y	Y

Biol = only biological treatment   Biol + C = biol. treatment + coagulant
Biol + F = biol. treatment. + filtration   Biol + C + F: = biol treatment. + coagulant + filtration
N = No: does not comply with the P standard   V = meets P standard in a variable or marginal form
Y = Yes: complies with the P standard   Y* = meets P standard with a highly efficient clarification
Source: Adapted from EPA (1987b)

Table 7.5. Typical concentrations of total nitrogen and ammonia in the effluent, and sensitivity to low BOD_5:P ratio values in the influent

Process	Ammonia (mg/L)	N total (mg/L)	Sensitivity to low BOD_5:P values (*)
A^2O/3-stage Phoredox.	<5	6–12	High
5-stage Bardenpho/Phoredox	<5	<6	High
UCT/VIP/Modif. UCT	<5	6–12	Low
Sequencing batch reactor	<5	6–12	Variable with the cycle

(*): desirable: values of the BOD_5:P ratio higher than 20
Source: Adapted from Sedlak (1991)

supernatant withdrawal phases. Further details on sequencing batch reactors are found in Chapter 8.

7.2.2 Selection among the biological nitrogen and phosphorus removal processes

Tables 7.4 and 7.5 present a comparison among the main processes used for biological phosphorus removal. Effluent polishing is also included, in case very high quality standards for the effluent are necessary. The effluent polishing processes considered are:

- *addition of coagulant agents* (metallic ions): phosphorus precipitation
- *filtration of the effluent*: removal of the phosphorus present in the suspended solids
- *combined addition of coagulants and filtration*

Table 7.6. Design criterion for biological nitrogen and phosphorus removal

Parameter	A^{2}O/3-stage Phoredox	UCT/VIP	5-stage Bardenpho/ Phoredox
MLVSS (mg/L)	2000–4000	1500–3500	1500–4000
Total θ_c (d)	5–10	5–10	10–30
Aerobic θ_c (d)	$\geq$5	$\geq$5	$\geq$8
HDT – anaerobic zone (hour)	0.5–1.5	1.0–2.0	1.0–2.0
HDT – 1st anoxic zone (hour)	0.5–1.0	2.0–4.0	2.0–4.0
HDT – aerobic zone (hour)	3.5–6.0	2.5–4.0	4.0–12.0
HDT – 2nd anoxic zone (hour)	–	–	2.0–4.0
HDT – final aerobic zone (hour)	–	–	0.5–1.0
BOD removal ratio – anoxic zone/ aerobic zone	0.7	0.7	0.7
Sludge recirculation ratio R$_{sludge}$ (Q$_r$/Q) (%)	20–50	50–100	50–100
Aerobic to anoxic recirculation ratio (Q$_{aer}$/Q) (%)	100–300	100–200	300–500 ratio
Anoxic to anaerobic recirculation ratio (Q$_{anox}$/Q) (%)	–	100–200	–
Power level in the anoxic and anaerobic zones (W/m^3)	5–10	5–10	5–10
Average DO in the aerobic zone	2.0	2.0	2.0

Source: Adapted from IAWPRC (1987), Metcalf and Eddy (1991), Randall *et al.* (1992), EPA (1987b, 1993)

7.2.3 Design criterion for the biological removal of nitrogen and phosphorus

The main design criteria and parameters for the design of activated sludge systems with biological removal of nitrogen and phosphorus are presented in Tables 7.6 and 7.7. The coefficients and rates related to nitrification and denitrification are listed in Table 7.3. The design of sequencing batch reactors is presented in Chapter 8.

Aspects of importance in the design and operation, which can affect the performance of the plant, are found in Section 6.5.2. Several considerations about design, covered in the section related to biological nitrogen removal (Section 7.1.4), are also valid for biological phosphorus removal.

7.2.4 Design example of a reactor for biological phosphorus removal

Design the anaerobic zone of the reactor from the example in Section 7.1.5, so that the system can also remove phosphorus biologically. The data of interest are:

Influent:

- Average influent flow: Q = 9,820 m^3/d
- Phosphorus concentration in the raw sewage: P$_{inf}$ = 12 mg/L

Table 7.7. Typical values of the rates and kinetic and stoichiometric coefficients for the modelling of biological phosphorus removal

Coefficient or rate	Unit	Range or typical values
Fraction of rapidly biodegradable influent COD (f_{rb})	–	0.15–0.30 (raw sewage) 0.20–0.35 (settled sewage)
COD/BOD$_5$ ratio in the influent	mgCOD/mgBOD$_5$	1.7–2.4
Yield coefficient (Y)	mgVSS/mgBOD$_5$	0.4–0.8
Coefficient of endogenous respiration of the biodegradable SS (K_d)	d^{-1}	0.08–0.09
Biodegradable fraction of the VSS (X_b/X_v) (f_b)	mgSS$_b$/mgVSS	0.55–0.70 (conventional activated sludge) 0.40–0.65 (extended aeration)
VSS/SS (X_v/X) ratio	mgVSS/mgSS	0.70–0.85 (conventional activated sludge) 0.60–0.75 (extended aeration)

Use of the coefficients and rates: see Item 6.5.3
Source: WRC (1994), von Sperling (1996a, 1996b). See also Tables 3.1 and 3.2

- Phosphorus removal efficiency in the primary sedimentation: 20% (adopted)
- BOD concentration in the settled sewage: BOD = 239 mg/L (calculated in Section 5.3.2)
- COD/BOD ratio in the influent = 1.8 (adopted)
- Rapidly biodegradable fraction of the influent COD: $f_{rb} = 0.25$ (Table 7.7, system with primary sedimentation)

Coefficients and ratios:

- $Y = 0.6$ mgVSS/mgBOD (adopted in Section 5.2.a)
- $K_d = 0.08$ d^{-1} (adopted in Section 5.2.a)
- SS$_b$/VSS ratio: $f_b = 0.73$ mgSS$_b$/VSS (calculated in Section 5.3.3)
- VSS/SS ratio in the reactor: VSS/SS = 0.77 (calculated in Section 5.3.6.c)

Reactor:

- Sludge age: $\theta_c = 6$ d (adopted in Section 5.3.1.b)

Effluent:

- Effluent soluble BOD: S = 4 mg/L (calculated in Section 5.3.3)
- Suspended solids: SS$_{eff}$ = 30 mg/L (adopted in Section 5.1)

Solution:

(a)   Removal of P in the primary sedimentation tank

The concentration of P in the effluent from the primary sedimentation tank is:

$$P_{eff\ prim} = P_{inf\ prim} \cdot (100 - E)/100 = 12 \times (100 - 20)/100 = 9.6\ mg/L$$

(b)   Volume of the anaerobic zone

The volume of the reactor with nitrogen removal, determined in Example 7.1.5.c, is 2,215 m^3, with a total hydraulic detention time of 5.4 hours (pre-anoxic and aerobic zones).

From Table 7.6, a hydraulic detention time in the anaerobic zone of 1.2 hours may be adopted. The total detention time will then be:

$$t_{tot} = 5.4 + 1.2 = 6.6 \text{ hours}$$

The f_{an} ratio between the volume of the anaerobic zone and the total volume is proportional to the ratio between the detention times:

$$f_{an} = V_{anaer}/V_{tot} = 1.2/6.6 = 0.18$$

The volume of the anaerobic zone is:

$$V = t \cdot Q = 1.2 \times 9,820/24 = 491 \text{ m}^3$$

(c)   Fraction of P in the suspended solids

- Influent COD:

$$COD = BOD \times (COD/BOD \text{ ratio}) = 239 \times 1.8 = 430 \text{ mg/L}$$

- Propensity factor for phosphorus removal (Equation 6.30):

$$P_f = (f_{rb} \cdot COD - 25) \cdot f_{an} = (0.25 \times 430 - 25) \times 0.18 = 14.9$$

- Phosphorus fraction in the active biomass (Equation 6.31):

$$P/X_a = 0.35 - 0.29 \cdot e^{-0.242 \cdot P_f} = 0.35 - 0.29 \cdot e^{-0.242 \times 14.9}$$
$$= 0.34 \text{ mg P/mg X}_a$$

- Ratio between active SS and volatile SS (Equation 6.32):

$$f_a = \frac{1}{1 + 0.2 \cdot K_d \cdot \theta_c} = \frac{1}{1 + 0.2 \times 0.08 \times 6} = 0.91 \text{ mgX}_a/\text{mgX}_v$$

- Fraction of P in the volatile suspended solids (Equation 6.35):

$$P/X_v = f_a \cdot (P/X_a) = 0.91 \times 0.34 = 0.31 \text{mgP/mgVSS}$$

- Fraction of P in the total suspended solids (Equation 6.36):

$$P/X = \left(\frac{VSS}{SS}\right) \cdot f_a \cdot (P/X_a) = 0.77 \times 0.91 \times 0.34 = 0.24 \text{ mgP/mgSS}$$

This result indicates that the system is able to allow a high accumulation of P in the suspended solids of the excess sludge, representing 24% of the mass of the SS. In terms of design, it is more suitable to work with a safety factor. A maximum value of 7% is suggested in Section 6.5.4.a, which is usual in a large number of wastewater treatment plants with biological phosphorus removal. Therefore, the P/X and P/X_v ratios should be corrected in view of this maximum suggested value of 0.07.

- Correction of the fraction of P in the total suspended solids, for the maximum limit of 7%

$$P/X = 0.07 \text{ mgP/mgSS}$$

- Correction of the fraction of P in the volatile suspended solids, for the maximum limit of 7% in the P/X ratio:

$$P/X_v = (P/X)/(VSS/SS) = 0.07/0.77 = 0.09 \text{ mgP/mgVSS}$$

(d)   Removal of P with the excess sludge

- Concentration of P removed with the excess sludge (Equation 6.40):

$$P_{rem} = \frac{Y}{1 + f_b \cdot K_d \cdot \theta_c} \cdot (P/X_v) \cdot (S_o - S)$$

$$= \frac{0.6}{1 + 0.73 \times 0.08 \times 6} \times 0.09 \times (239 - 4)$$

$$= 0.44 \times 0.09 \times 235 = 9.3 \text{ mgP/L}$$

If this removal value (9.3 mg/L) were higher than the concentration of influent P to the biological stage (in this example, 9.6 mg/L), it should be assumed that the removal is equal to the influent concentration, that is, generating a concentration of effluent soluble P equal to zero.

(e)   Effluent P concentrations

- Effluent soluble P (Equation 6.41):

$$P_{sol\ eff} = P_{tot\ inf} - P_{rem} = 9.6 - 9.3 = \mathbf{0.3\ mgP/L}$$

- Effluent particulate P (present in the effluent SS) (Equation 6.42):

$$P_{part\ eff} = SS_{eff} \cdot (P/X) = 30 \times 0.07 = \mathbf{2.1\ mgP/L}$$

- Effluent total P (Equation 6.43):

$$P_{tot\ eff} = P_{sol\ eff} + P_{part\ eff} = 0.3 + 2.1 = \mathbf{2.4\ mgP/L}$$

It is observed that most of the effluent phosphorus is associated with the effluent SS. If lower concentrations of P are desired, around 1 mg/L, a very efficient secondary sedimentation should be adopted, or the SS removal should be supplemented by polishing with dissolved air flotation or sand filtration.

(f)   P removal efficiency

- Total efficiency:

$$E = \frac{(P_{inf} - P_{eff})}{P_{inf}} \times 100 = \frac{(12.0 - 2.4)}{12.0} \times 100 = \mathbf{80\%}$$

8

Intermittent operation systems (sequencing batch reactors)

8.1 INTRODUCTION

Although use of intermittent operation reactors (sequencing batch reactors – SBR) started many decades ago, it was from the early 1980s that this technology became more widespread and used in the treatment of a larger diversity of effluents. This is partially due to a better knowledge of the system, to the use of more reliable effluent withdrawal devices, to the development of a more robust instrumentation and to the use of automated control by microprocessors. In the past few years, in view of the growing concern with the discharge of nutrients in watercourses, sequencing batch reactors have been modified to accomplish nitrification, denitrification and biological phosphorus removal.

8.2 PRINCIPLES OF THE PROCESS

The principle of the intermittent operation activated sludge process consists in the incorporation of all the unit operation and processes usually associated with the conventional treatment by activated sludge (primary sedimentation, biological oxidation and secondary sedimentation) in a single tank. Using a single tank, these processes and operations simply become sequences in time, and not separate units as in the conventional continuous-flow processes. The intermittent flow activated sludge process can also be used in the extended aeration mode, in which the single tank also incorporates sludge digestion.

Table 8.1. Stages in a typical operational cycle of sequencing batch reactor for carbon removal

Stage	Scheme	Aeration	Description
Fill		*on/off*	• The fill operation consists of the addition of sewage and substrate for microbial activity. • The fill cycle can be controlled by float valves to a pre-established volume or by timers for systems with more than one reactor. A simple method that is ordinarily applied to control the fill cycle is based on the volume of the reactor, resulting in fill times inversely related to the influent flow. • The fill phase can include several operational phases, and is subject to several control modes, named *static fill, fill with mixing,* and *fill with reaction.* • The *static fill* involves the introduction of the influent without mixing or aeration. This type of filling is more common in plants for nutrient removal. In these applications, the static fill is followed by a *fill with mixing,* so that the microorganisms are exposed to a sufficient amount of substrate, while anoxic or anaerobic conditions are maintained. Both mixing and aeration are performed in the *fill with reaction* stage. • The system can alternate among *static fill, fill with mixing* and *fill with reaction* throughout the operational cycle.
React		*on*	• The objective of the reaction stage is to complete the reactions started during the fill stage. • The reaction stage can comprise mixing, aeration or both. As in the case of the fill phase, the desired processes can require alternated aeration cycles. • The duration of the reaction phase can be controlled by timers, by the level of the liquid or by the degree of treatment, through the monitoring of the reactor. • Depending on the amount and duration of the aeration during the fill phase, there may or may not be a dedicated reaction phase.
Settle		*off*	• The solids–liquid separation occurs during the sedimentation phase, similar to the operation of a secondary sedimentation tank in a conventional plant. • The sedimentation in an intermittent system can be more efficient than in a continuous-flow sedimentation tank, due to more quiescent conditions of the liquid in a sequencing batch tank, with no interference of liquids entering and leaving.

Table 8.1 (*Continued*)

Stage	Scheme	Aeration	Description
Draw		*off*	• The clarified effluent (supernatant) is removed during the draw phase. • Drawing can be carried out by several mechanisms, the most frequently used ones being floating or adjustable weirs.
Idle		*on/off*	• The final phase is named idle, and is only used in applications with several tanks. • The main objective is to adjust the operational cycle of one reactor with the operational cycle of another reactor. • The time intended for the idle phase depends on the time required by the preceding tank to complete its cycle. • Wastage of excess sludge usually happens in this phase.

Source: Adapted from EPA (1993)

The process consists of complete-mix reactors where all treatment stages occur. This is attained by the establishment of operational cycles with defined duration. The biological mass remains in the reactor during all the cycles, thus eliminating the need for separate sedimentation tanks and sludge recirculation pumping stations. This is the essence of a sequencing batch reactor: biomass retention without the need for sludge recirculation by pumping. By preserving the biomass in the system, the sludge age becomes higher than the hydraulic detention time, which is a fundamental feature of the activated sludge process. The usual stages in the treatment cycle are summarised in Table 8.1.

The usual duration of each cycle can be altered in view of the variations of the influent flow, the treatment requirements and the characteristics of the sewage and biomass in the system.

The excess sludge is generally wasted during the last phase (*idle*). However, since this phase is optional, because its purpose is to allow an adjustment among the operational cycles of each reactor, the wastage may occur in other phases of the process. The quantity and frequency of the sludge wastage are established according to the performance requirements, as in conventional continuous-flow processes.

The plant usually has two or more sequencing batch reactors operating in parallel, each one in different stages of the operational cycle. This need is compulsory in systems that receive inflow during all day (such as domestic sewage), because a reactor in the sedimentation stage, for example, is not able to receive influent. At this time, the influent is being directed to another reactor, which is in the fill phase. In plants receiving wastewater intermittently, such as in industries that work only 8 hours per day, there may be just one reactor, that works in fill (and possibly react) phase for 8 hours, and carries out the other stages of the cycle in the subsequent 16 hours. Figure 8.1 shows schematically a plant with three sequencing batch reactors in parallel.

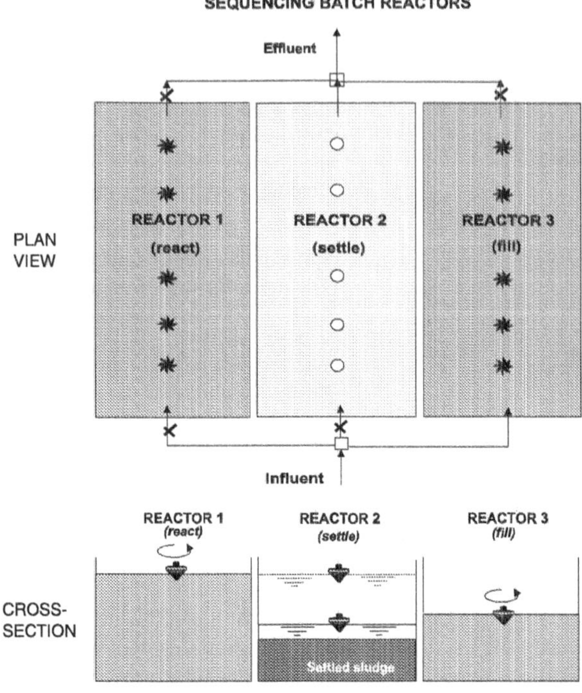

Figure 8.1.   Arrangement with three sequencing batch reactors in parallel

8.3 PROCESS VARIANTS

Several modifications have been incorporated into the process, in order to achieve different objectives in the wastewater treatment. These changes refer both to the form of operation of the system (continuous feeding and discontinuous drawing) and to the sequence and duration of the cycles associated with each phase of the process. The variations presented can also be used for the treatment of industrial effluents (Goronszy, 1997). Examples of process variants are presented next, some of them being protected by patent.

(a)   Sequencing batch reactor for biological nitrogen removal

Biological nitrogen removal can be reached by the incorporation of an anoxic stage after the aerobic reaction stage (Figure 8.2). In this case, there is a post-denitrification stage, which can be easily accomplished, although it occurs under endogenous respiration conditions, that is, at lower denitrification rates, due to the smaller availability of organic carbon.

If very low nitrogen values are not required, then a post-anoxic stage will not be necessary. In this case, a substantial amount of nitrate can be removed in a

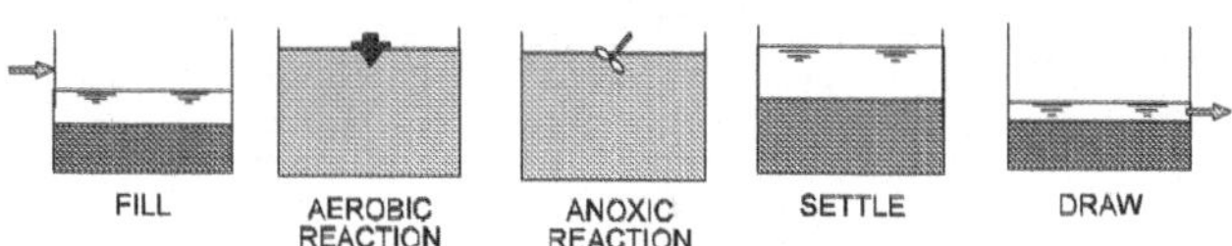

Figure 8.2.   Sequencing batch reactor for removal of carbon and nitrogen (post-denitrification)

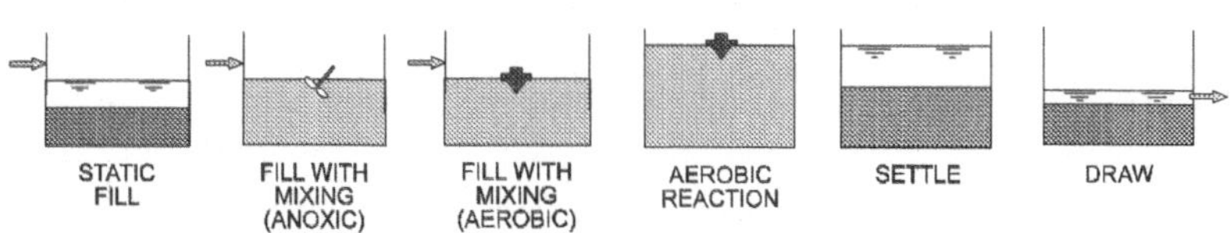

Figure 8.3.   Sequencing batch reactor for the removal of carbon and nitrogen (pre-denitrification)

pre-anoxic period during fill, and the carbon from the raw sewage will be used for pre-denitrification (Figure 8.3). The ratio between the fill volume and the total volume of the reactor determines the maximum level that nitrogen removal can be reached. The lower the ratio between the fill volume and the total volume, the larger the nitrogen removal, assuming that all the nitrate is reduced prior to the beginning of aeration (Randall *et al.*, 1992).

(b)   Sequencing batch reactor for biological phosphorus removal

The adaptation of the process for biological removal of phosphorus is made by the creation of a sequence of anaerobic conditions followed by aerobic conditions, provided that there is sufficient rapidly biodegradable organic matter during the anaerobic phase. Thus, the basic configuration of the operational cycles for the removal of BOD and suspended solids, as presented in Table 8.1, is changed in order to incorporate an anaerobic period. In this configuration (Figure 8.4), the incorporation of BOD and the release of phosphorus occur during the anaerobic reaction phase, with subsequent excess phosphorus incorporation and carbon oxidation occurring during the aerobic reaction phase. The operation of the system under these conditions is able to reduce the total phosphorus levels to less than 1 mg/L in the effluent, with no need of supplementary addition of chemical products (WEF/ASCE, 1992).

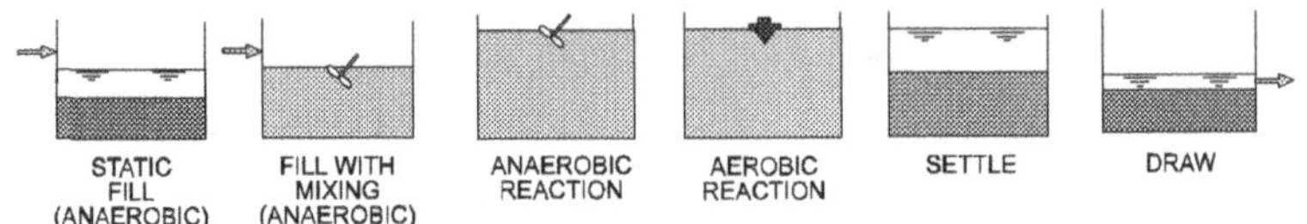

Figure 8.4.   Sequencing batch reactor for the removal of BOD and phosphorus

Figure 8.5.   Sequencing batch reactor for the removal of BOD, nitrogen and phosphorus

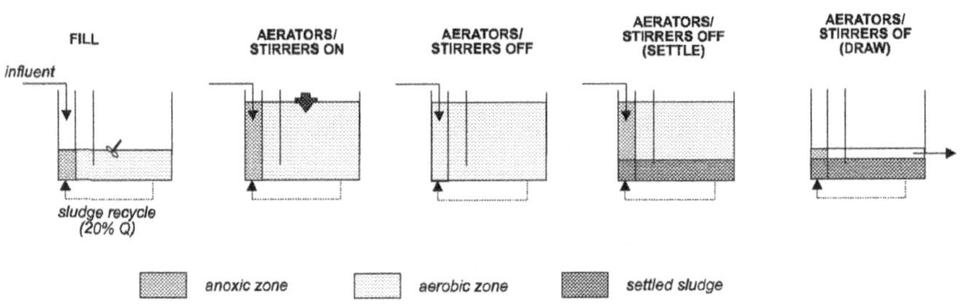

Figure 8.6.   Cyclic Activated Sludge System (CASS)

(c)   Sequencing batch reactor for biological removal of phosphorus and nitrogen

The operational cycles of the process can be modified to reach the combined oxidation of carbon and nitrogen and the removal of nitrate and phosphorus, as illustrated in Figure 8.5. The main difference is the incorporation of an anoxic phase after the aerobic reaction phase. Simultaneous removal of N and P is advantageous: if the system nitrifies but is not able to denitrify, the remaining nitrates will affect the conditions for creating a truly anaerobic environment during the anaerobic phase.

(d)   Cyclic Activated Sludge System

The Cyclic Activated Sludge System (CASS) is patented. Its operation is similar to that of other intermittent systems (see Figure 8.6). The differentiating element is

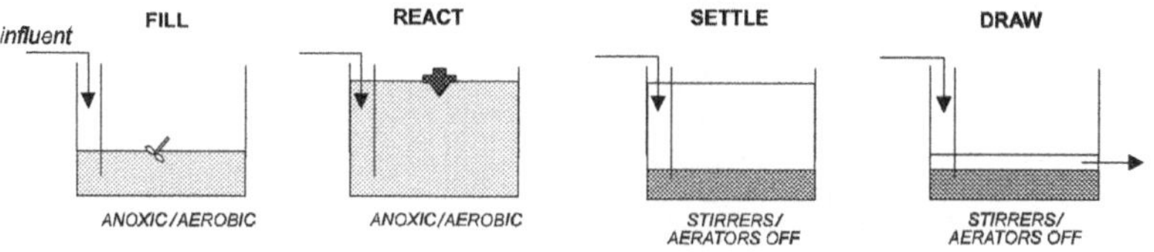

Figure 8.7.   Intermittent Cycle Extended Aeration System (ICEAS)

the incorporation of a selector (see Chapter 10 for the concept of selectors), which can receive a continuous inflow. The selector is a baffled compartment, where the raw or settled sewage is mixed with return sludge (non-existent in most of the SBR versions). The liquid effluent from the selector enters the reaction zone. When limiting or eliminating the aeration in the selector, the organic matter concentration is high and oxygen becomes deficient. These conditions apparently favour the growth of floc-forming bacteria and the inhibition of filamentous bacteria, which improves the settleability of the sludge (EPA, 1993).

(e)   Intermittent Cycle Extended Aeration System

The *Intermittent Cycle Extended Aeration System* (ICEAS) is patented (see Figure 8.7). Its main characteristic is that there is entrance of influent in all the stages of the cycle, differently from other variable volume variants. The inlet compartment aims at ensuring that the flow and load variations are evenly distributed among the reactors, preventing peak flows or shock loads from continuously overloading a tank. Another advantage of the continuous-flow regime of the ICEAS is the simplified control of the inflow, compared with other intermittent inflow variants. As there is influent entrance all the time, the ICEAS does not provide total quiescence during the sedimentation phase, differently from the intermittent flow versions. The ICEAS also uses an anoxic selector to allow denitrification and promote the growth of floc-forming bacteria, inhibiting the filamentous bacteria (EPA, 1993).

(f)   Alternated aeration activated sludge system

This variant has been patented by the Federal University of Minas Gerais, Brazil, and further details of the process can be found in von Sperling (2002). The inflow and outflow are continuous and the water level is constant, which are advantages of the continuous-flow systems. There is an increase in the total reactor volume from 33 to 50% (compared with reactors from continuous-flow activated sludge systems) to account for the volume of sedimentation. Figures 8.8 and 8.9 illustrate the conception and the operating principle of the system.

In this system, the reactor is divided into, say, three reactors, with communicating openings among them, which guarantees the constant water level in all chambers. The reactors have a high length/breadth ratio, with the influent entering

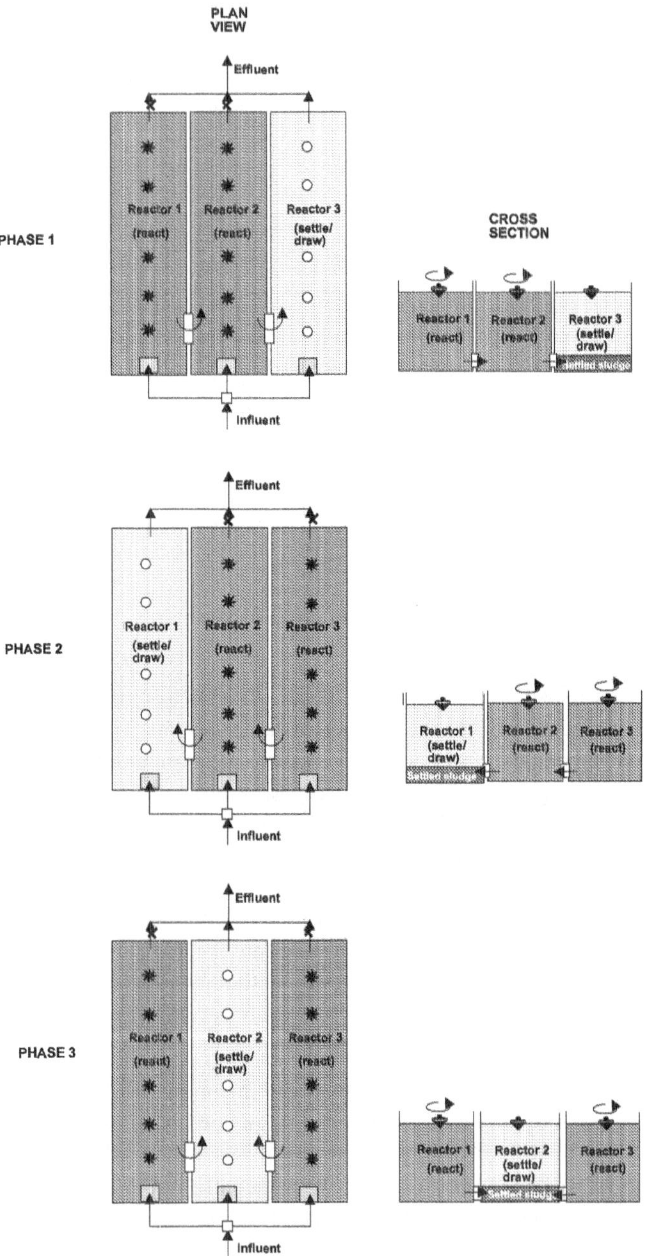

Figure 8.8.   Alternated aeration activated sludge, composed of three reactors

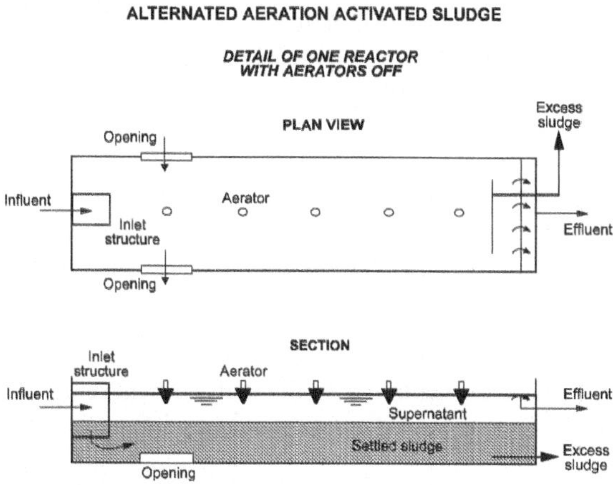

Figure 8.9.   Schematics of a reactor from the alternated aeration activated sludge system

simultaneously in the inlet end of all reactors. The effluent leaves from one reactor at a time (alternately), at the opposite end of the tank. The aeration system in the three reactors operates in an intermittent and alternated basis. In one reactor the aerators are switched off, in which occasional solids settlement takes place, followed by the supernatant (effluent) withdrawal. In the other reactors the aerators are switched on, the biomass is in suspension, and the biochemical reactions take place. In these reactors, in which the biomass is suspended, there is no effluent withdrawal. After a certain period, the reactors in sedimentation and in reaction alternate in such a way that at the end of the cycle, all reactors have performed the roles of reaction and sedimentation/withdrawal.

Because there is only one reactor in sedimentation, while the other two are under reaction, it may be assumed that this reactor corresponds to the secondary sedimentation tank, with an increase of 50% (or 1 in sedimentation/2 in reaction) of the total volume required for reaction. In case the system has a total of four reactors, with only one under sedimentation, the increase will be of only 33% (1 in sedimentation/3 in reaction).

Depending on the treatment objectives, other phases can be incorporated, such as anoxic and anaerobic, for biological nutrient removal.

8.4 DESIGN CRITERIA FOR SEQUENCING BATCH REACTORS

The design criteria for the traditional **sequencing batch reactor** (intermittent inflow and outflow), as described in Section 8.2, are presented below. The design of the reactor uses the basic criteria and parameters of the continuous-flow activated sludge systems, with special considerations on the hydraulic and organic loading aspects. Thus, the determination of the volume of the reactor should meet the following aspects: (a) kinetic criteria for carbon (and nitrogen and phosphorus, if applicable) removal, and (b) need to adapt the operating cycles to the influent flows (Chernicharo and von Sperling, 1993).

(a) Sludge age

The sludge age can be adopted according to the wide range available for the continuous-flow systems, that is, covering the conventional and extended aeration modes. The desirability to remove nutrients or not should also be taken into consideration. Examples of different sludge ages can be:

- $\theta_c = 4$–6 days: conventional mode, with no nutrient removal
- $\theta_c = 8$–10 days: conventional mode, with nutrient removal
- $\theta_c = 20$–25 days: extended aeration mode, with nutrient removal

However, due to the pursuit of operational simplicity in the sequencing batch reactors, a more convenient design of small- and medium-sized plants should adopt an extended aeration sludge age. EPA (1993) suggests 20 days to 40 days. However, in warm climate regions it is not necessary to adopt sludge ages higher than 30 days, in order to achieve sludge stabilisation. In the extended aeration mode, the whole treatment system may consist of only preliminary treatment, reactor, and sludge dewatering. Should nutrient removal be required, the conventional sludge ages provide higher removal efficiency.

(b) MLVSS concentration

The concentration of suspended solids during the reaction phase can be adopted similarly to the concentration of MLVSS in continuous-flow systems. If a safe positioning is desired, a slightly lower concentration can be adopted. EPA (1993) suggests concentrations of MLVSS between 1500 mg/L and 3500 mg/L.

(c) Operational cycles

According to WEF/ASCE (1992), the operational cycles vary widely, from approximately 6 to 48 hours. Generally, older systems have more conservative design criteria (longer cycles), while the systems designed more recently have shorter cycles, ranging from 6 to 12 hours. This optimisation has been achieved due to a deeper knowledge and greater control of the process, as well as to the automation of the system. Total operational cycles recommended by EPA are as follows (1993):

- conventional system: 4 to 6 hours
- system with biological nutrient removal: 6 to 8 hours

EPA (1993) proposes the division of the operational cycle according to the stages listed in Table 8.2.

Depending on the diurnal variations of the influent flows to the system, which can sometimes increase (minimum flow periods) and sometimes decrease (maximum flow periods) the reactor fill time, the operational cycles can have durations longer than those recommended. The automation level of the system also interferes with the duration of the operational cycles.

Table 8.2. Duration of each stage of the cycle, according to different removal purposes

Stage	Extended aeration BOD removal		Extended aeration BOD and N removal	
	Duration (hour)	% of the total	Duration (hour)	% of the total
Fill	1.0	*23.8*	1.0	*21.3*
Fill with mix	0.5	*11.9*	0.5	*10.6*
Fill with aeration	0.5	*11.9*	0.5	*10.6*
Aerobic/anoxic react	0.5	*11.9*	1.0	*21.3*
Settle	0.7	*16.7*	0.7	*14.8*
Draw	0.5	*11.9*	0.5	*10.6*
Idle	0.5	*11.9*	0.5	*10.6*
Total	**4.2**	***100.0***	**4.7**	***100.0***

Source: EPA (1993)

(d)   Mathematical model

In the design of a *continuous-flow* activated sludge system, the mathematical model of the reactor uses the detention time values in the anaerobic, anoxic and aerobic zones to estimate the quality of the effluent and the oxygen requirements. In these conditions, several mathematical models available in the literature can be adopted.

In *intermittent flow* systems, the mathematical model can use the time allocated for each stage of the cycle (anaerobic, anoxic and aerobic) to do the same estimates. The degree of uncertainty in the application of the model is higher in intermittent systems because the reactions do not occur in physically different zones, but in different periods of time. Thus, some reactions may be overlapped within the same period. However, it is believed that the order of magnitude of the results achieved by using a generic model can be maintained. Thus, the effluent quality can be estimated by using the C, N and P removal models described in this book, and by making the adaptations mentioned above, that is, converting the detention times in the reactor zones into times for each stage in the cycle. The design should be flexible enough to allow operation to tune the cycles in order to achieve the best effluent quality.

(e)   Aeration equipment

Aeration in sequencing batch reactors can be achieved by means of diffusers, floating aerators, jet aerators and aspirating aerators. The systems provided with diffusers should not allow clogging during settle, draw and idle periods. The mechanical aerators should be floating because of the variation of the water level throughout the operational cycle (fill and draw). For design purposes, it should be considered that the whole oxygen demand for stabilisation of the organic matter should be satisfied during the reaction phase. Thus, the power of the equipment installed in each reactor should be enough to supply the whole oxygen mass required during a shorter time interval (aeration phase).

Consequently, the installed power is higher than the consumed power. The consumed power can be estimated by means of the usual calculation methods of oxygen requirements, while the installed power should take into consideration the ratio between the total cycle time and the time with the aerators turned on. For example, in a system with a 12-hour cycle, in which 6 hours are with the aerators turned on (aerobic fill + aerobic reaction), the ratio between total time/time with aerators turned on will be $12/6 = 2$. In these conditions, the installed power should be twice higher than the consumed power.

(f)   Supernatant removal device

The removal of the clarified supernatant, without causing the suspension of the settled solids, is an item of great importance in the operational performance of a sequencing batch reactor. Fixed and floating outlet structures have been used, but the latter ones are more appropriate, as they can follow the water level, extracting the most superficial and, therefore, the most clarified layer (baffles may be installed for scum retention). Several floating mechanisms have been used, provided with flexible hoses or articulated mechanisms connected with the floating weirs.

8.5 DESIGN METHODOLOGY FOR SEQUENCING BATCH REACTORS

A sequence of calculations proposed by the author for estimating the volume of the reactor and the duration of the operational cycles of sequencing batch reactors (conventional reactor, with intermittent flow and variable level) is presented below. This methodology has been proposed by von Sperling (1998). Other methodologies are presented and exemplified in Eckenfelder Jr. (1989), Metcalf and Eddy (1991), Randall *et al.* (1992), Orhon and Artan (1994), and Artan *et al.* (2001). All these latter methodologies adopt the SVI (Sludge Volume Index) for estimation of the concentration and volume of the settled sludge, while the methodology proposed by von Sperling (1998) uses the concept of the zone settling velocity to estimate the sedimentation time, the concentration and the volume of the settled sludge. The methodology proposed focuses on an operational cycle intended for BOD removal (with no explicit removal of N and P), consisting of the following stages: *fill*, *react*, *settle*, *draw* and *idle*.

The height, volume and concentrations of interest in the design of sequencing batch reactors are presented in Figure 8.10.

(a)   Input data

Sludge age (θ_c). The sludge age is related to the active time of the cycle, which corresponds to the fill and reaction periods. The sludge age can be adopted according to the comments in Section 8.4.a.

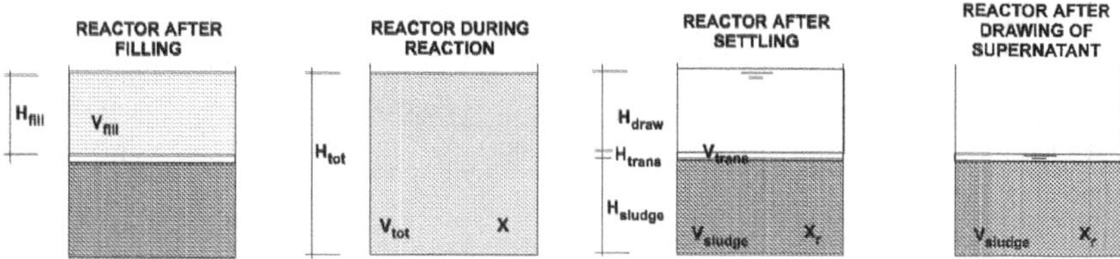

Figure 8.10.   Height, volume and concentrations of interest in the design of a sequencing batch reactor

MLVSS concentration (X_v). The MLVSS concentration refers to the reaction stage, in which all solids are dispersed in the reactor. The MLVSS values can be adopted according to Section 8.4.b. The MLSS concentration is obtained by the usual manner, dividing MLVSS by the VSS/SS ratio in the reactor.

Kinetic and stoichiometric coefficients. The kinetic and stoichiometric coefficients (Y, K_d, f_b, and MLVSS/MLSS) can be adopted similarly as those of the continuous-flow activated sludge systems (see Table 3.2).

Number of cycles per day. The number of cycles per day (**m**) depends on the total time desired for the cycle. Thus, in case a total time of 6 hours is desired for the cycle, the number of cycles per day will be (24 hours/d) / (6 hours/cycle) = 4 cycles/d.

Time of wastewater input to the plant. In the case of domestic sewage, the influent is assumed to arrive during 24 hours per day. In the case of industries that work only during one or two shifts per day, lower times can be adopted, in compatibility with the time of production of wastewater (e.g., 8 hours per day). In this case, 1 cycle/d can be adopted ($m = 1$), with a cycle duration of 8 hours.

Reactor height. The total height of the reactor (H_{tot}) (liquid depth) should be selected in view of the aeration equipment and the local conditions. According to Section 3.4, H_{tot} is usually within the following range: 3.5 to 4.5 m (mechanical aeration) and 4.5 to 6.0 m (diffused air).

(b)   Design sequence

The design sequence is presented in this section in a summary table (Table 8.3), including all equations (von Sperling, 1998). The application of the equations can be more clearly understood in the design example in Section 8.6.

8.6 DESIGN EXAMPLE OF A SEQUENCING BATCH REACTOR

Design an extended aeration sequencing-batch-reactor system for the treatment of the wastewater from the general example in Chapter 5.

Table 8.3. List of equations and summary of the design sequence

Item to be calculated	Unit	Equation	Equation number
Number of cycles per day	–	m (adopt)	–
Total cycle time	hour	$T_{total} = \frac{24}{m}$	(8.1)
Time of arrival of influent during the cycle	hour	$T_{arrival\ of\ influent\ during\ cycle} = T_{arrival\ influent\ during\ day}/m$	(8.2)
Biodegradable fraction of the MLVSS	–	$f_b = \frac{0.8}{1+0.2 \cdot K_d \cdot \theta_c}$	(8.3)
Volume for reaction	m³	$V_{react} = \frac{Y \cdot \theta_c \cdot Q \cdot (S_o - S)}{X_v \cdot (1 + f_b \cdot K_d \cdot \theta_c)}$	(8.4)
Fill volume	m³	$V_{fill} = \frac{Q}{m}$	(8.5)
Transition volume	m³	$V_{trans} = f_{Hfill} \cdot V_{fill}$	(8.6)
Sludge volume	m³	$V_{sludge} = V_{react}$	(8.7)
Total reactor volume	m³	$V_{tot} = V_{react} + V_{fill} + V_{trans}$	(8.8)
Total reactor height	m	H_{tot} (adopt)	–
Fill height	m	$H_{fill} = \frac{V_{fill}}{Area} = \frac{V_{fill}}{(V_{tot}/H_{tot})}$	(8.9)
Transition height	m	$H_{trans} = f_{Hfill} \cdot H_{fill}$	(8.10)
Sludge height	m	$H_{sludge} = H_{tot} - (H_{fill} + H_{trans})$	(8.11)
MLSS concentration	mg/L	$X = \frac{X_v}{(SSV/SS)}$	(8.12)
MLSS mass in the reactor	kg	$M_x = \frac{X \cdot V_{tot}}{1000}$	(8.13)
SS concentration in the settled sludge	mg/L	$X_r = \frac{M_x \cdot 1000}{V_{sludge}}$	(8.14)
Number of reactors	–	n (adopt)	–
Volume of each reactor	m³	$V_{reactor} = \frac{V_{tot}}{n}$	(8.15)
Fill time within cycle	hour	$T_{fill} = T_{arrival\ of\ influent\ during\ cycle}/n$	(8.16)
Active time within cycle (= fill time + react time)	hour	$T_{active} = T_{total} \cdot \frac{V_{react}}{V_{tot}}$	(8.17)
Reaction time within cycle	hour	$T_{react} = T_{active} - T_{fill}$	(8.18)
Settling velocity of the sludge interface	m/hour	$v = v_o \cdot e^{-K \cdot X}$	(8.19)
Settle time within cycle	hour	$T_{settle} = \frac{(H_{trans} + H_{fill})}{v}$	(8.20)
Supernatant withdrawal time within cycle	hour	T_{draw} (adopt; $\leq T_{total} - T_{fill} - T_{react} - T_{settle}$)	(8.21)
Idle time within cycle	hour	$T_{idle} = T_{total} - (T_{fill} + T_{react} + T_{settle} + T_{draw})$	(8.22)
Number of effluent removals per day	–	Number removals per day = m·n	(8.23)
Volume of effluent in each removal	m³	Vol. each removal = Q/(m·n)	(8.24)
Flow of effluent in each removal	m³/hour	Flow each removal = Vol. each rem./T_{draw}	(8.25)

Y = yield coefficient (gMLVSS/gBOD₅ removed)
θ_c = sludge age (d)
Q = inflow (m³d⁻¹)
S_o = total influent BOD (mgL⁻¹)
S = total effluent soluble BOD (mgL⁻¹)
X_v = MLVSS concentration (mgL⁻¹)
K_d = decay coefficient (d⁻¹)
v_o, K = settling velocity equation coefficients (see Tables 4.2 and 4.3)

Source: von Sperling (1998)

Input data for the example

Influent and effluent characteristics:

- Average inflow: $Q = 9,820$ m^3/d
- Influent BOD concentration: $S_o = 341$ mg/L
- Effluent soluble BOD concentration (desired): $S = 9$ mg/L

Coefficients:

- Yield coefficient: $Y = 0.6$ mg/mg
- Endogenous respiration coefficient: $K_d = 0.08$ d^{-1}
- SSV/SS ratio in the reactor: SSV/SS $= 0.69$

Design criteria:

- Sludge age: $\theta_c = 25$ d (extended aeration)
- MLVSS concentration (during reaction): $X_v = 2,415$ mg/L
- Sludge settleability: between *fair* and *poor*
- Number of cycles per day: $m = 3$ (adopted)
- Number of reactors: $n = 3$ (adopted)
- Time with arrival of incoming sewage (per day) $= 24$ hours/d
- Total height of the reactor: $H_{tot} = 4.00$ m

(a)   Total cycle time

Equation 8.1:

$$T_{total} = \frac{24}{m} = \frac{24}{3} = 8 \text{ hours/cycle}$$

(b)   Time of arrival of influent during each cycle

The time of wastewater input is not necessarily 24 hours/d, because there might
be some cases in which the influent is generated during less than 24 hours/d (e.g.,
8 hours/d), as is the case with some industries. From Equation 8.2, and considering
the inflow for 24 hours/d (domestic sewage):

$$T_{\text{arrival of influent during cycle}} = T_{\text{arrival of influent during day}}/m = 24/3 = 8 \text{ hours}$$

(c)   Volume of the reactor

- Volume for reaction (Equations 8.3 and 8.4):

  The volume for reaction can be calculated using any suitable mathematical
  steady-state model for the continuous-flow activated sludge process. From

Equations 8.3 and 8.4:

$$f_b = \frac{0.8}{1 + 0.2 \cdot K_d \cdot \theta_c} = \frac{0.8}{1 + 0.2 \times 0.08 \times 25} = 0.57$$

$$V_{react} = \frac{Y \cdot \theta_c \cdot Q \cdot (S_o - S)}{X_v \cdot (1 + f_b \cdot K_d \cdot \theta_c)} = \frac{0.6 \times 25 \times 9{,}820 \times (341 - 9)}{2{,}415 \times (1 + 0.57 \times 0.08 \times 25)}$$

$$= 9{,}463 \text{ m}^3$$

- Fill volume (Equation 8.5):

$$V_{fill} = \frac{Q}{m} = \frac{9820}{3} = 3{,}273 \text{ m}^3$$

- Transition volume (Equation 8.6):

Before starting the supernatant withdrawal, the sludge must settle a distance equal to the height of fill plus a transition height. This transition height is routinely included for safety in other design sequences available in the literature, and aims at avoiding the situation whereby the weir level coincides with the level of the settled sludge. By doing so, there will be a clarified transition layer, which will remain even after the supernatant withdrawal. The transition height H_{trans} is normally fixed as a fraction (f_{Hfill}) of the total fill height. The value of f_{Hfill} is usually adopted around 0.1 (H_{trans} is equal to 10% of H_{fill}).

$$V_{trans} = f_{Hfill} \cdot V_{fill} = 0.1 \times 3{,}273 = 327 \text{ m}^3$$

- Sludge volume (Equation 8.7):

$$V_{sludge} = V_{react} = 9{,}463 \text{ m}^3$$

- Total reactor volume (Equation 8.8):

$$V_{tot} = V_{react} + V_{fill} + V_{trans} = 9{,}463 + 3{,}273 + 327 = 13{,}063 \text{ m}^3$$

(d) Heights in the reactor

- Fill height (Equation 8.9):

$$H_{fill} = \frac{V_{fill}}{(V_{tot}/H_{tot})} = \frac{3{,}273}{(13{,}063 \, / \, 4.00)} = 1.00 \text{ m}$$

- Transition height (Equation 8.10):

$$H_{trans} = f_{Hfill} \times H_{fill} = 0.1 \times 1.00 \text{ m} = 0.10 \text{ m}$$

- Sludge height (Equation 8.11):

$$H_{sludge} = H_{tot} - (H_{fill} + H_{trans}) = 4.00 - (1.00 + 0.10) = 2.90 \text{ m}$$

(e) MLSS mass and concentration

- MLSS concentration in the reactor, during the *react* stage (Equation 8.12):

$$X = \frac{X_V}{(SSV/SS)} = \frac{2,415}{0.69} = 3,500 \text{ mg/L}$$

- MLSS mass in the reactor (Equation 8.13):

$$M_x = \frac{X \cdot V_{tot}}{1,000} = \frac{3,500 \times 13,063}{1,000} = 45,721 \text{ kgSS}$$

(f) Average SS concentration in the settled sludge

SS concentration in the layer of settled sludge (Equation 8.14):

$$X_r = \frac{M_x \cdot 1,000}{V_{lodo}} = \frac{V_{tot} \cdot X}{V_{tot} \cdot (H_{lodo}/H_{tot})}$$

$$= \frac{X}{(H_{lodo}/H_{tot})} = \frac{3,500}{(2,90 / 4,00)} = 4,828 \text{ mg/L}$$

This concentration corresponds to the concentration of excess sludge, if it is removed during the idle stage.

(g) Times within the cycle

- Fill time (Equation 8.16):

$$T_{fill} = T_{\text{arrival of influent during cycle}}/n = 8/3 = 2.7 \text{ hours}$$

- Active time (Equation 8.17):

$$T_{active} = T_{total} \cdot \frac{V_{react}}{V_{tot}} = \frac{9,463}{13,063} = 5.8 \text{ hours}$$

- Reaction time (Equation 8.18):

$$T_{react} = T_{active} - T_{fill} = 5.8 - 2.7 = 3.1 \text{ hours}$$

- Settle time

  Initially, the settling velocity of the sludge-liquid interface must be calculated. Assuming a settleability between *fair* and *poor*, as specified in the example, the coefficients v_o and K from Table 4.3 must be interpolated between the values given for *fair* and *poor* settleability, resulting in:

$$v_o = (8.6 + 6.2)/2 = 7.40 \text{ m/hour}$$

$$K = (0.50 + 0.67)/2 = 0.59 \text{ m}^3/\text{kg}$$

The hindered settling velocity is a function of the sludge concentration, being thus given by (Equation 8.19):

$$v = 7.4 \cdot e^{(-0.59 \cdot X/1000)} = 7.4 \cdot e^{(-0.59 \times 3500/1000)} = 0.94 \text{ m/hour}$$

Time spent by the sludge-liquid interface to settle the distance $H_{fill} + H_{trans}$ (Equation 8.20):

$$T_{settle} = \frac{(H_{trans} + H_{fill})}{v} = \frac{0.10 + 1.00}{0.94} = 1.2 \text{ hours}$$

- Supernatant withdrawal time

  The supernatant withdrawal time is adopted at this stage. The following constraint applies (Equation 8.21):

$$T_{draw} \leq T_{total} - T_{fill} - T_{react} - T_{settle}$$

$$T_{draw} = 0.5 \text{ hour (adopted)}$$

- Idle time (Equation 8.22):

  The idle time is the time left to complete the cycle.

$$T_{idle} = T_{total} - (T_{fill} + T_{react} + T_{settle} + T_{draw})$$

$$= 8.0 - (2.7 + 3.1 + 1.2 + 0.5) = 0.5 \text{ hour}$$

(h)   Summary of the duration of each phase in the cycle

Stage	Nomenclature	Duration (hours)	Percentage of the total cycle (%)
Fill	T_{fill}	2.7	33.8
React	T_{reac}	3.1	38.8
Settle	T_{settle}	1.2	15.0
Draw	T_{draw}	0.5	6.2
Idle	T_{idle}	0.5	6.2
Total	–	8.0	100.0

(i)   Effluent flow from each reactor

The effluent (supernatant) flow is different from the inflow to the reactor, because the effluent removal is concentrated on a shorter period. This larger instantaneous flow affects the dimensioning of the outlet structures and pipes.

The number of effluent removals per day is equal to the product of the number of cycles per day (m) and the number of reactors (n) (Equation 8.23):

$$\text{Number of removals per day} = m{\cdot}n = 3 \times 3 = 9 \text{ removals/d}$$

The average volume of each removal (m^3) corresponds to the average daily influent flow (Q) divided by the number of removals per day (Equation 8.24):

$$\text{Volume of each removal} = Q/(m{\cdot}n) = 9820/(3 \times 3) = 1{,}091 m^3/\text{removal}$$

The flow in each removal (m^3/h) is given by the quotient between the volume of each removal and the removal time (T_{draw}) (Equation 8.25):

$$\text{Flow of each removal} = \text{Volume of each removal}/T_{draw} = 1{,}091/0.5$$

$$= 2{,}182 \ m^3/\text{hour} = 606 \ L/s$$

(j)   Oxygen requirements and sludge production

Refer to the calculation methodology presented in the example of the continuous-flow activated sludge system (Chapter 5).

When calculating the power requirements, it should be noted that the installed power should be greater than the consumed power. This is because the aerators have to transfer the oxygen required by the biomass during the time when they are switched on. Therefore, the required power must be multiplied by a factor equal to time with aerators on/total time. In this example, if the aerators are turned on only during the react phase, the time with aerators on will be 3.1 hours/cycle, and the total cycle time will be 8.0 hours/cycle. The correction factor is, therefore: 8.0/3.1 = 2.6. The installed power needs to be 2.6 times greater than the consumed power.

9

Activated sludge for the post-treatment of the effluents from anaerobic reactors

9.1 DESIGN CRITERIA AND PARAMETERS

The main characteristics, applications, advantages and disadvantages of the systems composed of upflow anaerobic sludge blanket (UASB) reactors followed by the activated sludge system were presented in Chapter 1.

The present chapter, based on von Sperling *et al.* (2001) and on the results from the Brazilian Research Programme on Basic Sanitation (PROSAB), lists the main criteria and parameters used in the design of the post-treatment stage. The approach used here is simpler and more direct than that adopted in the previous chapters on this section on activated sludge. However, the results are not substantially different from those obtained using the more complete design sequences presented earlier. The mathematical model described in this book for BOD removal and determination of the required reactor volume, required power and sludge production can be applied to the present situation.

The main design parameters, which determine the behaviour of the system and the required volumes and areas, are (a) *reactor*: sludge age (θ_c) and mixed liquor volatile suspended solids (MLVSS) concentration; and (b) *secondary sedimentation tank*: hydraulic loading rate (HLR) and solids loading rate (SLR).

Table 9.1. Design parameters of activated sludge systems for the post-treatment of effluents from anaerobic reactors

Item	Parameter	Value
Aeration tank	Sludge age (d)	6–10
	F/M ratio (kg BOD/kgMLVSS·d)	0.25–0.40
	Hydraulic detention time (hour)	3–5
	MLVSS concentration (mg/L)	1,100–1,500
	MLSS concentration (mg/L)	1,500–2,000
	VSS/SS ratio in the reactor (−)	0.73–0.77
	Biodegradable fraction of the VSS ($f_b = SS_b/VSS$)	0.68–0.74
Aeration system	Average O_2 requirements – carbonaceous demand (kgO_2/kgBOD rem)	0.80–0.94
	Average O_2 requirements–nitrogenous demand (kgO_2/kgTKN applied)	3.8–4.3
	Average O_2 requirements–nitrogenous demand (kgO_2/kgN available)*	4.6
	Maximum O_2 consumption/average O_2 consumption ratio	1.2–1.5
	Standard oxygenation efficiency (kgO_2/kW·hour)	1.5–2.2
	Correction factor: standard oxygen. efficiency/ field oxygen. efficiency	1.5–1.8
Sludge production	Product. excess AS sludge (returned to UASB) (kgSS/kgBODrem from AS)	0.78–0.90
	Per capita product. of excess AS sludge (returned to UASB) (gSS/inhabitant·d)	8–14
	Concentration of SS in the AS sludge returned to the UASB (mg/L)	3,000–5,000
	Removal efficiency of VSS from the AS sludge in the UASB reactor	0.25–0.45
	Production of anaerobic sludge (kgSS/kgBOD applied to the UASB)	0.28–0.36
	Per capita production of anaerobic sludge (gSS/inhabitant·d)	14–18
	Production of total mixed sludge (to be dewatered) (kgSS/kgBOD applied)	0.40–0.60
	Per capita production of total mixed sludge (to be dewatered) (gSS/inhabitant·d)	20–30
	Per capita volumetric product. total mixed sludge (to be dewatered) (L/inhabitant·d)	0.5–1.0
	Concent. mixed sludge (AS + anaerobic) removed from the UASB (%)	3.0–4.0
Secondary sediment. tank	Hydraulic loading rate (Q/A) (m^3/m^2·d)	24–36
	Solids loading rate [$(Q + Q_r)$·X/A] ($kgSS/m^2$·d)	100–140
	Sidewater depth (m)	3.0–4.0
	Recirculation ratio (Q_r/Q)	0.6–1.0
	Concentration of SS in the sludge recirculated to the aeration tank (mg/L)	3,000–5,000
Sludge treatment	Per capita production of SS in the sludge to be disposed of (gSS/inhabitant·d)	20–30
	Per capita volum. production of sludge to be disposed of (L sludge/inhabitant·d)	0.05–0.15
	Solids content (centrifuge, belt press) (%)	20–30
	Solids content (filter press) (%)	25–40
	Solids content (sludge drying bed) (%)	30–45

* N available for nitrification = influent TKN − N in excess sludge (10% of the excess VSS is N)
AS: activated sludge

Table 9.1 lists the main design parameters used for the activated sludge process as post-treatment of effluents from anaerobic reactors.

Kinetic and stoichiometric coefficients. Regarding the values of the kinetic and stoichiometric coefficients (mainly Y and K_d), the same usual values of the classical configurations of the activated sludge system are used in this chapter. However, it should be highlighted that the values of these coefficients, applied to the specific case of activated sludge as post-treatment of effluents from anaerobic reactors, should be further investigated, due to the possible influence of the previous anaerobic treatment on the process kinetics.

Design parameters for the reactor. The design parameters for the activated sludge reactor as post-treatment are similar to those for the *conventional activated sludge* systems. The main difference lies in the lower MLSS concentration usually assumed for the post-treatment activated sludge. If higher values are adopted, the volume of the aeration tank will be very reduced (detention time shorter than 2.0 hours; no full-scale operational experience so far to demonstrate the process stability of such small tanks).

Design parameters for the secondary sedimentation tanks. The loading rates in the secondary sedimentation tanks of post-treatment activated sludge systems are presumably different from those in conventional activated sludge systems, since the former work with lower MLSS concentrations and with a sludge of slightly different characteristics. Besides that, the UASB reactors provide a certain smoothing in the flow to be treated, reducing the Q_{max}/Q_{av} ratio in the influent to the sedimentation tanks. These are items that deserve continued investigations, with experience in full-scale wastewater treatment plants, to get specific design parameters for this configuration.

Nitrification. With respect to the removal of ammonia in the UASB-activated sludge system, it should be mentioned that there have been operational difficulties in the maintenance of full nitrification in the aerobic reactor. This fact is apparently associated with toxicity problems to the nitrifying bacteria, possibly caused by the presence of sulphides. For this reason, even in warm-climate regions, sludge ages equal to or greater than 8 days should be adopted, if nitrification is desired.

Biological nutrient removal. Post-treatment activated sludge systems are not particularly efficient in the removal of nitrogen, since there is little availability of organic carbon for the denitrifying bacteria, as a large fraction of the organic matter has been previously removed in the UASB reactor. A means of supplying organic carbon to the activated sludge reactor is by a partial by-pass to the UASB reactor, supplying raw sewage to the anoxic zone in the aeration tank. A similar comment can be made for the biological removal of phosphorus: the previous removal of a large fraction of the organic carbon in the UASB reactor hinders the biological P removal process. Similarly, a partial by-pass of the raw sewage may be helpful.

Sequencing batch activated sludge reactor. The design of a sequencing batch activated sludge reactor after an anaerobic reactor should propose an operational cycle that is suitable for the condition of low organic load in the influent to the aerobic stage. Designs that do not pursue an optimisation may lead to large fill volumes, compared with the reaction volumes, which may result in an uneconomical, large volume of the aerobic reactor.

9.2 DESIGN EXAMPLE OF AN ACTIVATED SLUDGE SYSTEM FOR THE POST-TREATMENT OF THE EFFLUENT FROM A UASB REACTOR

Undertake a simplified design of a continuous-flow activated sludge system acting as post-treatment of the effluent from a UASB reactor. Determine the volume of the reactor, the oxygen consumption, the power of the aerators and the production and removal of excess sludge. Use the same input data as those in the general example in Chapter 5 (design of conventional activated sludge system) and the design parameters presented in Tables 9.1 and 1.3.

Input data:

- population equivalent: 67,000 inhabitants
- average influent flow: $Q = 9,820$ m^3/d
- loads in the raw sewage:
  BOD: 3,350 kg/d
  SS: 3,720 kg/d
  TKN: 496 kg/d
- concentrations in the raw sewage:
  BOD: 341 mg/L
  SS: 379 mg/L
  TKN: 51 mg/L
- removal efficiencies in the UASB reactor (assumed):
  BOD: 70%
  TKN: 10%

The design of the UASB reactor is not presented here.

Solution:

(a) Characteristics of the influent to the activated sludge (AS) stage

The influent to the activated sludge system is the effluent from the UASB reactor. Considering the removal efficiencies provided in the input data, one has:

- Influent BOD load AS = raw sewage BOD load × (1 – UASB Efficiency) = 3,350 kg/d × (1–0.70) = **1,005 kgBOD/d**
- Influent BOD concentration AS = raw sewage BOD concentration × (1 – UASB Efficiency) = 341 mg/L × (1–0.70) = **102 mgBOD/L**
- Influent TKN load AS = raw sewage TKN load × (1 – UASB Efficiency) = 496 kg/d × (1 – 0.10) = **446 kgTKN/d**

- Influent TKN concentration AS = raw sewage TKN concentration × (1 − UASB Efficiency) = 51 mg/L × (1–0.10) = **46 mgTKN/L**

(b)　Characteristics of the final effluent from the treatment plant

By adopting overall typical removal efficiencies for the UASB-activated sludge system presented in Table 1.3, the estimated concentrations in the final effluent of the treatment plant are as follows:

Parameter	Overall removal efficiency (%)	Concentration in the raw sewage (mg/L)	Estimated concentration in the final effluent (mg/L)
BOD	85–95	341	16–47
SS	85–95	379	19–57
TKN	75–90	51	5–13

Effluent concentration = Influent concentration × (100 − Efficiency)/100

(c)　Design of the reactor

Design parameters adopted (see Table 9.1):

- Sludge age: $\theta_c =$ **8 d**
- Mixed liquor volatile suspended solids: MLVSS = $X_v =$ **1,500 mg/L**
- Effluent soluble BOD: S = 10 mg/L (adopted)

Coefficients adopted (see Table 3.2):

- $Y = 0.6$ gVSS/gBOD
- $K_d = 0.08$ gVSS/gVSS·d

The biodegradable fraction of mixed liquor volatile suspended solids is given by (Equation 2.2):

$$f_b = \frac{0.8}{1 + 0.2 \cdot K_d \cdot \theta_c} = \frac{0.8}{1 + 0.2 \times 0.08 \times 8} = 0.71$$

The volume of the reactor is given by (Equation 2.4):

$$V = \frac{Y \cdot \theta_c \cdot Q \cdot (S_o - S)}{X_v \cdot (1 + f_b \cdot K_d \cdot \theta_c)} = \frac{0.60 \times 8 \times 9{,}820 \times (102 - 10)}{1{,}500 \times (1 + 0.71 \times 0.08 \times 6)} = 1{,}988 \text{ m}^3$$

The volume of the reactor can also be calculated based on the F/M ratio concept, which does not require the knowledge of coefficients Y and K_d. By adopting an F/M value equal to 0.35 kgBOD/kgMLVSS·d (Table 9.1), the resulting reactor volume is:

$$V = \frac{Q \cdot DBO_{\text{influent AS}}}{X_v \cdot (F/M)} = \frac{9{,}820 \times 102}{1{,}500 \times 0.35} = 1{,}908 \text{ m}^3$$

It is observed that the volumes resulting from both calculations are very similar. In the remainder of the design, the value obtained from the calculation using the sludge age ($V = 1,988$ m^3) is used.

Two tanks can be adopted, each one with a volume of $(1,988$ m$^3)/2 = 994$ m^3.

By adopting a depth of **3.5 m**, the surface area of each tank is 994 m^3/3.5 m $= 284$ m^2.

The length/breadth ratio can vary according to the layout and to the arrangement of the aerators (in case of mechanical aeration). For the purposes of this example, adopt:

Length **L = 30.0 m** and breadth **B = 9.5 m** (length/breadth ratio: **L/B = 3.2**)

The resulting total volume is **1,995 m^3**.

The resulting HDT in the aeration tank is:

$$HDT = V/Q = 1,995 \text{ m}^3/9,820 \text{ m}^3/d = 0.20 \text{ d}$$

$$= \textbf{4.8 hours} \text{ (appropriate, according to Table 9.1)}$$

The MLVSS/MLSS ratio ($= $ VSS/SS $= X_v/X$) adopted in the aeration tank is 0.75 (see Table 9.1).

The MLSS concentration (X) in the aeration tank is:

$$MLSS = MLVSS/(VSS/SS) = (1,500 \text{ mg/L})/(0.75) = \textbf{2,000mg/L}$$

(d)   Production and removal of excess sludge

Coefficient of sludge production: 0.84 kgSS/kgBOD removed in the activated sludge (see Table 9.1 or Table 2.6 – sludge age of 8 days, with solids in the influent, with primary sedimentation tank, which, in this case, is replaced by the UASB reactor).

The BOD load removed from the aeration tank is:

$$\text{BOD load rem} = Q \cdot (S_o - S) = 9,820 \text{ m}^3/d \times (102 - 10) \text{ g/m}^3$$

$$= 903,440 \text{ gBOD/d} = 903 \text{ kgBOD/d}$$

The production of excess aerobic activated sludge is, therefore:

$$P_X = 0.84 \text{ kgSS/kgBOD} \times 903 \text{ kgBOD/d} = 759 \text{ kgSS/d}$$

In the activated sludge system as post-treatment for anaerobic effluents, the production of solids is low, due to the fact that the anaerobic reactor removes previously a large part of the substrate (BOD) required for biomass growth. In these conditions, the loss of solids in the final effluent should be taken into consideration when estimating the amount of solids to be wasted. Assuming an average concentration of SS in the final effluent equal to 20 mg/L, the loss corresponds to:

$$\text{Loss of SS in the final effluent} = 9{,}820 \text{ m}^3/\text{d} \times 20 \text{ g/m}^3 = 196{,}400 \text{ gSS/d}$$

$$= 196 \text{ kgSS/d}$$

The SS load to be intentionally wasted from the aerobic reactor and returned to the UASB reactor is, therefore:

$$\text{Production of SS} = P_x - \text{SS loss} = 759 - 196 = \textbf{563 kgSS/d}$$

The per capita production of aerobic activated sludge is:

$$\text{Per capita } P_X = 563 \text{ kgSS/d}/67{,}000 \text{ inhabitants} = 0.008 \text{ kgSS/inhabitant·d}$$

$$= 8 \text{ gSS/inhabitant·d (appropriate, according to Table 9.1).}$$

The distribution of the excess sludge in terms of volatile solids and fixed solids is a function of the VSS/SS ratio (equal to 0.75 in this example). Thus, the distribution is:

- Total suspended solids: $P_X = 563$ kgSS/d
- Volatile suspended solids: $P_{XV} = (\text{VSS/SS}) \times P_X = 0.75 \times 563 = 422$ kgVSS/d
- Fixed suspended solids: $P_{XF} = (1 - \text{VSS/SS}) \times P_X = (1{-}0.75) \times 563 = 141$ kgFSS/d

The concentration of the excess aerobic activated sludge (AS) is the same as that of the return sludge, since the excess sludge is removed from the recirculation line. This concentration is a function of the MLSS concentration and the recirculation ratio R ($=Q_r/Q$). In the example, MLSS = 2000 mg/L and R is adopted as **0.8** (see Table 9.1). The SS concentration in the excess aerobic sludge and in the return sludge (X_r) is:

$$X_r = X \cdot (1 + R)/R = 2{,}000 \text{ mg/L} \times (1 + 0.8)/0.8 = \textbf{4,500 mgSS/L}$$

$$= 4{,}500 \text{ gSS/m}^3 = 4.5 \text{ kgSS/m}^3$$

The flow of excess aerobic activated sludge (AS) returned to the UASB reactor is:

$$\text{flow} = \text{load/concentration} =$$

$$Q_{\text{ex aerobic}} = (563 \text{ kgSS/d})/(4.5 \text{ kgSS/m}^3) = \textbf{125 m}^3\textbf{/d}$$

This flow is very low in comparison with the influent flow to the UASB reactor, representing only approximately **1.3%** (125/9,820 = 0.013), that is, the hydraulic impact of the return of the excess aerobic sludge to the UASB reactor is non-significant. On the other hand, the organic load in the excess sludge is estimated to be 282 kgBOD/d (1 kg of SS generates approximately 0.5 kgBOD, that is, 563 kgSS/d $\times$ 0.5 kgBOD/kgSS $=$ 282 kgBOD/d). Hence, the BOD load from the aerobic sludge returned to the UASB reactor is (282 kg/d)/(3,350 kg/d) $=$ **8%** of the BOD load in the influent. This increased load should not affect significantly the performance of the UASB reactor.

(e)   Oxygen consumption and required power for the aerators

The average O_2 consumption for the carbonaceous demand (oxidation of BOD) is 0.87 kgO$_2$/kgBOD removed in the aeration tank (see Table 9.1 or Table 2.6). The BOD load removed in the activated sludge system is 903 kgBOD/d (calculated in item (d)). The O_2 consumption is:

$$\text{Average } O_2 \text{ consumption (carbonaceous demand)}$$

$$= 0.87 \text{ kgO}_2/\text{kgBOD} \times 903 \text{ kgBOD/d} = 786 \text{ kgO}_2/\text{d}$$

The average O_2 consumption adopted for the nitrogenous demand (oxidation of the ammonia) is 4.6 kgO$_2$/kg N available (see Table 9.1). The TKN load available corresponds to the applied load minus the N load incorporated into the excess sludge (10% of the VSS production). In this example, the VSS load produced was calculated as 452 kgVSS/d. The N load available is:

$$N \text{ load available} = N \text{ load applied} - N \text{ load excess sludge}$$

$$= 446 - 0.1 \times 452 = 401 \text{ kgN/d}$$

The O_2 consumption for the nitrogenous demand is:

$$\text{Average } O_2 \text{ consumption (nitrogenous demand)}$$

$$= 4.6 \text{ kgO}_2/\text{kgTKN} \times 401 \text{ kgTKN/d} = 1,845 \text{ kgO}_2/\text{d}$$

This value corresponds to (1,845 kgO$_2$/d)/(446 TKN applied) $=$ 4.1 kgO$_2$/kgTKN applied (matches with value in Table 9.1).

The *total* average consumption is:

$$\text{Total average } O_2 \text{ consumption} = \text{carbonaceous demand} + \text{nitrogenous demand}$$

$$= 786 + 1,845 = \mathbf{2{,}631 \ kgO_2/d}$$

It can be observed that, differently from the conventional activated sludge system, the oxygen consumption in this case is controlled by the nitrogenous demand (1,845/2,631 $=$ 70% of the total), as most of the BOD was previously removed in the UASB reactor.

The oxygen consumption necessary to meet the demand in peak conditions is a function of the ratio between the maximum O_2 consumption and the average O_2 consumption. In this example, a value of 1.3 was adopted, considering the presence of the UASB reactor upstream and the fact that the plant is of medium size (see Table 9.1):

$$\text{Maximum } O_2 \text{ consumption} = (\text{maximum consumption/average consumption ratio}) \times \text{average consumption} = 1.3 \times 2{,}631 \text{ kgO}_2/\text{d}$$

$$= \textbf{3{,}420 kgO}_2\textbf{/d}$$

This oxygen consumption is the field demand (actual consumption in the treatment plant). The production of oxygen to be specified for standard conditions (clean water, 20 °C, sea level) is greater, so that, in the field, the reduced value of the oxygen production equals the field oxygen demand. The standard/field oxygenation efficiency correction factor adopted is 1.6 (see Table 9.1). The required O_2 in standard conditions is:

$$O_2 \text{ required in standard conditions} = (\text{standard/field oxygenation efficiency ratio}) \times \text{field } O_2 \text{ consumption} = 1.6 \times 3{,}420 \text{ kgO}_2/\text{d}$$

$$= \textbf{5{,}472 kgO}_2\textbf{/d} = 228 \text{ kgO}_2/\text{hour}$$

By adopting a standard oxygenation efficiency of 1.8 $kgO_2/kW{\cdot}hour$ (see Table 9.1), the power requirement is:

$$\textit{Required power} = O_2 \text{ consumption/oxygenation efficiency}$$

$$= (228 \text{ kgO}_2/\text{hour})/(1.8 \text{ kgO}_2/\text{kW}{\cdot}\text{hour}) = 127 \text{ kW} = 173 \text{ HP}$$

As there are two aeration tanks, and the length/breadth ratio in each one is 3, **three** aerators can be adopted in each tank, making up a total of **six aerators.** The power of each aerator is:

$$\text{Power required for each aerator} = \text{total power / number of aerators}$$

$$= 173 \text{ HP}/6 = 29 \text{ HP}.$$

A commercial value higher than that required should be adopted for the installed power, so that the oxygenation capacity is sufficient when there is a by-pass of the raw sewage to the UASB reactor (supply of organic carbon to the aerobic reactor, if applicable). In this example, **40 HP** aerators should be used.

$$\text{The total } \textit{installed} \text{ power is: } 40 \text{ HP} \times 6 = \textbf{240 HP} = \textbf{176 kW}$$

The per capita installed power is 176,000 W / 67,000 inhabitants = 2.63 W/inhabitant (appropriate, according to Table 1.3).

If aeration is controlled by switching on/off the aerators, by changing the submergence of the aerators, or by other methods, and taking into account that the by-pass of the raw sewage will be only occasional, the average *consumed* power will be lower than the installed power. The calculation of the average consumption should be based on the ratio between the maximum and average consumptions. However, the ratio between maximum O_2 consumption/average O_2 consumption adopted (1.3) is not high, in view of the smoothing provided by the UASB reactor. It may be difficult to make the production of oxygen equal to the average consumption throughout the day (this practice would be easier if the ratio between maximum and average consumption were larger, such as in the conventional activated sludge system). Therefore, adopt, in this example, the consumed power as equal to the required power, which is calculated according to the maximum O_2 consumption:

Consumed power = 127 kW × 24 hours/d × 365 d/year = **1,112,520 kW·hour/year** (18 kW·hour/inhabitant·year, appropriate, according to Table 1.3).

The average power level, a parameter that expresses the mixing capacity of the aerators, is calculated as:

Power level = average power/reactor volume = $(127,000$ W)/$(1,995$ m$^3)$ = 64 W/m^3 (sufficient to maintain the sludge in suspension).

(f)   Design of the secondary sedimentation tank

Design parameters adopted (see Table 9.1):

- Hydraulic Loading Rate: HLR = 30 m^3/m^2·d
- Solids Loading Rate: SLR = 120 kgSS/m^2·d

The required surface area, according to the concept of the hydraulic loading rate (HLR adopted = 30 m^3/m^2·d), is:

$$\text{Area} = Q/\text{HLR} = (9{,}820 \text{ m}^3/\text{d})/(30\text{m}^3/\text{m}^2 \cdot \text{d}) = 327 \text{ m}^2$$

The required surface area, according to the concept of the solids loading rate, depends on the load of influent solids to the sedimentation tanks. For the calculation of the solids load, the sludge return flow is $Q_r = R \times Q$. In item (d) of the example, the recirculation ratio R adopted was $(=Q_r/Q)$ 0.8. Therefore, the sludge return flow is $Q_r = 0.8 \times 9{,}820$ m^3/d = **7,856 m^3/d**. The MLSS concentration, calculated in item (c), is 2,000 mg/L = 2,000 g/m^3 = 2.0 kg/m^3. For the solids loading rate of 120 kgSS/m^2·d, the required surface area is:

$$\text{Area} = \text{SS load}/\text{SLR} = (Q + Q_r) \times \text{MLSS} / \text{SLR}$$

$$= [(9{,}820 + 7{,}856) \text{ m}^3/\text{d} \times 2.0 \text{ kgSS/m}^3]/(120 \text{ kgSS/m}^2 \cdot \text{d}) = 295 \text{ m}^2$$

In this case, HLR was more restrictive, because the concentration of SS in the aeration tank is low, which results in low solids loads to the sedimentation tank. Adopt the highest value between the two calculated values (327 m^2 and 295 m^2), that is, **327 m^2**.

By adopting two sedimentation tanks, the surface area of each one is: 327 m^2/2 = 164 m^2.

By adopting circular sedimentation tanks, the diameter of each sedimentation tank is:

$$\text{Diameter} = (\text{Area} \times 4/\pi)^{1/2} = (164\,\text{m}^2 \times 4/3.14)^{0.5} = \mathbf{14.5\ m}$$

By adopting a sidewater depth $\mathbf{H = 3.5\ m}$, the total volume of the sedimentation tanks is 3.5 m $\times$ 327 m^2 = **1,145 m^3**.

The slope of the bottom of the sedimentation tanks depends on the type of sludge removal device: scrapers require a slope of approximately 1:12 (vertical/horizontal), while suction removers are suitable for a flat bottom. Dortmund-type sedimentation tanks have a much higher slope and a lower sidewater depth. If there is slope, the volume of the conical part can be included in the calculation of the total volume.

The hydraulic detention time in the secondary sedimentation tanks is:

$$\text{HDT} = \text{V/Q} = (1{,}145\,\text{m}^3)/(9{,}820\,\text{m}^3/\text{d}) = 0.12\,\text{d} = \mathbf{2.9\ hours}$$

(g)   Sludge processing

According to item (d), the load of aerobic sludge generated in the activated sludge system and returned to the UASB reactor is:

Aerobic sludge, before digestion in the UASB reactor:
- Volatile solids: P_{XV} = 422 kgVSS/d
- Fixed solids: P_{XF} = 141 kgFSS/d
- Total solids: P_X = 563 kgSS/d

Assuming a removal of 35% of VSS from the aerobic sludge during digestion in the UASB reactor (Table 9.1: values between 25 and 45%), and knowing that the load of fixed solids remains unchanged, the load of aerobic sludge wasted from the UASB reactor is:

Aerobic sludge, after digestion in the UASB reactor:
- Volatile solids: P_{XV} = 422 kgVSS/d $\times$ (1–0.35) = 274 kgVSS/d
- Fixed solids: P_{XF} = 141 kgFSS/d
- Total solids: P_X = 274 + 141 = **415 kgSS/d**

The sludge to be removed from the UASB reactor also includes the anaerobic sludge, which is usually produced in the UASB reactor. The production of anaerobic sludge is between 0.40 and 0.50 kgSS/kgBOD *removed* in the UASB reactor, or between 0.28 and 0.36 kgSS/kgBOD *applied* to the UASB reactor (see Table 9.1). By adopting a coefficient of anaerobic sludge production of 0.30 kgSS/kgBOD *applied* to the UASB reactor, the production of anaerobic sludge is:
Anaerobic sludge:

$$\text{Total solids: } P_X = \text{coefficient of sludge production} \times \text{load of BOD}$$

$$\text{in the raw sewage} = 0.30\,\text{kgSS/kgBOD} \times 3{,}350\,\text{kgBOD/d} = \mathbf{1{,}005\ kgSS/d}$$

The total amount of sludge to be wasted from the UASB reactor (digested anaerobic sludge + digested aerobic sludge) is:

$$\text{Total production of sludge} = \text{anaerobic sludge} + \text{aerobic sludge}$$

$$= 1{,}005 + 415 = \mathbf{1{,}420\ kgSS/d}$$

The per capita sludge production, expressed as dry solids, is: 1,420 kgSS/d/ 67,000 inhabitants $= 0.021$ kgSS/inhabitant·d $= 21$ gSS/inhabitant·d (matches with Tables 9.1 and 1.3)

Assuming a concentration of SS of **3.0%** in the sludge removed from the UASB reactor (see Table 9.1), which is equivalent to approximately 30,000 mgSS/L or 30 kgSS/m^3, the flow of sludge removed from the UASB reactor and directed to the sludge processing is:

$$Q_{ex\ UASB} = \text{load/concentration} = (1{,}420\ kgSS/d)/(30\ kgSS/m^3)$$

$$= \mathbf{47 m^3/d}\ (0.76L\ /\ \text{inhabitant·d},\ -\text{matches with Table 9.1})$$

The sludge removed from the UASB reactor is usually already digested and thickened, requiring just a dewatering stage. Assuming, for simplicity, a solids capture efficiency of 100% in the dewatering and a density of 1.0 for the dewatered sludge, and adopting a solids content of **25%** (approximately 250,000 mgSS/L $= 250{,}000$ gSS/m^3 $= 250$ kgSS/m^3) for the dewatered sludge (mechanical dewatering, see Table 9.1), the characteristics of the sludge for final disposal are:

Sludge to be disposed of (cake):

- Load of solids $= \mathbf{1{,}420\ kgSS/d}$ (equal to the influent load to dewatering)
- Daily volume $=$ load/concentration $= (1{,}420\ kgSS/d)/(250\ kgSS/m^3) = $ **5.7 m^3/d**

The per capita production of sludge to be disposed of is:

- Per capita load of SS $= 1{,}420$ kgSS/d/67,000 inhabitants $= 0.021$ kgSS/ inhabitant·d $= 21$ gSS/inhabitant·d (matches with Tables 9.1 and 1.3)
- Per capita volume of sludge $= 5.7$ m^3/d/67,000 inhabitants $= 5{,}700$ L/d/67,000 inhabitants $= 0.09$ L sludge/inhabitant·d (matches with Tables 9.1 and 1.3)

(h)   Comparison with the conventional and extended aeration activated sludge systems

Chapter 5 presents a full design example of a conventional activated sludge, using the same input data. Example 2.11 presents a simplified design of an extended aeration system, also using the same input data. For the sake of comparison, the main values resulting from the three designs are listed below.

Item	Conventional activated sludge	Extended aeration	Activated sludge after UASB reactor
Sludge age (d)	6	25	8
Volume of aeration tank (m^3)	2,051	6,366	1,995
Volume of secondary sedimentation tanks (m^3)	2,128	4,416	1,145
Production of sludge to be treated (kgSS/d)	1,659 (*)	3,119 (**)	1,420 (***)
Installed power for aeration (HP)	400	600	240

(*) Add production of primary sludge. Treatment of mixed sludge by thickening, digestion and dewatering
(**) Treatment of the aerobic sludge by thickening and dewatering
(***) Aerobic and anaerobic sludge after digestion in the UASB reactor. Treatment by dewatering

Therefore, the wide range of advantages of the combined UASB reactor-activated sludge system is noticed, mainly in terms of sludge production and power consumption. In terms of unit volumes, the volume of the UASB reactor should still be added to this alternative, while the volumes of the units associated with the sludge treatment should be added to the other alternatives. The total volume of all the units in the UASB-activated sludge alternative is still a little smaller than the total volume from the other two alternatives.

10

Biological selectors

10.1 INTRODUCTION

The successful operation of an activated sludge plant depends on an efficient solids-liquid separation in the secondary sedimentation tank, with the following main objectives (a) produce a clarified effluent and (b) thicken the sludge on the bottom of the sedimentation tank to a satisfactory concentration for its recirculation to the reactor.

Both functions can be harmed in case the sludge presents poor settleability and thickening capacity. There are several types of deterioration of the sludge characteristics, but the most frequent one is **sludge bulking**, which is caused by an imbalance between the populations of microorganisms that make up the activated sludge floc. In a simplified manner, the floc consists of:

- *Floc-forming bacteria*. These bacteria have a gelatinous matrix, which facilitates the gathering of new microorganisms, producing a floc of larger dimensions and, as a consequence, with a higher settling velocity.
- *Filamentous bacteria*. These bacteria, which have a predominantly elongated morphology, are responsible for the floc structure, when present in a suitable number.

The balance between the filamentous and the floc-forming organisms is delicate, and a good part of the operational success of the activated sludge plant depends on it. Three conditions can occur (Horan, 1990) (see Figure 10.1).

- *Balance between filamentous and floc-forming organisms*. Good settling and thickening capacity of the sludge.

**INFLUENCE OF THE FILAMENTOUS ORGANISMS
ON THE FLOC STRUCTURE**

IDEAL, NON-BULKING FLOC

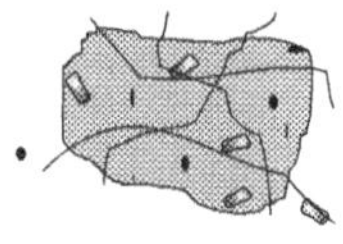

1. Filamentous and floc-forming organisms
   in equilibrium
2. Strong and large floc
3. Filaments do not interfere
4. Clear supernatant
5. Low SVI

PIN-POINT FLOC

1. Few or no filamentous organisms
2. Weak and small floc
3. Turbid supernatant
4. Low SVI

BULKING SLUDGE

1. Filamentous organisms prevail
2. Large and strong floc
3. Filaments interfere with settling and thickening
4. Clear supernatant
5. High SVI

Figure 10.1.   Effect of the filamentous organisms on the structure of the activated sludge floc (EPA, 1987)

- *Predominance of floc-forming organisms*. The floc is insufficiently rigid, which generates a small, weak floc, with poor settleability. This condition is named *pin-point floc*.
- *Predominance of filamentous organisms*. The filaments are projected outside the floc, preventing the adherence of other flocs. Thus, after sedimentation, the flocs occupy a large volume (represented by a high value of the SVI – Sludge Volume Index), which causes an increased level of the sludge blanket in the secondary sedimentation tank. This increase can lead to loss of solids, causing the deterioration of the quality of the final effluent. This condition is named *sludge bulking*.

There are several possible causes for sludge bulking, all of them associated with the environmental conditions to which the bacteria are submitted. Among them, the following can be mentioned:

- low dissolved oxygen (DO)
- low F/M ratio
- septic influent wastewater
- nutrient deficiency
- low pH

Until recently, this phenomenon was controlled only at the operational level, such as with manipulation of the return sludge flow, supply of the necessary amount of oxygen, addition of chemical products and chlorination. However, the recent progresses in the understanding of the dynamics of microbial populations in the reactor has allowed, in the design stage, the incorporation of preventive measures against sludge bulking.

The essence of this mechanism lies in the creation of environmental conditions that favour the predominance of floc-forming bacteria over filamentous bacteria. The most desirable microorganisms in the reactor are then selected by the incorporation of special reactors, named **selectors,** in the design of the biological reactor.

The subject of biological selectors is very broad and complex. Many researches are being carried out worldwide, and a substantial progress is being made in the understanding of the phenomenon. This chapter intends just to give an introductory view on the subject. Further details can be obtained in specific books on the theme, such as Jenkins *et al.* (1993) and Wanner (1994), besides recent technical papers.

Chapter 12 presents several possible forms of controlling sludge-bulking problem in existing wastewater treatment plants.

10.2 TYPES OF SELECTORS

10.2.1 Classification concerning the physical configuration

In terms of configuration of the selectors, there are basically the following types (see Figure 10.2):

- *plug-flow reactors*
- *separate, sequential compartments in plug-flow reactors*
- *separate selector tanks upstream of complete-mix reactors*

The three types are based on the principle that a *high F/M ratio favours the predominance of floc-forming organisms.* This is due to the fact that, in the zone of large food availability (high F/M), the floc-forming bacteria have better conditions to assimilate the high load of substrate than the filamentous bacteria (Metcalf and Eddy, 1991).

In *plug-flow reactors* (Fig. 10.2.a), the inlet end of the reactor has a high F/M ratio, due to the higher BOD concentrations caused by the entrance of the influent wastewater. In fact, studies in several activated sludge plants with plug-flow reactors have indicated a better sludge settleability and lower SVI values than in plants with complete-mix reactors (WRC, 1990). A plug-flow reactor is predominantly longitudinal, either by means of a long, unidirectional tank, or by means of a tank with several U or baffle walls (see Figure 7.4 of a U-shaped reactor). The U-shape, which is also frequently used in biological nutrient removal plants, enables the allocation of a reactor approaching plug flow in a not predominantly longitudinal area. Besides that, the length of some piping can be reduced, especially that of the internal recirculation line.

PHYSICAL CONFIGURATIONS OF SELECTORS

PLUG-FLOW REACTOR

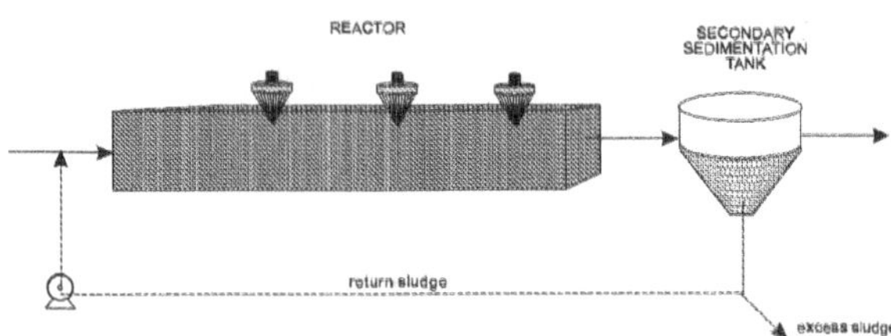

PLUG-FLOW REACTOR WITH INCORPORATED SELECTORS

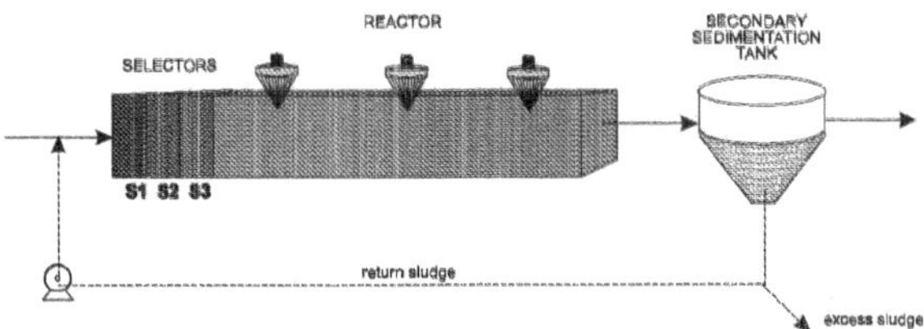

COMPLETE-MIX REACTOR WITH SELECTOR UPSTREAM

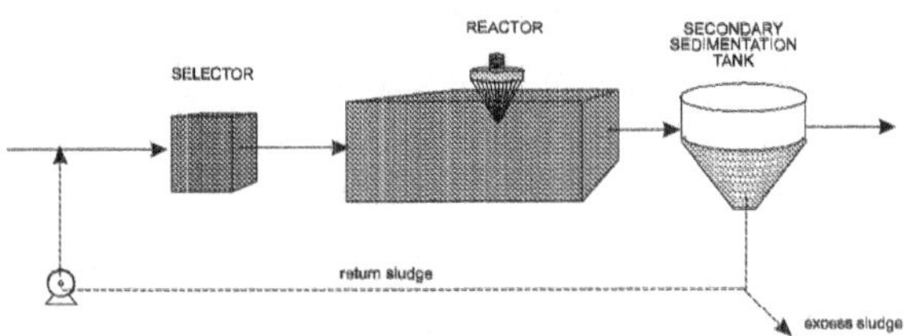

Figure 10.2. Types of configurations of biological selectors

In plug-flow reactors, the initial part can be divided into *compartments* by dividing walls, creating one or more selectors physically separated from the main part of the reactor (Figure 10.2.b). Each of these compartments has a high F/M ratio, a small volume, and a short detention time.

In the case of *complete-mix reactors*, the selector should comprise a separate tank (Figure 10.2.c), also with a high F/M ratio and a short detention time. A complete-mix reactor is predominantly square or not very elongated in plan.

The design of a plug-flow reactor can still incorporate an additional flexibility relating to the influent addition point. If the influent is distributed at several points along the tank, the system is named **step feeding**. This configuration is also used for the control of solids in the system (Keinath, 1981; EPA, 1987; Copp *et al.*, 2002). When the secondary sedimentation tank can no longer accommodate the

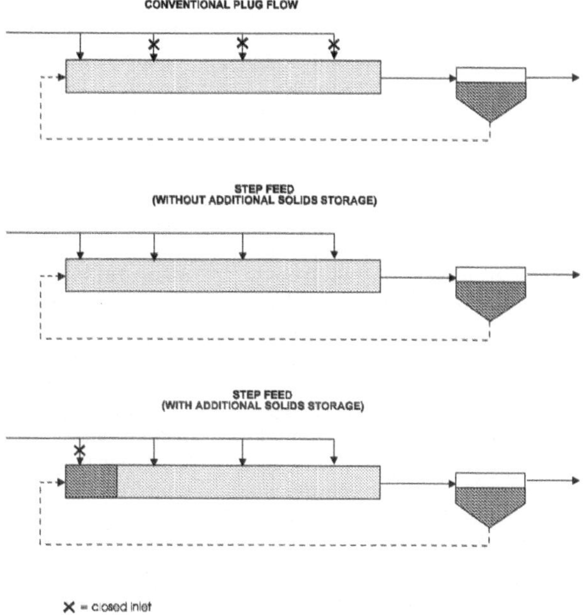

Figure 10.3.   Variations of the plug-flow reactor. Conventional reactor and step feeding

solids, and the sludge blanket begins to rise (due, for instance, to sludge bulking), the solids can be temporarily stored at the entrance of the reactor, provided that the influent is diverted further downstream (Figure 10.3). This constitutes a measure to control the *effect* of the bulking, and not its *cause*. However, it is effective, being an additional resource available for the operator in the important aspect of the control of solids.

See Section 3.3 for a specific analysis of activated-sludge reactors.

10.2.2 Classification concerning the availability of oxygen

Regarding the presence or absence of oxygen, the selectors can be of either one of the three types below:

- *aerobic*
- *anoxic*
- *anaerobic*

The purpose of having different types of selectors is that, by recognising the different environmental requirements of the several organisms, it is possible to create environmental conditions that favour the growth of selected organisms.

Table 10.1. Comparison between the selector types

Type of selector	Advantage	Disadvantage
Aerobic	• Simple process • Does not need internal recirculations, besides the sludge return • Depends on the tank geometry, and not on nitrification	• Does not reduce the oxygen requirements • Requires a complex aeration system to supply the maximum oxygen demand in the initial zone of high F/M ratio
Anoxic	• Allows savings in the oxygen requirements • Allows savings in alkalinity consumption (increases the resistance to pH reduction) • Reduces the denitrification possibilities in the secondary sedimentation tank and the occurrence of rising sludge • The initial zone of high F/M ratio occurs in the anoxic zone (the high oxygen demand is supplied by nitrate, instead of oxygen)	• Cannot be used in a non-nitrifying process • Requires an additional internal recirculation line • Requires care in the design and operation, to reduce the introduction of oxygen into the anoxic zone • A poor design can cause sludge bulking due to low DO levels • Operational problems can generate bad odours
Anaerobic	• Simple design • Does not need internal recirculations, besides the sludge return • Selector of simpler operation • Can be used for biological phosphorus removal	• Does not reduce the oxygen requirements • It may not be compatible with high sludge ages • Requires care in the design and operation, to reduce the introduction of oxygen into the anaerobic zone • A poor design can cause sludge bulking due to low DO levels

Source: partly adapted from WEF/ASCE (1992)

The design of the selector-reactor system should be compatible with a broader view of the treatment plant as a whole. Aspects to be taken into consideration include (a) the nitrification capacity of the system (function of the sludge age) and (b) the desirability to encourage the denitrification in the reactor (function of the reactor configuration and of the recirculations). It is interesting that the selector is provided with an additional flexibility, allowing it to work as either anoxic or aerobic (Sampaio and Vilela, 1993)

The anoxic and anaerobic zones should be provided with *stirrers*, to ensure that the biomass remains in suspension. In the aerobic zones, there should be either mechanical or diffused air aeration.

Table 10.1 presents a balance between the advantages and disadvantages of the three types of selectors, related to the availability of oxygen.

11

Process control

11.1 INTRODUCTION

The main purposes of the implementation of operational control in a wastewater treatment plant can be (Andrews, 1972, 1974; Lumbers, 1982; Markantonatos, 1988; von Sperling and Lumbers, 1988; Olsson, 1989, von Sperling, 1990):

- produce a final effluent with a quality that complies with the discharge standards
- reduce the variability of the effluent quality
- avoid large process failures
- reduce operational costs
- increase the treatment capacity without physical expansion of the system
- implement an operation with variable efficiency to accommodate seasonal variations
- reduce labour requirements
- allow faster start-up

Being highly variable, the influent loads to a sewage treatment plant represent an incentive for the adoption of operational control but, at the same time, they introduce a great difficulty in its implementation. The control of a sewage treatment plant differs from the control of an industrial process, mainly regarding the great variability in the characteristics of the influent. In industrial processes, where control techniques have been traditionally used, the characteristics of the influent are deterministic, or have minor variations around the reference value, being usually directly controllable. An additional complexity of biological treatment systems results from its own dynamics, which contains (a) non-linearities, (b) very wide ranges of time constants, (c) a heterogeneous culture of microorganisms metabolising a heterogeneous substrate, (d) inaccuracy and (e) stability interrupted by abrupt failures (Beck, 1986).

In terms of automated operational control, additional difficulties that have reduced its application in a broader way have been (Lumbers, 1982; Beck, 1986; Markantonatos, 1988; von Sperling, 1990):

- the characteristics of the influent are of a dynamic, stochastic nature, with unknown disturbances and measurement noises superposed to variations in the process
- the effect of the control actions varies for the different process variables, in terms of time lag and magnitude of the response
- there is a lack of reliable on-line sensors for some process variables
- not all the process variables can be directly measured
- the control actions are usually limited by the physical restrictions of the system
- in several plants, the possibility of control is limited due to a design with little flexibility
- there are difficulties in incorporating complex process models in the control algorithms and, conversely, there are limitations in the control strategies based on very simple process models

However, several of these problems have been recently reduced by the development of more robust sensors, cheaper and more accessible information technology, more reliable mathematical models, new control algorithms, and designs that are more flexible and adaptable to automated strategies. The automated, advanced control of real-scale activated sludge plants is covered in several IWA publications and scientific and technical reports (e.g., Copp *et al.*, 2002), von Sperling (1989a, 1990, 1992, 1994d), von Sperling and Lumbers (1988, 1991a, 1991b) and Olsson and Newell (1999)).

Because the advanced control algorithms depend on dynamic models of the system, which are not covered in this book, they are not dealt with in this chapter. The objective of the chapter is to provide the control principles of the activated sludge process, without going into detail into the control algorithms and the principles of control engineering. Therefore, this text presents only the classical or conventional control strategies.

Special attention is given in this chapter to two process variables: dissolved oxygen (DO) and mixed liquor suspended solids (MLSS). These two variables play an important role in the efficiency and in the operational costs of the activated sludge plants, as it has been already described in several sections of this book.

The last section in this chapter covers the important topic of monitoring, which is an essential requirement for process control and the evaluation of the performance of the plant.

11.2 BASIC CONCEPTS OF PROCESS CONTROL

11.2.1 Variables involved

Some basic concepts of control engineering applied to wastewater treatment plants are briefly described here. The operational control of a treatment plant can be classified according to the degree of automation, as follows (Andrews, 1972):

- manual operation, with (a) evaluation of the performance by human senses and (b) manual process control
- manual operation, with (a) performance evaluation by analyses or indicating or recording instruments and (b) manual process control
- automatic control, with (a) evaluation of the performance by automated sensors and (b) automated process control

In this chapter, emphasis is given to the second operational form, which is more frequently practised in developing regions.

In a control system, an important step is the identification of the variables involved in the process. Four types can be distinguished (von Sperling and Lumbers, 1988; von Sperling, 1990) (see Figure 11.1):

- input variables
- control variables (state variables and/or output variables)
- measured variables (input variables and/or control variables)
- manipulated variables

The *input variables* are those that force the system (forcing functions) and that cannot be directly controlled in most of the treatment plants. Examples are the influent characteristics, such as flow, BOD, SS and TKN.

The *control variables* are those that need to be controlled. They include the *state variables*, such as MLSS, DO and the sludge blanket level. A particular case

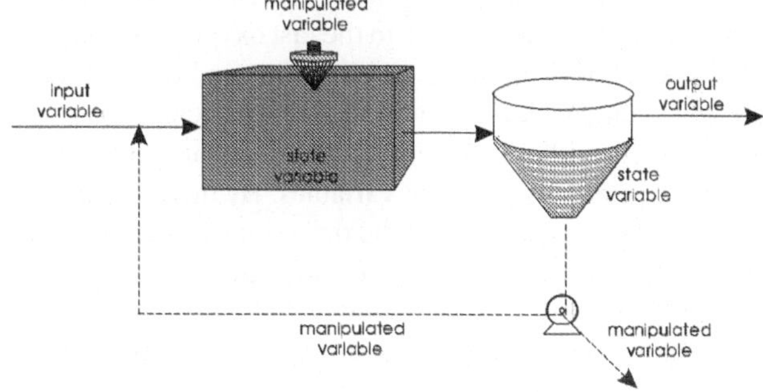

Figure 11.1.   Variables involved in the control of the activated sludge process

is represented by the *output variables*, which define the effluent quality, such as effluent BOD, SS and N.

The *measured variables* are the input, control or other variables, which provide information for the definition of the control action. The selection of the variables depends on the control algorithm and on the suitability for either direct or on-line measurement.

The *manipulated variables* are those that are altered to maintain the control variables at the desired level, as determined by the control algorithm. The activated sludge process is relatively poor in terms of availability of manipulated variables, compared with industrial production lines, but it is one of the most flexible processes in comparison with other wastewater treatment processes. The main manipulated variables in the activated sludge systems are:

- aeration level (oxygen transfer coefficient – $\mathbf{K_L a}$)
- return sludge flow ($\mathbf{Q_r}$)
- excess sludge flow ($\mathbf{Q_{ex}}$)

Other manipulated variables can be the influent flow (if there are equalisation tanks), storage of the return sludge (requiring an additional tank), and variation of the inlet point in step-feed reactors. As they are more specific, these control forms are not covered in this chapter.

11.2.2 Control algorithms

There are several algorithms in the control-engineering field that can be used for activated sludge systems. The most common ones are the *feedback* and the *feedforward* controls.

The *feedback control* measures the output variable and takes a corrective action based on the deviation with relation to the set point. A common example is the control of DO, which is measured at each pre-established time interval, either increasing or decreasing the $K_L a$ (manipulated variable) according to the comparison between the current and the desired concentration. To guarantee a quick response, the dynamics of the control variable should be fast, as it is the case with DO, in which the variations occur in a relatively short time. This is due to the fast oxygen consumption by the microorganisms and to the fast oxygen transfer by the aerators. In the *feedback control*, it is not necessary to know and model the system, since the actions are based on deviations that have already occurred.

The other control algorithm is the *feedforward*, in which the corrective actions are based on measurements of the input variables. By means of a dynamic model of the system, the control variables and the deviations from the set point are estimated, finally leading to the adjustment of the manipulated variables. An example is the control of MLSS by the manipulation of the excess sludge flow (Q_{ex}). As the response of the system to variations in Q_{ex} is slow, the use of a *feedback* controller would not be adequate, and a *feedforward* process could be applied. In fact, several changes in activated sludge are slow, especially those based on biochemical reactions. In contrast to the *feedback* control, in *feedforward* control, a considerable knowledge of the process is necessary, so that the output variables can be estimated. Unfortunately, this is not the case with wastewater treatment systems, and the incorporation of a significant portion of *feedback* control is frequently necessary (Andrews, 1974). This statement, made in 1974, remains true until today, in spite of the deeper knowledge of the process acquired in the past years.

Other control approaches that can be adopted are (a) optimal control and (b) control by expert systems and variants. The *optimal control* implies the existence of an objective function (e.g., cost or performance) to be optimised (either minimised or maximised) by using appropriate mathematical techniques. Constraints are established to the variables, to conform them to the physical limitations of the system and also to specified criteria, such as those related to performance or cost. The values of the manipulated variables are determined by an optimisation algorithm (von Sperling, 1990; von Sperling and Lumbers, 1991a, 1992).

The *expert systems*, a branch of artificial intelligence, incorporate the knowledge of experts, and apply this knowledge to solve problems for the users, whose capacity to interpret information and to take control decisions is not the same as that of an expert (Berthouex *et al.*, 1989). The expert systems can be used for process control or for diagnosis and correction of process failures (von Sperling, 1990; von Sperling and Lumbers, 1991b).

11.3 DISSOLVED OXYGEN CONTROL

Due to the diurnal variations of the influent BOD and ammonia loads, the oxygen demand varies with time following a certain diurnal pattern and also incorporating unpredictable or random components. If oxygen is supplied at a constant rate, equal to the average oxygen demand, there will be periods of either overaeration or underaeration during the day. To avoid this, an oxygen transfer rate corresponding to the peak demand is frequently adopted, naturally leading to overaeration periods during the day. The control of the dissolved oxygen aims at equalling the supply of oxygen to its consumption.

The methods used to regulate the aeration level vary according to the type of aeration (Flanagan *et al.*, 1977; WPCF/ASCE, 1988):

- mechanical aeration
    - switching on-off of aerators
    - variation of the rotational speed of the aerators (two speeds or variable speeds)
    - variation of the level of the aerator (variation of the submergence of the aerators by acting on the shaft)
    - variation of the water level (variation of the submergence of the aerators by adjusting the outlet weir)
- aeration by diffused air
    - variation of the speed of the blowers
    - variation of the inlet vanes

- adjustment of the suction valves of all operating blowers to maintain a constant pressure on the air feeding piping

In terms of DO control, the conventional solutions are:

- variation of $K_L a$ according to the time of the day
- variation of $K_L a$ according to the influent flow
- variation of $K_L a$ by feedback control of DO

The first method represents a simple solution, in which $K_L a$ is varied during some pre-established hours of the day (Schlegel, 1977). This is a control form that is a function of time. However, this solution assumes that the diurnal variations are the same everyday, which is improbable, especially if the influent contains a representative portion of industrial wastewaters.

The second method, which assumes the variation of $K_L a$ according to the measured influent flow, can also lead to some distortions. The first reason may result from the fact that the BOD concentration does not necessarily vary proportionally to the flow. The second reason is associated with the lag between the arrival of the BOD load and the associated oxygen consumption, due to the time necessary for the intracellular assimilation of the particulate carbonaceous material, which is not directly available like the soluble form (Clifft and Andrews, 1981b). However, both control forms represent an evolution compared with the option of no control, allowing energy savings with no need of installation of DO sensors.

The third conventional system is the *feedback* control, in which $K_L a$ varies according to the need to either increase or decrease the DO concentration in relation to the set point. As commented, the DO dynamics are fast and, consequently, suitable for *feedback* control.

An additional stage in the control of DO considers the optimum use of aeration, which involves several monitoring locations, variable set points, and manipulations in the oxygen demand itself (Lumbers, 1982). In this line, alternative or complementary approaches are:

- DO profile
- respirometry-based control – SCOUR / SNOUR (Specific Carbonaceous Oxygen Utilisation Rate / Specific Nitrogenous Oxygen Utilisation Rate)
- feedforward control
- self-adjustable control

The analysis of these advanced items is outside the scope of this text. A complementary discussion can be found in von Sperling and Lumbers (1988), von Sperling (1990) and Copp *et al.* (2002).

11.4 SOLIDS CONTROL

11.4.1 Manipulation of the variables

The main manipulated variables for the control of solids in the activated sludge process are the return sludge flow (Q_r) and the excess sludge flow (Q_{ex}). From a practical point of view, their importance can be understood as (Takase and Miura, 1985):

- Q_{ex} controls the total SS mass in the system, and maintains it at a specified value
- Q_r controls the balance between the SS mass in the reactor and in the secondary sedimentation tanks, maintaining it at a specified ratio

The solids control methods based on Q_r and Q_{ex} are analysed separately here for an easier understanding, although both are interconnected.

(a)   Return sludge flow (Q_r)

Strategies commonly used for manipulation of Q_r are (von Sperling and Lumbers, 1988):

- constant Q_r
- Q_r proportional to the influent flow Q
- Q_r function of SVI
- Q_r function of the sludge blanket level in the secondary sedimentation tanks

The return sludge flow maintained constant corresponds to a non-control strategy, which is very simple and adopted in several wastewater treatment plants. However, the return sludge flow should be large enough to accommodate the fluctuations in the solids load entering the sedimentation tanks (Lohmann and Schlegel, 1981), in terms of both flow and MLSS, especially the diurnal variations. To achieve this objective, a large flow is usually adopted, which generally recirculates more sludge than necessary.

Another very common strategy is the maintenance of Q_r proportional to Q, by adopting a fixed Q_r /Q ratio. This reduces the total quantity of sludge to be pumped (Lohmann and Schlegel, 1981) and provides a better balancing of the loads onto the sedimentation tanks.

The third method corresponds to controlling the return sludge flow by measuring the Sludge Volume Index (SVI or variants). A high value of this parameter indicates poor sludge settleability and the possible need to increase Q_r. The SVI tests are not usually performed on-line, and the manipulations are made based on the operator's experience.

The fourth method provides the largest guarantee against the loss of solids in the effluent. Its principle is the control of the return sludge flow according to the height of the sludge blanket in the secondary sedimentation tanks. Adopting Q_r as a continuous function of the sludge blanket level can present some difficulties, but either increasing or decreasing Q_r as soon as the sludge blanket level reaches a certain height is a practical solution. For example, if the blanket reaches a high specific height, the sensor located in this position detects it and sends a signal so that the sludge outlet valve in the sedimentation tanks opens more, thus increasing the sludge flow. This control can also be done manually by the operator, in a less intensive manner.

In the short term, the MLSS (X) and RASS (X_r) concentrations are ruled by purely hydraulic phenomena, and the bacterial growth reactions are irrelevant. As a result, a change in Q_r causes a rapid effect on both variables (especially RASS). If Q_r increases, X also increases, within certain limits, due to the larger solids load taken to the reactor. However, an increased Q_r usually results in a decreased X_r, which at last limits the increase in X, until the system reaches a state of equilibrium. The reverse happens if Q_r is decreased. Thus, it can be understood that the variations in MLSS due to the manipulations in Q_r are somehow limited.

(b)   Excess sludge flow (Q_{ex})

Manual control of the excess sludge removal rate is practised in almost all activated sludge systems. Some commonly used strategies are:

- control of MLSS (constant MLSS)
- control of the sludge load (constant F/M ratio)
- control of the sludge age (constant θ_c)

Due to the importance of these three procedures, they are covered separately in the following section.

11.4.2 Control of process indicators

The classical methods, traditionally used for the solids control in activated sludge systems, are:

- control of MLSS (constant MLSS)
- control of the sludge load (constant F/M ratio)
- control of the sludge age (constant θ_c)

There are other methods, at an intermediate level, which are not covered in this book. They include (a) control of the Oxygen Utilisation Rate, (b) feedback control of the effluent BOD, (c) feedback control of the effluent nitrogen, (d) control of the sludge blanket level in the secondary sedimentation tank (mentioned above, but also subject to control by Q_{ex}).

(a)   Control of MLSS

This is probably the strategy most commonly used by operators, though intuitively. Its purpose is to maintain MLSS constant. If an appropriate level of MLSS is maintained, a good quality of the effluent is usually expected. In terms of soluble BOD, the control of the MLSS concentration by the removal of excess sludge is equivalent to the control of the F/M ratio (Item b below) and sludge age (Item c below), under steady-state conditions. However, in the operation of a wastewater treatment plant, steady-state conditions rarely occur, and the system usually operates in the dynamic state (continuous variation of the flows and influent concentrations, causing continuous changes in the state variables).

The manipulation of the excess sludge flow is more frequently used for the control of MLSS, although the manipulation of the return sludge flow can be used within certain limits. The response of the system to Q_{ex} variations is slow (reduced mass of solids wasted per day, compared with the existing total solids mass). Regarding Q_r, the response in the secondary sedimentation tank is fast (smaller mass of solids present in the sedimentation tank), while the response is slow in the reactor (larger mass of solids and, as a consequence, higher inertia).

The selection of the desired MLSS level is essential for a successful control. The critical aspects include:

- A constant MLSS implies a variable solids load to the sedimentation tank, since the influent flow is usually variable. Depending on the MLSS level, this variability can be harmful to the performance of the system in terms of effluent suspended solids.
- The MLSS level affects the removal of carbonaceous matter (BOD) and the nitrification and denitrification. Higher MLSS values can increase the BOD removal efficiency, but they can, in parallel, cause a higher consumption of dissolved oxygen, which can lead to a possible reduction in the DO concentration in the reactor, to the point of affecting nitrification.

(b)   Control of the F/M ratio

The sludge load, or food/microorganism ratio (F/M), is a practical design and operational parameter. F/M represents the substrate load per unit sludge mass, according to the formula:

$$\frac{F}{M} = \frac{Q \cdot S_o}{V \cdot X_v} \tag{11.1}$$

where:

F/M = food/microorganism ratio (d^{-1})
Q = influent flow (m^3/d)
S_o = concentration of influent substrate (BOD$_5$ or COD) (g/m^3)
X = biomass concentration (total – MLSS, volatile – MLVSS or active) (g/m^3)
V = volume of the reactor (m^3)

The purpose of the control is usually to maintain a constant F/M ratio to ensure a uniform substrate removal. The F/M value to be adopted is usually a design data, but it is frequently adjusted by experience during the operation. The procedure to control the F/M ratio is by adjusting the solids concentration X (by manipulating Q_{ex} or Q_r) according to the influent substrate load to maintain the F/M ratio constant (see Equation 11.1).

However, some problems are related to the F/M control (von Sperling, 1992, 1994d):

- BOD$_5$ cannot be used in the control as substrate indicator, since laboratory results take 5 days to be obtained.

- The unit day^{-1} is usually confusing for operators.
- The F/M ratio is an essentially steady-state parameter, and its association with the quality of the effluent is not valid under dynamic conditions, which prevail in wastewater treatment plants.
- The possibilities of instantaneous control of the F/M ratio by using Q_r to change the MLSS concentration are limited, due to the large mass of solids in the reactor. The manipulation of Q_{ex} leads to effects only in the medium term, being therefore unable to absorb transients and diurnal variations of the influent BOD load.
- The F/M ratio is quantitatively related to the quality of the final effluent only in terms of soluble BOD. However, soluble BOD is usually low, especially in extended aeration systems (von Sperling and Lumbers, 1989a). The biggest problem regarding the effluent BOD is usually the particulate BOD, caused by the suspended solids in the effluent. Increased values of MLSS to maintain the F/M ratio constant can cause an overload of solids into the secondary sedimentation tank, with possible deterioration of the particulate BOD of the effluent.

(c)   Control of the sludge age

Solids Retention Time (SRT), Mean Cell Residence Time (MCRT) and Sludge Age (θ_c) are designations used to express the average time the biomass remains in the system. Under steady-state conditions, the growth rate of the cells should be compensated by their removal via the excess sludge, to maintain the biomass concentration constant. Under these conditions, in which the biomass production is equal to its wastage, the sludge age can be defined as:

$$\theta_c = \text{(mass of solids in the system)/(mass of solids produced per day)}$$
$$= \text{(mass of solids in the system)/(mass of solids wasted per day)} \qquad (11.2)$$

As commented in Section 2.15, there are two classical methods to control the sludge age, with the purpose of keeping it at a constant value:

- wastage of solids from the return sludge line (the concentration of excess sludge is equal to the concentration of the return sludge RASS)
- wastage of solids from the aeration tank or from its effluent (the concentration of the excess sludge is equal to MLSS). This method is named hydraulic control

The hydraulic control is conceptually simpler, without the need for the measurement of the solids concentration. A fraction of the volume of the reactor equal to $1/\theta_c$ should be removed daily. Thus, if a 20-day sludge age is desired, a volume equal to 1/20 of the reactor should be discarded per day. If this fraction is removed daily, the sludge age will remain theoretically constant, independent of the influent flow. If the influent BOD load is constant, the concentration of solids will remain constant, and the θ_c control is equivalent to the control of MLSS. If the influent substrate load increases, the concentration of solids will also increase. Hence, both the mass of solids present in the system and the mass of solids being discarded will increase proportionally, and the sludge age will remain constant.

However, these considerations are only valid in the steady state or in each hypothetical steady state of the operation, not covering the transients between one stage and another. This concept is consequently limited because, in the real operation of a plant, the transients occur more frequently than the occasional steady states. In the dynamic state, the two relations in Equation 11.2 are no longer the same, and the sludge production is different from the sludge wastage, generating either positive or negative mass accumulations in the reactor. Under the steady-state assumption, a sudden increase of substrate concentration is assumed as being immediately followed by an increase of the biomass concentration. However, the bacterial growth takes time, and a deterioration of the effluent will not be noticed until a new steady state is achieved (if it is at all achieved).

Other problems of the control by the sludge age are (von Sperling, 1992, 1994d):

- The sludge age concept comprises only the soluble substrate in the reactor, not covering the usually more important component related to the effluent particulate BOD from the system.
- The sludge age concept was mainly developed for the removal of carbonaceous matter. However, the sludge age of the nitrifying bacteria, whose growth rate is very slow, is usually different from the sludge age of the heterotrophic bacteria responsible for the BOD removal (under dynamic conditions and modifications of the environmental conditions, such as dissolved oxygen concentration). Therefore, there is no general sludge age for all bacteria.
- The control by sludge age does not take into consideration the contribution of the influent inert SS to the biological stage, which can change the balance between production (including influent) and wastage of solids.
- The control of the sludge age focus only on the reactor, and does not take into consideration the important stage of final sedimentation and its implications on the quality of the final effluent, in terms of suspended solids and particulate BOD.

(d) Discussion on the classical methods

A general evaluation of the classical methods leads to the following main points:

- The classical strategies do not integrate the simultaneous control of the reactor and the secondary sedimentation tank, and do not recognise the fundamental importance of the secondary sedimentation tank to the overall quality of the effluent.
- The classical strategies do not focus simultaneously on the purposes of removing the carbonaceous and nitrogenous matter.
- The classical strategies are based mainly on the separate manipulation of the return sludge and the excess sludge. Therefore, its potential for an integrated and simultaneous manipulation is not used.

- There is an inherent difficulty regarding the choice of the MLSS, θ_c or F/M value to be maintained. A certain value can be satisfactory under certain conditions, but unsatisfactory under others.
- The classical strategies are based on steady-state operating conditions, which rarely occur in real practice.

Based on the considerations mentioned above, an impression might have been created that there are no adequate strategies for the control of activated sludge systems, which are, paradoxically, the most flexible wastewater treatment process. This impression should not be true, and the point discussed herein is that an integrated management should be adopted, instead of the usual approach to control the system according to a single variable only. Even a simple combination of two control variables, such as MLSS and sludge blanket level, has better chances of being successful in terms of the overall performance than each of the separate strategies.

Besides that, it is believed that a dynamic model for the reactor – secondary sedimentation tank system can be directly used to evaluate a control strategy. Even though there is still a certain reluctance by many operators in using mathematical models, it should be remembered that strategies conceptually as simple as the control by sludge age or by F/M ratio have as a support a model (although simplified) of the kinetics of bacterial growth and substrate removal.

The ideal approach is the adoption of a dynamic model that, even with a simplified structure, covers the reactor and the secondary sedimentation tank, and simulates the removal of carbonaceous and nitrogenous matter. The simultaneous consideration of the units and processes is considered a minimum requirement for any control strategy to be adopted. However, due to their complexity, dynamic models are not included in the scope of this book.

In summary, it is believed that the control strategies to be adopted should have the following characteristics (von Sperling, 1992, 1994d):

- Integrated control of the system, by (a) simultaneous actuation on the manipulated variables (Q_r and Q_{ex}), (b) consideration of the interactions between the reactor and the secondary sedimentation tank, (c) consideration of the simultaneous purposes of BOD, SS and ammonia (sometimes N and P) removal and (d) incorporation of the minimisation of the operational costs as some of the purposes.
- Operation not directed to the control of certain variables (e.g., MLSS, F/M or θ_c) to fixed (questionable) set points, but to the output variables (e.g., BOD, SS, ammonia), which should explicitly comply with quality standards for the effluent.
- Non-use of a single process indicator or a single variable (e.g., MLSS, F/M or θ_c), but use of an integrated dynamic model of the system, covering the reactor and the secondary sedimentation tank, and with all the important input, state and manipulated variables interacting simultaneously. The model, and not just a single variable, should be used to drive the control strategy (von Sperling, 1990).

Table 11.1. Monitoring programme for activated sludge systems (liquid phase)

Place	Parameter	Use	Sample Frequency	Type
Raw sewage	BOD	PE	Weekly	composite
	COD	PE	Weekly	composite
	SS	PE	Weekly	composite
	VSS	PE	Weekly	composite
	TKN	PE	Weekly	composite
	pH	PC	Daily	simple
	Alkalinity	PC	Weekly	simple
	Coliforms	PE	Weekly	simple
Primary effluent	BOD	PE	Weekly	composite
	COD	PE	Weekly	composite
	SS	PE	Weekly	composite
Reactor	Temperature	PC	Daily	simple
	DO	PC	Daily or continuous	simple or sensor
	SS	PC	Daily or continuous	simple or sensor
	VSS	PC	Weekly	simple
	NO_3^-	PC	Weekly	simple
	SVI	PC	Daily	simple
Return sludge	SS	PC	Daily	composite
Final effluent	BOD	PE	Weekly	composite
	COD	PE	Weekly	composite
	SS	PE	Weekly	composite
	SSV	PE	Weekly	composite
	TKN	PE	Weekly	composite
	NH_3	PE	Weekly	composite
	NO_2^-	PE	Weekly	composite
	NO_3^-	PE	Weekly	composite
	pH	PE	Daily	simple
	Coliforms	PE	Weekly	simple

PE = performance evaluation; PC = process control
Other wastewater characterisation parameters can be included, depending on the need
The programme can vary according to the size and relative importance of the plant
Source: adapted from WEF (1990)

11.5 MONITORING THE SYSTEM

Process monitoring is essential for its adequate performance. Table 11.1 proposes a programme for typical activated sludge plants, without automated process control. Naturally, depending on the size and the degree of relative importance of the treatment plant, the frequency and the number of parameters can be either increased or reduced.

12

Identification and correction
of operational problems

12.1 INTRODUCTION

This chapter presents a synthesis of the main operational problems liable to occur in activated sludge systems, including their causes and control measures. Due to the large variety of problems, the list does not intend to be exhaustive and deep, but only an initial guide for the operator. The items focused refer to the increase in the concentration of the following parameters in the final effluent: (a) suspended solids, (b) BOD and (c) ammonia.

The structure of the presentation is in terms of a knowledge basis, which can be used for the development of expert systems for guiding the operator in the solution of operational problems.

The classification of the problems, their detection, causes and control forms are based on a review of several references, including Adelusi (1989), Gall and Patry (1989), WRC (1990), Kwan (1990), WEF (1990), Gray (1991), Metcalf and Eddy (1991) and Wanner (1994).

12.2 HIGH CONCENTRATIONS OF SUSPENDED SOLIDS IN THE EFFLUENT

12.2.1 Causes

> - *Rising sludge*
> - *Bulking sludge*
> - *Pin-point floc*
> - *Dispersed sludge*
> - *Overload of solids in the secondary sedimentation tanks (non-bulking sludge)*
> - *Hydraulic overload in the secondary sedimentation tanks*
> - *Foam and scum*
> - *Other operational problems of the secondary sedimentation tanks*

12.2.1.1 Rising sludge

Detection:

> - Sludge clumps floating on the secondary sedimentation tank surface
> - Gas bubbles entrapped in the floc
> - Supernatant possibly clarified (except for the clumps); low turbidity
> - Possibly high SVI
> - Non-significant presence of filamentous bacteria on microscopic examination

Causes:

> - Denitrification in the secondary sedimentation tank (with release of nitrogen gas bubbles)
> - Gas bubbles adhered to the floc
> - Septic sludge (with release of gas bubbles from anaerobic decomposition)
> - Emulsified grease and oil

Detailing and correction of the causes:

(a)   Denitrification in the secondary sedimentation tank

Cause 1: Insufficient denitrification in the preceding units	
Secondary causes	• Lack of anoxic zones in the preceding units • Insufficient anoxic zones in the preceding zones • Insufficient organic carbon in the anoxic zone • Insufficient amount of nitrified effluent in contact with the anoxic zone • pH in the anoxic zone outside the range from 6.5 to 8.0

Control	1. Create or increase the anoxic zone in the reactor, and ensure that the denitrifying bacteria are supplied with enough organic carbon:
	• Anoxic zone downstream the aerated zone (post-denitrification)
	  – Introduce/increase bypass of raw sewage to the post-anoxic zone (raw sewage as source of organic carbon)
	  – Increase volume of the anoxic zone
	  – If it is not possible to add raw sewage to the anoxic zone, complement the organic carbon requirements with methanol or other similar product
	• Anoxic zone upstream the aerated zone (pre-denitrification)
	  – Introduce/increase internal recirculation from the aerated zone (nitrified liquid) to the anoxic zone
	  – Avoid recirculation containing oxygen
	  – Increase volume of the anoxic zone
	• Anoxic zones upstream and downstream the aerated zone
	  – Introduce/increase internal recirculation from the aerated zone (nitrified liquid) to the anoxic zone
	  – Avoid recirculation containing oxygen
	  – Increase volume of the anoxic zone
	• Simultaneous nitrification/denitrification (oxidation ditches)
	  – control aeration to maintain a balance between the aerobic/anoxic zones (nitrification/denitrification) (for pH control: see Section 12.2.1.2.b)
	2. If the pH is out of range, wait a certain time, since nitrification will also be affected, thus reducing denitrification itself
	3. Reseed with active denitrifying biomass

Cause 2: Long detention time of the sludge in the secondary sedimentation tank	
Secondary causes	• Low return sludge flow • Low velocity of the sludge removal mechanism • Problems with the sludge removal mechanism

Control	1. Reduce the sludge detention time in the secondary sedimentation tank • Increase the return sludge flow • Increase the velocity of the sludge scraping or collection mechanism • Repair the sludge scraping or collection mechanism, if defective • If the problem is in only one or in some tanks, reduce the influent flow to the defective tanks

Cause 3: Undesired nitrification in the reactor, leading to denitrification in the secondary sedimentation tank	
Secondary cause	• If the activated sludge system has not been designed to nitrify and denitrify (effluent ammonia is not an important item in this plant) and if nitrification is occurring, it may lead to denitrification in the secondary sedimentation tank
Control	1. Either reduce or eliminate nitrification in the reactor • Reduce the DO concentration in the reactor • Reduce the sludge age by increasing the excess sludge flow

(b)   Gas bubbles attached to the floc

Causes	• If diffused air is used, an excessive aeration can cause bubbles adhered to the floc • If there is a post-anoxic zone, nitrogen gas bubbles may remain adhered to the floc directed to the secondary sedimentation tank
Control	1. Reduce the aeration level 2. Introduce a reaeration stage after the anoxic zone, to release the gas bubbles prior to the secondary sedimentation tank

(c)   Septic sludge

Detection	• Odour • Analyse sewage in terms of sulphides and volatile organic acids
Causes	• Low return sludge flow • Problems with the mechanical scrapers • Presence of highly concentrated industrial wastes

Control	1. Reduce the sludge detention time in the secondary sedimentation tank • Increase the return sludge flow • Increase the velocity of the sludge scraping or collection mechanism • Repair the sludge scraping or collection mechanism, if defective • If the problem is in only one or in some tanks, reduce the influent flow to the defective tanks 2. Increase the removal efficiency of the highly concentrated industrial wastes • Reduce the excess sludge flow to increase the MLVSS concentration • Increase the aeration level (see Section 12.2.1.2.a)

(d)   Emulsified grease and oil

Cause	• Industrial wastes
Control	1. Use hose jets to direct oil and grease to the scum remover 2. Verify whether the scum removal equipment in the primary and secondary sedimentation tanks are working well 3. Increase the frequency and duration of the surface scraping to assure an appropriate removal of oil and grease 4. Verify, in the primary sedimentation tank, whether the effluent baffle is deep enough to prevent oil and grease from passing underneath 5. Remove the oil and grease at the source

12.2.1.2 Bulking sludge

Detection:

• Cloudy mass in the secondary sedimentation tank • High SVI value • Low concentration of SS in the return sludge • High sludge blanket level • Clear supernatant • Filamentous bacteria present in the microscopic examination

Causes:

- Low concentrations of DO in the reactor
- pH lower than 6.5
- Low floc load in the entrance of the reactor
- Nutrient deficiency
- Septic sewage
- Presence of large amounts of rapidly degradable carbohydrates

Control based on the causes of the problem:

(a)   Low DO concentrations in the reactor

Cause 1: Insufficient oxygen supply due to problems in the aeration system	
• Mechanical aeration	
Secondary causes	• Defective aerators • Defective DO control system • Accidental switching-off of the aerators • Power failure
Control	• Repair or replace defective aerators • Lubricate bearings and motors of the aerators • Repair defective DO control system • In case of frequent power failures, install stand-by generator
• Diffused-air aeration	
Secondary causes	• Clogged diffusers • Dirty blowers • Defective blowers • Defective DO control system • Power failure
Control	1. Clean clogging in the diffusers • Fixed porous dome diffusers: empty the tank and scrub with detergent or mild muriatic acid • Tube diffusers: remove the tubes from the tanks and replace them, allowing the aeration to continue. Clean the tube with running water, and leave it in a strong detergent solution. Rinse it, and test its permeability under pressure 2. Increase temporarily the air flow to clean clogged coarse bubble diffusers 3. Install air purification system before the air enters the diffusers

	4. Use solvents to clean blowers 5. Lubricate/replace bearings when necessary 6. Repair defective DO control system 7. In case of frequent power failures, install a stand-by generator

Cause 2: Insufficient oxygen supply due to inadequate control of the aeration rate (for mechanical aeration)	
• *Manual control by switching on/off the aerators*	
Secondary causes	• Selection of an excessive switching-off time of the aerators • Selection of an excessive number of switched-off aerators • Incorrect selection of the switching on/off times of the aerators • Incorrect selection of the aerators to be turned off • Stepwise variation of the aeration rate, leading to periods with insufficient aeration • Insufficient submergence of the aerators
Control	1. Reduce the duration of certain switching-off periods of the aerators 2. Reduce the number of aerators turned off 3. Change the selection of the switching-off times of the aerators 4. Change the selection of the aerators to be turned off 5. Increase submergence of the aerators
• *Manual control by two rotation-speed aerators*	
Secondary causes	• Incorrect selection of the rotation reduction times • Incorrect selection of the aerators to have their rotation reduced • Stepwise variation of the aeration rate, leading to periods with insufficient aeration • Defective rotation variation mechanism
Control	• Change the selection of the rotation reduction times • Change the selection of the aerators to have their rotation reduced • Install more aerators • Verify the rotation variation control mechanism
• *Switching on/off control by timer*	
Secondary cause	• Refer to "Manual control by switching on/off" above
Control	1. Refer to "Manual control by switching on/off" above

• *Manual switching on/off control by continuous DO measurement and limit value alarm*	
Secondary causes	• Refer to "Manual control by switching on/off" above • Incorrect switching off alarm set point • Poor operation of the DO sensor due to (a) defective instrument, (b) insufficient calibration, (c) incorrect calibration or (d) foul sensor

Control	1. Refer to "Manual control by switching on/off"above
	2. Adjust the switching off alarm set point
	3. Verify the DO sensor:
	  • Replace defective parts
	  • Recalibrate
	  • Clean sensor

• *Manual control by two rotation-speed aerators, by continuous DO measurement and limit value alarm*	
Secondary causes	• Refer to "Manual control by two rotation-speed aerators" above • Incorrect switching off alarm set point • Poor operation of the DO sensor due to (a) defective instrument, (b) insufficient calibration, (c) incorrect calibration or (d) foul sensor

Control	1. Refer to "Manual control by two rotation-speed aerators" above
	2. Adjust the switching off alarm set point
	3. Verify the DO sensor:
	  • Replace defective parts
	  • Recalibrate
	  • Clean sensor

• *Automatic switching on/off control of aerators, based on continuous DO measurements*	
Secondary causes	• Refer to "Manual switching on/off control by continuous DO measurement and limit value alarm" above • Excessively low "switch on" point • Excessively low "switch off" point
Control	1. Refer to "Manual switching on/off control by continuous DO measurement and limit value alarm" above 2. Raise "switch on" point 3. Raise "switch off" point
• *Automatic control of two rotation-speed aerators, based on continuous DO measurements*	
Secondary causes	• Refer to "Manual control by two rotation-speed aerators, by continuous DO measurement and limit value alarm" above

Control	1. Refer to "Manual control by two rotation-speed aerators, by continuous DO measurement and limit value alarm" above
• *Automatic control of multiple-rotation speed aerators, based on continuous DO measurements*	
Secondary causes	• Refer to "Automatic control of two rotation-speed aerators, based on continuous DO measurements" above • Low DO set point • Insufficient range for variation of the rotation speed • Inadequate relation between the aerator rotation and the DO (gains from the feedback control) • Increased rotation does not lead to increased oxygen transfer rate
Control	1. Refer to "Automatic control of two rotation-speed aerators, based on continuous DO measurements" above 2. Raise DO set point 3. Change parameters in the relation between rotation speed and DO (gains from the feedback control) 4. Verify rotation variation mechanism and repair/replace defective parts
• *Automatic control of the aeration level by variation of the outlet weir level, based on continuous DO measurements*	
Secondary causes	• Low DO set point • Insufficient weir level variation range • Inadequate relation between the weir level and the DO (gains from the feedback control) • Increased weir level does not lead to increased oxygen transfer rate • Defective weir level variation mechanism • Poor operation of the DO sensor due to (a) defective instrument, (b) insufficient calibration, (c) incorrect calibration or (d) foul sensor

Control	1. Raise DO set point
	2. Change parameters in the relation between weir level and DO (gains from the feedback control)
	3. Verify weir level variation mechanism and repair/replace defective parts
	4. Verify the DO sensor:
	    • Replace defective parts
	    • Recalibrate
	    • Clean sensor

• *Automatic control of the aeration level by variation of the level of the vertical shaft of the aerator, based on continuous DO measurements*	
Secondary causes	• Low DO set point • Insufficient range of variation of aerator shaft • Inadequate relation between the shaft level and the DO (gains from the feedback control) • Lowering the shaft level of the aerator not leading to increased oxygen transfer rate • Defective mechanism for variation of the shaft level • Poor operation of the DO sensor due to (a) defective instrument, (b) insufficient calibration, (c) incorrect calibration or (d) foul sensor
Control	1. Raise DO set point 2. Change parameters in the relation between shaft level and DO (gains from the feedback control) 3. Verify mechanism for variation of shaft level, and repair/replace defective parts 4. Verify the DO sensor: • Replace defective parts • Recalibrate • Clean sensor

Cause 2: Insufficient oxygen supply due to inadequate control of the aeration rate (for diffused air aeration)	
• *Manual control by switching on/off blowers*	
Secondary causes	• Selection of an excessive switching-off time of the blowers • Selection of an excessive number of blowers turned off • Incorrect selection of the switching on/off times of the blowers • Stepwise variation of the aeration rate, leading to periods with insufficient aeration
Control	1. Reduce the duration of certain switching-off periods of the blowers 2. Reduce the number of blowers turned off 3. Change the selection of the switching-off times of the blowers
• *Manual control by variation of the opening of the inlet vanes*	
Secondary causes	• Insufficient opening of the inlet vanes • Incorrect selection of the opening/closing times of the vanes
Control	1. Open the inlet vanes more 2. Change the selection of the opening/closing times of the vanes

• *Switching on–off control of the blowers by timer*	
Secondary causes	• Refer to "Manual control by switching on/off blowers" above
Control	1. Refer to "Manual control by switching on/off blowers" above
• *Control of the variation of the opening of the inlet vanes by timer*	
Secondary causes	• Refer to "Manual control of the variation of the opening of the inlet vanes" above
Control	1. Refer to "Manual control of the variation of the opening of the inlet vanes" above
• *Manual switching on/off control of blowers by continuous DO measurement and limit value alarm*	
Secondary causes	• Refer to "Manual switching on/off control of blowers" above • Incorrect switching-off alarm set point • Poor operation of the DO sensor due to (a) defective instrument, (b) insufficient calibration, (c) incorrect calibration or (d) foul sensor
Control	1. Refer to "Manual switching on/off control of blowers" above 2. Adjust the switching-off alarm set point 3. Verify the DO sensor: • Replace defective parts • Recalibrate • Clean sensor
• *Manual control for opening of the inlet vanes by continuous DO measurement and limit value alarm*	
Secondary causes	1. Insufficient opening of the inlet vanes 2. Incorrect switching-off alarm set point 3. Poor operation of the DO sensor due to (a) defective instrument, (b) insufficient calibration, (c) incorrect calibration or (d) foul sensor
Control	1. Open the inlet vanes more 2. Adjust the switching-off alarm set point 3. Verify the DO sensor: • Replace defective parts • Recalibrate • Clean sensor

	• *Automatic control of the aeration level by switching on/off the blowers, based on continuous DO measurements*
Secondary causes	• Excessively low "switching on" point • Excessively low "switching off" point • Stepwise variation of the aeration rate, leading to periods with insufficient aeration • Poor operation of the DO sensor due to (a) defective instrument, (b) insufficient calibration, (c) incorrect calibration or (d) foul sensor

Control	1. Raise the "switching on" point
	2. Raise the "switching off" point
	3. Verify the DO sensor:
	  • Replace defective parts
	  • Recalibrate
	  • Clean sensor

	• *Automatic control of the aeration level by variation of the opening of the vanes, based on continuous DO measurements*
Secondary causes	• Low DO set point • Inadequate relation between the opening of the vanes and the DO (gains from the feedback control) • Poor operation of the DO sensor due to (a) defective instrument, (b) insufficient calibration, (c) incorrect calibration or (d) foul sensor

Control	1. Raise DO set point
	2. Change parameters in the relation between opening of the vanes and DO (gains from the feedback control)
	3. Verify the DO sensor:
	  • Replace defective parts
	  • Recalibrate
	  • Clean sensor

	• *Automatic control of the aeration level by variation of the rotation of the blowers, based on continuous DO measurements*
Secondary causes	• Low DO set point • Inadequate relation between rotation of the blowers and DO (gains from the feedback control) • Poor operation of the DO sensor due to (a) defective instrument, (b) insufficient calibration, (c) incorrect calibration or (d) foul sensor

Control	1. Raise DO set point
	2. Change parameters in the relation between rotation of the blowers and DO (gains from the feedback control)
	3. Verify the DO sensor:
	  • Replace defective parts
	  • Recalibrate
	  • Clean sensor

Cause 3: Insufficient aeration capacity	
Control	1. Mechanical aeration: • Investigate the cost–benefit relation for installation of more aerators • Investigate the cost–benefit relation for a local supplementation of oxygen for peak periods 2. Diffused air aeration: • Investigate the cost–benefit relation for installation of more diffusers and blowers • Investigate the cost–benefit relation for a local supplementation of oxygen for peak periods

Cause 4: Excessive oxygen consumption	
• Consumption for BOD oxidation (synthesis)	
Secondary causes	• High influent BOD load • High load of solids and BOD returned from the supernatant of sludge thickeners • High load of BOD returned from the supernatant of sludge digesters
Control	1. Regulate the influent flow • Use stormwater storage tanks to reduce peaks (in combined sewerage systems) • Introduce/use equalisation tanks 2. Improve the operation of the thickeners • Remove thickened sludge more frequently • Reverse operation from continuous to batch (or vice-versa) • Add coagulants or coagulant aids to improve sludge thickening 3. Improve the operation of the digesters • Prevent the entrance of excessive volumes of highly organic sludge in the digester • Prevent the entrance of toxic materials in the digesters, which can inhibit the methanogenic organisms • Ensure adequate mixing in the digesters • Suspend temporarily the removal of supernatant from the digesters 4. Return supernatant from the thickeners or digesters during periods of low influent flow

• Consumption for biomass respiration (endogenous respiration)	
Secondary causes	• High MLSS concentrations • Low excess sludge flow • Low frequency of removal of excess sludge • Limited thickening, digestion, dewatering, storage and disposal capacity for the sludge • High return sludge flow • High influent organic load leading to a high growth of the biomass • Problems with the excess sludge removal pumps

Control	1. Reduce the MLSS concentration
	   • Increase the excess sludge removal flow
	   • Increase the removal frequency of the excess sludge
	   • Analyse the need/feasibility to expand the sludge treatment units
	   • Reduce return sludge flow
	   • Repair/replace defective excess sludge removal pumps

• Consumption for ammonia oxidation (nitrification)	
Secondary cause	• High influent ammonia load

Control	1. Regulate the influent flow
	   • Use stormwater storage tanks to reduce peaks (in combined sewerage systems)
	   • Introduce/use equalisation tanks

(b) pH concentrations in the reactor lower than 6.5

Cause 1: Oxidation of the carbonaceous and nitrogenous matter	

Control	1. Temporary change in the pH
	   • Add alkaline agents to increase buffer capacity in the reactor
	   • Produce temporary anoxic zone by the intermittent switching off of aerators to encourage denitrification, whilst saving alkalinity
	2. Permanent change in the pH
	   • Create permanent anoxic zones to encourage denitrification, whilst saving alkalinity

Cause 2: Presence of low-pH industrial wastes	

Control	1. Temporary change in the pH
	   • Add alkaline agents to increase buffer capacity in the reactor
	   • Eliminate problem at the source

	2. Permanent change in the pH • Isolate the source of acidity, demanding some form of control of the industrial wastes (either neutralisation or separate treatment)

Cause 3: Return of inadequately digested supernatant from the digesters	

Control	1. Temporary change in the pH
	  • Improve the operation of the digesters
	  • Prevent the entrance of excessive volumes of highly organic sludge in the digester
	  • Prevent the entrance of toxic materials in the digesters, which can inhibit the methanogenic organisms
	  • Ensure adequate mixing in the digesters
	  • Suspend temporarily the removal of supernatant from the digesters
	2. Permanent change in the pH
	  • Improve the operation of the digesters (see above)
	  • Consider heating the digesters
	  • Expand the digesters

(c) Low floc load in the inlet end of the reactor

Detection	• Floc load = $[(COD_{inf} - COD_{eff}) \cdot Q]/(X_r \cdot Q_r)$ (mgCOD/gMLSS)
Causes	• Low load of influent BOD • High concentration of MLSS in the inlet end of the reactor
Control	1. Reduce the return sludge flow 2. Increase the excess sludge flow 3. In step-feed reactors, concentrate the entrance of influent on the inlet end of the reactor

(d) Nutrient deficiency

Detection	• Analyse influent and determine the BOD_5: N:P ratio • Conventional activated sludge – approximate ratio: 100:5:1 • Extended aeration – approximate ratio: 200:5:1
Causes	• Presence of industrial wastes deficient in N and/or P • Activated sludge operating to remove N and/or P
Control	1. Add nitrogen or phosphorus in immediately available forms

(e) Septicity

Detection	• Odour • Analyse the influent for sulphides or volatile organic acids • Gas bubbles on the surface of the primary sedimentation tank
Causes	• Influent with long detention time in the collection and transport system • Long periods between each sludge removal in the primary sedimentation tank • Problems with the sludge scraper of the primary sedimentation tank • Influent containing wastes with high concentration of organic matter
Control	1. Increase the removal frequency of the sludge from primary sedimentation tanks 2. Reduce the number of primary sedimentation tanks in operation 3. Increase the velocity of the sludge scraper in the primary sedimentation tank 4. Repair defective sludge scrapers in the primary sedimentation tank 5. Reduce the influent flow to the defective primary sedimentation tanks 6. Introduce pre-aeration to the influent 7. Add oxidising agents to the sewage collection and transportation system

(f) Presence of large amounts of rapidly biodegradable carbohydrates

Cause	• Presence of industrial wastes, such as those from dairies, breweries, sugar refineries
Control	1. Introduce biological pre-treatment upstream the activated sludge system, if the problem is permanent

Control based on the operation of the secondary sedimentation tank:

Objective	• Prevent/reduce the expansion of the sludge blanket
Control	1. Increase the return sludge flow 2. Reduce the MLSS concentration by increasing the excess sludge flow 3. Equalise the influent flow to the secondary treatment

	4. Direct the influent to the second and/or subsequent entrances in step-feed reactors
	5. Store the return sludge in sludge storage tanks, if there are any

Control based on rearrangement of the reactor (if feasible):

Objective	• Configure the reactor to induce conditions for better sludge settleability
Control	1. Introduce anoxic zones in the initial end of the reactor 　• Turn off the initial aerators intermittently, aiming at producing a temporary anoxic zone (for a short time) 　• Create an anoxic zone by introducing a dividing wall (without aeration, but with stirrers) 2. Induce plug-flow characteristics 　• Operate with cells in series 　• Introduce dividing walls in the reactor

Control based on the addition of chemical products:

Objective	• Temporarily control the filamentous organisms
Precautions	• Chemical products should be added carefully and under constant monitoring. Add the product starting with small doses, and examine the floc after a reasonable period of time. Continue increasing the dosage until the filamentous organisms start to decrease
Control	1. Toxic compounds (selectively eliminates the filamentous organisms, due to their larger surface area; not effective if bulking is due to nutrient deficiency) 　• Add chlorine or chlorine compounds at the entrance to the reactor or in the return sludge if bulking is severe, to kill the filamentous organisms 　• Add hydrogen peroxide to the return sludge (decay products are not harmful) 2. Flocculation agents (to increase the strength of the flocs) 　• Add metallic salts (aluminium, iron) to the reactor 　• Add polymers to the effluent from the reactor (influent to the secondary sedimentation tank)

Control based on the rearrangement or expansion of the plant:

Objective	• Undertake permanent physical rearrangement measures in the plant, to prevent the growth of filamentous organisms
Control	• Incorporate an anoxic zone upstream the reactor • Reduce dispersion in the reactor • Incorporate a selector tank

12.2.1.3 Pin-point floc

Detection:

• Small, spherical, discreet flocs • The larger flocs settle easily, leaving the small flocs, which generate a turbid effluent • Low SVI • Non-significant presence of filamentous bacteria, under microscopic examination

Causes:

• Insufficient number of filamentous organisms (affecting the structure of the floc, which becomes fragile) • Excessive aeration • Composition of the influent (unbalanced nutrients) • Excessive floc load at the entrance to the reactor

Detailing and correction of the causes:

Cause 1: Insufficient number of filamentous organisms	
Detection	• Microscopic examination
Secondary cause	• High sludge age (low F/M ratio)
Control	1. Increase the removal of excess sludge

Cause 2: Excessive aeration	
Detection	• Determination of DO in the reactor
Control	1. Reduce the aeration level

Cause 3: Composition of the influent (unbalanced nutrients)	
Detection	• Analyse influent and determine the BOD_5:N:P ratio • Conventional activated sludge – approximate ratio: 100:5:1 • Extended aeration – approximate ratio: 200:5:1

Secondary causes	• Presence of industrial wastes deficient in N and/or P • Activated sludge operating to remove N and/or P
Control	1. Add nitrogen or phosphorus in forms immediately available

Cause 4: Excessive floc load at the entrance to the reactor	
Detection	• Floc load = $[(COD_{inf} - COD_{eff}) \cdot Q]/(X_r \cdot Q_r)$ (mgCOD/gMLSS)
Secondary causes	• High load of influent BOD • Low concentration of MLSS at the inlet end of the reactor
Control	1. Increase the return sludge flow, mixing it well with the influent 2. Decrease the excess sludge flow 3. In step-feed reactors, direct the influent to the points after the inlet end of the reactor

12.2.1.4 Dispersed sludge

Detection:

• Turbid effluent • Undefined sedimentation zone • Variable SVI

Causes:

• Excessive shearing caused by hydraulic turbulence • Bacteria unable to aggregate themselves into flocs • Use of centrifugal pumps to pump the sludge and of centrifuges to dewater the sludge

**Detailing and correction of the causes:**

(a)   Excessive shearing caused by hydraulic turbulence

Cause	• Excessively vigorous aeration (mechanical aeration)
Control	1. Reduce the aeration level 2. Verify the size of the aerator and the rotation speed according to the tank dimensions

(b) Bacteria unable to aggregate themselves into flocs

Cause 1: Shock organic loads	
Control	1. Control the influent flow • Use stormwater storage tanks to reduce peaks (in combined sewerage systems) • Introduce/use equalisation tanks
Cause 2: Toxicity	
Detection	• Low oxygen utilisation rate (OUR), which suggests that toxic products are preventing the growth and respiration of the biomass and, consequently, the treatment level • Non-typical DO profile in plug-flow reactors • Reduction/loss of nitrification
Cause	• Presence of industrial effluents
Control	1. Increase sludge age (reduce the excess sludge flow) 2. Increase the MLSS concentration (reduce the excess sludge flow) 3. Increase the DO concentration 4. Consider the increase in the volume/number of reactors 5. Control toxicity at the source 6. Temporarily store toxic discharge, releasing it in small amounts, favouring dilution (if the biomass can be acclimatised to small amounts of the toxic agent) 7. Consider modification of the reactor to increase dispersion, leading to complete mix (if the toxic loads are frequent) 8. Divert the influent to other points further downstream in the reactor (in step-feed reactors) 9. Study the effect of toxicity on the biomass, to evaluate possible acclimatisation 10. Import biomass from other plants, for reseeding 11. Temporarily bypass the biological stage
Cause 3: Low concentrations of DO in the reactor	
Detection	• Measurement of DO in the reactor
Causes	• See Section 12.2.1.2.a
Control	1. See Section 12.2.1.2.a
Cause 4: Low pH values in the reactor	
Detection	• Measurement of pH in the reactor
Causes	• See Section 12.2.1.2.b
Control	1. See Section 12.2.1.2.b

Cause 4: Low sludge age (high F/M ratio)	
Detection	• Measurement of the influent BOD (COD) and MLSS for calculation of the F/M ratio • Measurement of MLSS and flow and SS concentration in the excess sludge
Causes	• High load of influent BOD • Low concentration of MLSS
Control	1. Decrease the excess sludge flow

(c)   Use of centrifugal pumps to pump the sludge and of centrifuges to dewater the sludge

Control	1. Change opening of the centrifugal pumps 2. Replace the centrifugal pumps with another type of pump 3. Add polymers to improve the solids capture in the centrifuge for thickening and/or dewatering (avoiding the return of fine solids to the system, which may eventually lead to dispersed sludge)

12.2.1.5 Overload of solids in the secondary sedimentation tanks (non-bulking sludge)

Detection:

• High sludge blanket level • Low SVI • Applied solids load higher than the maximum allowable solids load, given by the limiting solids flux

Causes:

• Insufficient capacity of the secondary sedimentation tanks in terms of surface area • Low sludge underflow removal from the secondary sedimentation tank • High MLSS • High influent flow • Large variation of the influent flow • Insufficient capacity of the secondary sedimentation tanks in terms of sludge storage (low sidewater depth) • Poor distribution of the influent flow to the secondary sedimentation tanks (overload in some units) • Low temperature, increasing the viscosity of the liquid and resulting in lower settling velocities

Control:

Control	<ol><li>Reduce the solids load applied per unit area of the sedimentation tanks<ul><li>Reduce the MLSS concentration (increase the excess sludge flow)</li><li>Equalise the variations of the influent flow</li><li>Increase the capacity of the secondary sedimentation tanks by building new units</li></ul></li><li>Increase the maximum allowable solids load per unit area of the sedimentation tanks<ul><li>Increase the underflow removal from the secondary sedimentation tanks</li></ul></li><li>Increase the sludge storage capacity<ul><li>Store sludge temporarily in tanks (if available)</li><li>Store sludge temporarily in the reactor, by directing the influent to points further downstream (in step-feed reactors)</li><li>Increase the sludge storage capacity by raising the sidewater wall, or by building new sedimentation tanks with higher sidewater depths</li></ul></li><li>Improve the flow distribution to the secondary sedimentation tanks, avoiding overload to some units</li></ol>

12.2.1.6 *Hydraulic overload of the secondary sedimentation tanks*

Detection:

• High sludge blanket level • Cloudy aspect of the effluent • Interface settling velocity lower than the hydraulic loading rate

Causes:

• High influent flow • Large variation of the influent flow • Poor distribution of the influent flow to the secondary sedimentation tanks (overload in some units) • Poor sludge settleability • Low temperature, increasing the viscosity of the liquid and resulting in lower settling velocities

Control:

Control	1. Equalise the influent flow 2. Increase the capacity of the secondary sedimentation tanks by building new units 3. Improve the settleability of the sludge (see Sections 12.2.1.2, 12.2.1.3 and 12.2.1.4)

12.2.1.7 Foam and scum

Detection:

• Visual observation of the reactors and/or secondary sedimentation tanks

Causes:

• Intense aeration • Filamentous organisms • Non-biodegradable detergents

Detailing and correction of the causes:

(a)   Intense aeration

Detection	• The foam disappears when the aerators are turned off
Control	1. Adjust the aeration, so that the foam is restricted to the reactor

(b)   Filamentous organisms

Detection	• The foam persists after the switching-off of the aerators • The foam has a brownish colour • The filamentous organisms incorporate air bubbles, forming a thick foam, which gets the brown colour due to the MLSS that gathers in it
Control	1. Remove the microorganisms by increasing the excess sludge flow 2. Allow the foam to go from the reactor to the secondary sedimentation tank 3. Remove the foam from the secondary sedimentation tank by scum removal equipment 4. Break the foam with high-pressure water jets

c)   Non-biodegradable detergents

Detection	• The foam persists after switching-off of the aerators • The foam is white
Control	1. Control at the source (replace the detergents with biodegradable products) 2. Break the foam with high-pressure water jets 3. Use products that prevent the formation of foams

12.2.1.8 Other operational problems of the secondary sedimentation tanks

Cause 1: Non-homogeneous distribution of the influent flow to the secondary sedimentation tanks	
Control	1. Adjust the distribution by changing the levels of the weirs in the flow split chamber 2. Improve the flow distribution by changing the hydraulic design of the flow division

Cause 2: Sidewater depth small to absorb variations in the level of the sludge blanket in the secondary sedimentation tanks	
Control	1. Store sludge temporarily in tanks (if available) 2. Store sludge temporarily in the reactor by directing the influent to points further downstream (in step-feed reactors) 3. Increase the sludge storage capacity by raising the sidewater depth, or by building new sedimentation tanks with higher side walls

Cause 3: Hydraulic short circuits caused by poor design or construction of the inlets and outlets of the sedimentation tanks	
Control	1. Improve the flow distribution and the energy dissipation in the entrance to the sedimentation tanks 2. Improve the levelling of the outlet weir 3. Introduce V-notch weirs 4. Reposition the effluent collection launder, if it is very close to the inlet 5. Reanalyse the hydraulic design of the sedimentation tanks (stability of the tank in terms of Froude Number)

Cause 4: High weir rate, leading to a high approaching velocity, which can resuspend the solids	
Control	1. Introduce more weirs and effluent launders in the sedimentation tanks
Cause 5: Resuspended solids on the external face of double-weir launders	
Control	1. Suppress the external weir by raising its level
Cause 6: Bottom outlet blocked in some sedimentation tanks	
Control	1. Unblock the sludge hoppers and the sludge lines
Cause 7: Poor operation of the return sludge pumps	
Control	1. Repair the sludge recirculation pumps 2. Direct the influent to the stormwater tanks (in combined sewerage systems) 3. Temporarily bypass the plant (emergency procedure)
Cause 8: Poor operation of the sludge removing mechanism	
Control	1. Repair the sludge removing mechanism

12.3 HIGH BOD CONCENTRATIONS IN THE EFFLUENT

The effluent BOD is present in two forms: particulate BOD and soluble BOD

12.3.1 High concentrations of particulate BOD

Detection:

• Determination of the SS and particulate BOD (total BOD – soluble BOD) concentrations in the final effluent

Cause:

• High SS concentrations in the final effluent (see Section 12.2)

Control:

• Control the effluent SS concentration (see Section 12.2)

12.3.2 High concentrations of soluble BOD

Detection:

> • Determination of the soluble BOD concentration in the final effluent

Causes:

> • Low DO concentrations in the reactor
> • Insufficient MLSS concentration
> • High load of influent BOD
> • Large variation of the influent BOD load
> • Inhibition by toxic substances
> • pH outside the range from 6.5 to 8.5
> • Unbalanced nutrients
> • Temperature variations

Detailing and correction of the causes:

Cause 1: Low DO concentrations in the reactor	
Detection	• See Section 12.2.1.2.a
Secondary causes	• See Section 12.2.1.2.a
Control	• See Section 12.2.1.2.a

Cause 2: Insufficient MLSS concentration	
Detection	• Measurement of the MLSS concentration in the reactor
Secondary causes	• High excess sludge flow • High influent flow, transferring the biomass to the secondary sedimentation tank • Loss of solids in the secondary sedimentation tank due to sedimentation problems • Insufficient return sludge flow • Problems in the return sludge pumping

Control	1. Reduce the excess sludge flow
	2. Control the influent flow
	  • Use stormwater storage tanks to reduce peaks (in combined sewerage systems)
	  • Introduce/use equalisation tanks
	3. Control the loss of solids in the secondary sedimentation tank (see Section 12.2)
	4. Increase the return sludge flow
	5. Repair the return sludge pumps

Cause 3: High load of influent BOD	
Detection	• Measurement of the influent flow and BOD concentration
Control	1. Reduce the excess sludge flow to increase the sludge age and the biomass

Cause 4: Large variation of the influent BOD load	
Detection	• Measurement of the influent flow and BOD concentration
Control	1. Reduce the excess sludge flow to increase the sludge age and the biomass 2. Increase the return sludge flow during peak periods to increase the MLSS concentration (limited to an instantaneous control) 3. Introduce/use equalisation tanks 4. Release sludge from the sludge tanks (if available) during peak loads

Cause 5: Inhibition by toxic products	
Detection	• See Section 12.2.1.4.b, Cause 2
Control	• See Section 12.2.1.4.b, Cause 2

Cause 6: pH outside the range from 6.5 to 8.5	
Detection	• See Section 12.2.1.2.b
Secondary causes	• See Section 12.2.1.2.b
Control	1. See Section 12.2.1.2.b

Cause 7: Unbalanced nutrients	
Detection	• See Section 12.2.1.3, Cause 3
Secondary causes	• See Section 12.2.1.3, Cause 3
Control	1. See Section 12.2.1.3 – Cause 3

Cause 8: Temperature variations	
Detection	• Measurement of temperature in the influent and/or reactor
Secondary causes	• Reduction in temperature • Increase in temperature

Control	1. Reduction in temperature
	   • Reduce the excess sludge flow, to increase MLSS and the sludge age
	   • Reduce heat losses
	2. Increase in temperature (if it is causing problems)
	   • Increase the excess sludge flow, to reduce MLSS
	   • Supplement aeration

12.4 HIGH AMMONIA CONCENTRATIONS IN THE EFFLUENT

12.4.1 Causes

> • *Inhibition of the growth of the nitrifying bacteria*
> • *Insufficient MLSS concentration*
> • *High loads of influent ammonia*

12.4.1.1 Inhibition of the growth of the nitrifying bacteria

Causes:

> • Low DO concentrations in the reactor
> • Low temperatures in the reactor
> • Low pH values in the reactor
> • Presence of inhibiting toxic substances

Detailing and correction of the causes:

Cause 1: Low DO concentrations in the reactor	
Detection	• See Section 12.2.1.2.a
Secondary causes	• See Section 12.2.1.2.a
Control	1. See Section 12.2.1.2.a

Cause 2: Low temperatures in the reactor	
Detection	• Measurement of the temperature in the influent and/or reactor
Control	1. Reduce the excess sludge flow, to increase MLSS and the sludge age 2. Increase the DO concentration 3. Reduce heat losses 4. Consider the increase in the volume/number of reactors

Cause 3: Low pH values in the reactor	
Detection	• See Section 12.2.1.2.b
Secondary causes	• See Section 12.2.1.2.b
Control	1. See Section 12.2.1.2.b

Cause 4: Presence of inhibiting toxic substances	
Detection	• See Section 12.2.1.4.b, Cause 2
Control	1. See Section 12.2.1.4.b, Cause 2

12.4.1.2 Insufficient MLSS concentration

Detection:

> • See Section 12.3.2, Cause 2

Causes:

> • See Section 12.3.2, Cause 2

Control:

> • See Section 12.3.2, Cause 2

12.4.1.3 High loads of influent ammonia

Detection:

> • Measurement of the influent flow and TKN concentration

Control:

> 1. Reduce the excess sludge flow to increase the sludge age and the biomass
> 2. Increase the return sludge flow during peak periods to increase the MLSS concentration (limited to an instantaneous control)
> 3. Introduce/use equalisation tanks
> 4. Release sludge from the sludge tanks (if available) during peak loads

13

Basic principles of aerobic biofilm reactors

R.F. Gonçalves

13.1 INTRODUCTION

New versions of wastewater treatment plants using biofilm reactors are compact, capable of being installed in urban areas with relatively low impacts (Rogalla *et al.*, 1992) and, above all, highly resistant to variations in temperature and to toxicity shock loads (Arvin and Harremöes, 1991). Operational stability is important in the case of small treatment plants, this being one of the reasons for the renewed interest in several locations for the "old" trickling filters and biodiscs (rotating biological contactors) for small-sized communities (Upton and Green, 1995). A similar interest to biofilm reactors applied to medium and large communities occurred in developed regions (e.g., USA), after the development of processes combining biomass in suspension with biomass attached to a support medium (Parker *et al.*, 1990). The process advantages renewed the interest for systems with attached biomass, stimulating the development of a great variety of processes.

The main concepts and technical aspects relative to biofilm reactors applied to wastewater treatment and the post-treatment of effluents from anaerobic reactors are presented in this chapter. The classification of the main types of biofilm reactors with relation to suspended-biomass reactors, as well as the behaviour of the biofilm and the influence of the transport phenomenon during reactions, is discussed. Usual configurations, as well as new configurations for the post-treatment of effluents from anaerobic reactors, including some design examples, main construction aspects and more common operational problems, mainly with the following processes:

- trickling filters
- rotating biological contactors (biodiscs)
- submerged aerated biofilters

Due to the great importance of UASB reactors in warm-climate regions, emphasis is given to aerobic biofilm reactors acting as post-treatment for anaerobic effluents.

13.2 CLASSIFICATION OF AEROBIC BIOFILM REACTORS

A better understanding of the mechanisms involved in the conversion processes taking place in biofilm reactors led to the development of new reactors from 1970 (Atkinson, 1981). Improvements concerning mixing of phases, oxygen transfer and separation of phases were incorporated, improving performance through an effective control of the biofilm thickness and an increment of the mass transfer.

Figure 13.1 (Lazarova and Manen, 1994) presents an alternative classification of aerobic reactors, based on the state of biomass fixation. The major difference with relation to old similar classifications is the group of hybrid reactors, which incorporate suspended biomass and fixed biomass in the same reaction volume. The processes with suspended biomass involve several variants of activated sludge. Among the hybrid processes, there are those with the support medium mechanically mixed (Oodegard *et al.*, 1993) and with structured supports inserted in the aeration tank (Bonhomme *et al.*, 1990). Both are variants of the activated sludge systems,

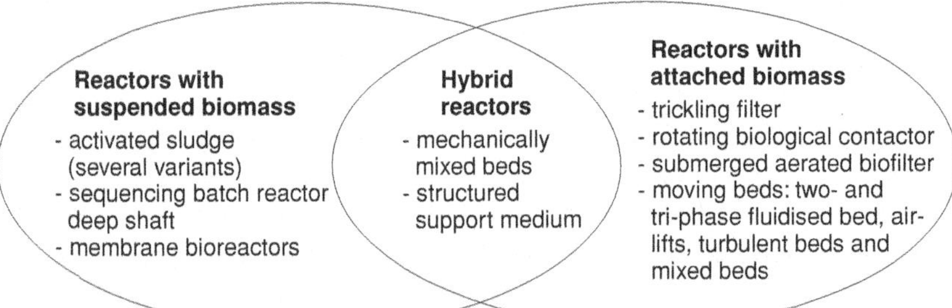

Figure 13.1.   Modern classification of mechanised aerobic treatment processes, with respect to the state of the biomass (adapted from Lazarova and Manen, 1994)

since they result from the incorporation of the support medium in the aeration tank. This technique has been used to upgrade overloaded treatment plants, because the applied organic load can even be three times higher compared to that in the conventional process (Lessel, 1993).

Reactors with attached biomass, or simply biofilm reactors, now include, besides trickling filters and rotating biological contactors, several other types of reactors with fixed or moving beds. Processes with moving beds have the support medium in permanent movement, hydraulically or mechanically driven. They generally use a material with large specific surface area for the attachment of the biomass, that can be grains of small diameter (0.2 to 2 mm) or a material with high porosity (e.g., sponges). High biomass concentrations are reached in these processes ($>20\,kgTSS/m^3$), resulting in a high treatment capacity. Their main advantage with relation to fixed bed processes is the absence of clogging of the filter medium, and their main disadvantages are the high operational costs (especially energy) and the sophisticated devices necessary for appropriate flow distribution and aeration. Among the main processes, the two-phase fluidised bed reactors stand out, counting with many full-scale treatment plants operating in the USA and in Europe (Lazarova and Manen, 1994). Indicated for the treatment of diluted effluents, their construction costs are reported to be lower than that of activated sludge systems, although the operation and maintenance costs may be higher (due to the saturation in oxygen and pumping).

13.3 FORMATION, STRUCTURE AND BEHAVIOUR OF BIOFILMS

The present item includes additional details.

In all reactors with attached biomass, the metabolic conversion processes take place inside the biofilm. Substrate transport occurs by diffusion processes, initially through the liquid film in the liquid/biofilm interface and later through the biofilm (Figures 13.2 and 13.3). The products of the oxidation and reduction reactions are transported in the opposite direction, to the exterior of the biofilm. The substrate donor as well as the electron acceptor must penetrate the biofilm for the biochemical reaction to take place.

The quantification of the limitations to the mass transfer is very important, so that better performance reactors can be designed. Improvement of performance is directly related with the reduction of these limitations, because the global reaction velocity in these heterogeneous systems may be lowered due to the mass transfer among the phases (Zaiat, 1996).

In many aerobic systems, the rate of oxygen transfer to the cells is the limiting factor that determines the biological conversion rate. Oxygen availability for microorganisms depends on the solubility and mass transfer, as well as on the rate at which dissolved oxygen is utilised. In biofilm reactors used for post-treatment of anaerobic effluents, the transport mechanisms involve oxygen and ammonia

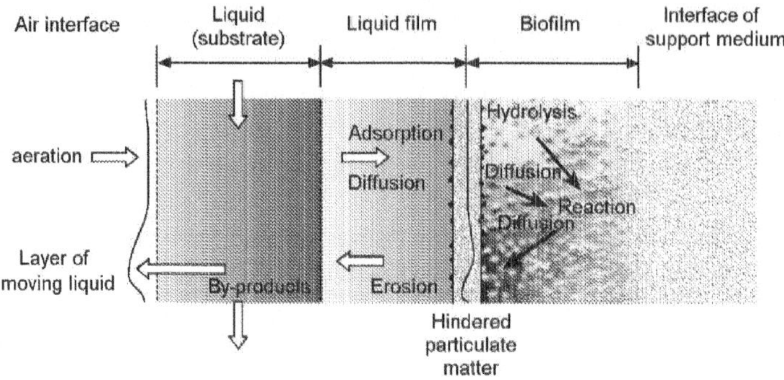

Figure 13.2.   Mechanisms and processes involved with the transport and substrate conversion in biofilms

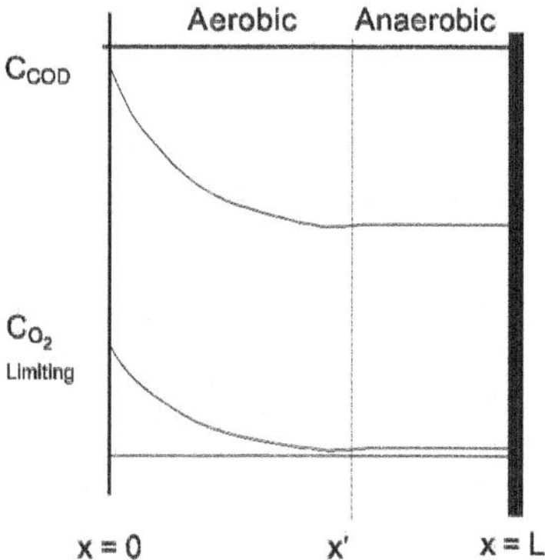

Figure 13.3.   Distribution of the concentration of two compounds involved in oxidation–reduction reaction in the biofilm (O_2 and COD)

nitrogen (O_2 and $N-NH_4^+$), besides the intermediate ($N-NO_2^-$) and final nitrogen product ($N-NO_3^-$). The main stages involved are:

- transfer of oxygen from the gaseous phase to the liquid medium
- transfer of oxygen, ammonia and nitrate from the liquid phase to the biofilm
- transfer of oxygen, ammonia and nitrite inside the biofilm
- transfer of the intermediary product ($N-NO_2^-$) and of the final product ($N-NO_3^-$) to the liquid medium

According to Chisti *et al.* (1989), oxygen, being poorly soluble in water, frequently becomes the limiting factor in aerobic biofilm processes. The main oxygen transport steps are illustrated in Figure 13.4, in which eight possible resistant structures to mass transfer are identified.

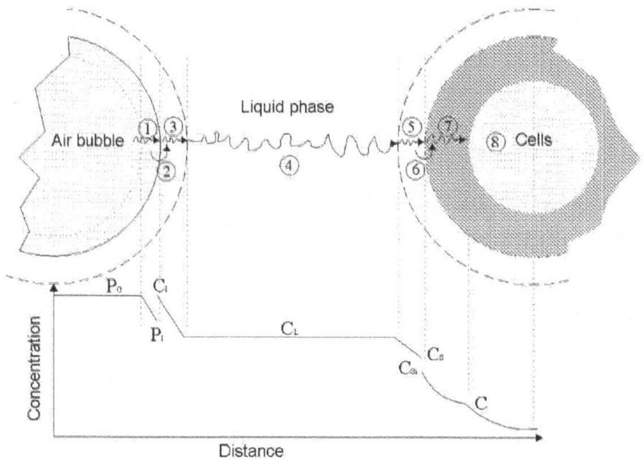

Figure 13.4.   Schematic diagram of the stages involved in the transport of oxygen (adapted from Bailey and Ollis, 1986, cited by Fazolo, 2000)

The resistances considered in the tri-phase systems are:

1. in the gaseous film inside the bubble, between the core of the gas in the bubble and the gas–liquid interface
2. in the gas–liquid interface
3. in the liquid film, close to the gas–liquid interface, between this interface and the liquid medium
4. in the liquid medium
5. in the liquid film, between the liquid medium and the liquid–solid interface (external resistance)
6. in the liquid–solid interface
7. in the solid phase (internal resistance)
8. in the sites of biochemical reaction (inside the microorganisms)

The relative magnitude of these resistances depends on the hydrodynamics of the bubble, solubility of oxygen, temperature, cellular activity, composition of the solution and interface phenomena (Bailey and Ollis, 1986). Therefore, the penetration depth of the substrates in the biofilm is of fundamental importance in the determination of the global conversion rate in the reactor. The ideal situation corresponds to a biofilm completely penetrated by the two substrates, resulting in a reaction limited only by the maximum rate of biochemical reaction.

However, the most common situation in the treatment of domestic sewage is the partial penetration of at least one of the two substrates in a thick biofilm layer, caused by an intrinsic volumetric high conversion rate and a great resistance to the diffusion in the biofilm (Figure 13.3). In this case, only the fine outer biofilm layer will be active with respect to the reaction in question, with the remaining biomass being inactive in the deepest layers. An intrinsically zero-order biochemical reaction may become half order, decreasing the overall surface conversion rate (Harremöes, 1982).

In the case of systems with nitrification, the critical ratio between the O_2 and NH_4^+ concentrations, that determines the limiting substrate, is between 0.3 and 0.4 (Gönenc and Harremöes, 1985). This makes oxygen the limiting substrate in most cases. Assuming, for example, a concentration of 2 mg/L of O_2 in the liquid phase of the reactor, the limiting ammonia concentration will be 0.6 mg/L. In the case of simultaneous oxidation of organic matter and nitrification, the competition between the heterotrophic and autotrophic (nitrifying) bacteria for oxygen determines the structure of the aerobic biofilm compartment. When the O_2/COD ratio is very small, the aerobic compartment is entirely dominated by the heterotrophic bacteria, and nitrification does not take place in the biofilm (Gönenc and Harremöes, 1990).

The understanding of these mass-transfer mechanisms is reflected in the configuration of the various new-generation biofilm reactors. In the case of submerged aerated biofilters, there prevail granular mediums with high specific surface that maximise the area for mass transfer and the amount of biomass in the reactor. With the use of granular mediums, high sludge ages are obtained without the need for clarification and biomass recirculation.

On the other hand, the severe hydrodynamic conditions in the biofilters propitiate the development of a fine and very active biofilm, especially in the bed layers that do not have contact with the settled wastewater. Hydraulic loads of 2 $m^3/m^2 \cdot$hour (wastewater) and 15 $m^3/m^2 \cdot$hour (air) are commonly practised in secondary treatment, resulting in a granular tri-phase medium submitted to a high turbulence. The association of the turbulence and the high velocity of the liquid controls the biofilm thickness and decreases the resistance to diffusion in the liquid film. Besides, high air flows increase the oxygen concentration in the liquid phase, facilitating its diffusion in the biofilm.

The stability of the process to temperature variations and toxic shock loads is also a consequence of the resistance to the diffusion in the biofilm (Arvin and Harremöes, 1991). The thickness of the active biofilm layer increases when the liquid temperature decreases, significantly reducing the sensitivity of the process to temperature variations (Okey and Albertson, 1987). Regarding nitrification, two factors resulting from the temperature drop contribute to alleviate the reduction in efficiency: increase in the DO concentration in the liquid (increasing diffusion) and decrease in biological activity (reducing conversion rates).

With respect to the resistance to toxicity shocks, the process behaves in a similar way to temperature drop. If the concentration of a certain toxic compound suddenly exceeds the inhibition threshold, the gradient of concentrations through the biofilm attenuates its impact on the treatment. Even if the outer biofilm layers are affected, the inner layers continue to degrade the concentrations reduced by the resistance to diffusion (Saez *et al.*, 1988).

The great capacity to tolerate shock loads, in spite of the low real hydraulic detention times in the granular medium of biofilters ($\approx$20 minutes), is due to the high biomass concentration in the reactors. Biomass concentrations higher than 20 gTSS/L are found in biofilters with granular mediums (specific surface > 600 m^2/m^3) applied to secondary treatment of domestic sewage (Gonçalves, 1993).

14

Trickling filters

C.A.L. Chernicharo, R.F. Gonçalves

14.1 DESCRIPTION OF THE TECHNOLOGY

14.1.1 Preliminary considerations

Trickling filters (TF) are wastewater treatment systems that can be widely used in developing countries, principally in view of their simplicity and low operational costs.

A trickling filter consists of a tank filled with a packing medium made of a material of high permeability, such as stones, wooden chips, plastic material or others, on top of which wastewater is applied in the form of drops or jets. After the application, the wastewater percolates in the direction of the drainage system located at the bottom of the tank. This downward percolation allows bacterial growth on the surface of the packing medium, in the form of a fixed film denominated biofilm. The wastewater passes over the biofilm, promoting contact between the microorganisms and the organic matter.

Trickling filters are aerobic systems, because air circulates in the empty spaces of the packing medium, supplying oxygen for the respiration of the microorganisms. Ventilation is usually natural. The application of wastewater on the medium is done frequently through rotating distributors, moved by the hydraulic head of the liquid. The wastewater quickly drains through the support medium. However, the organic matter is absorbed by the biofilm and is retained for a time sufficient for its stabilisation (see Figure 14.1).

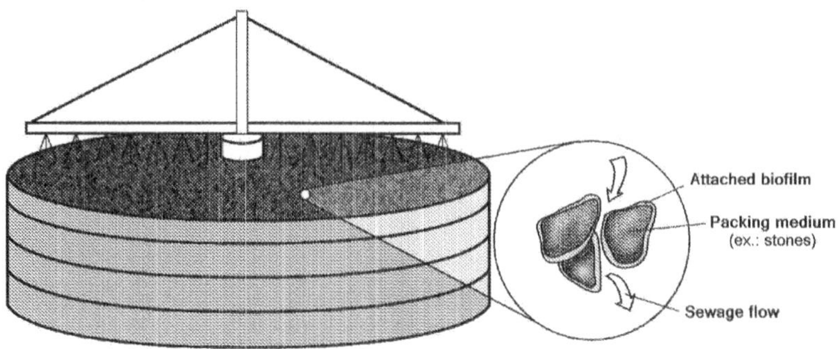

Figure 14.1.   Schematic representation of a trickling filter

The filters are usually circular, and can be several metres in diameter. Contrary to what the name indicates, the primary function is not filtering. The diameter of the stones (or other medium) used is in the order of centimetres, leaving a large space between them, which is inefficient for the act of filtering (sieving action). The function of the medium is only to supply a support for the formation of the microbial film. Synthetic mediums of several materials and forms also exist, having the advantage of being lighter than stones, besides presenting a greater surface area. However, the synthetic mediums are more expensive.

With the continued biomass growth on the surface of the stones, the empty spaces tend to decrease, thus increasing the downward velocity through the pores. When the velocity reaches a certain value, it causes a shearing stress that dislodges part of the attached material. This is a natural form of controlling the microbial population on the support medium. The dislodged sludge should be removed in the secondary settling tank to decrease the level of suspended solids in the final effluent.

14.1.2 Types of trickling filters

The trickling filters are generally classified according to the surface or the organic loading rate to which they are submitted, as described below. The main design criteria are presented in Table 14.1.

Low rate trickling filter

The low rate trickling filter is conceptually simple. Although its efficiency in the removal of BOD is comparable to that of the conventional activated sludge system, its operation is simpler, although less flexible. Trickling filters have lesser capacity in adjusting to influent flow variations, besides requiring a slightly higher total area. In terms of energy consumption, they have much lower consumption than activated sludge systems. Figure 14.2 presents the typical flowsheet of a low rate trickling filter.

Table 14.1. Typical characteristics of the different types of trickling filters

Operational conditions	Low rate	Intermediate rate	High rate	Super high rate	Roughing
Packing medium	Stone	Stone	Stone	Plastic	Stone/plastic
Hydraulic loading rate (m^3/m^2·d)	1.0 to 4.0	3.5 to 10.0	10.0 to 40.0	12.0 to 70.0	45.0 to 185.0
Organic loading rate (kgBOD/m^3·d)	0.1 to 0.4	0.2 to 0.5	0.5 to 1.0	0.5 to 1.6	Up to 8
Effluent recycle	Minimum	Occasional	Always*	Always	Always
Flies	Many	Variable	Variable	Few	Few
Biofilm loss	Intermitt.	Variable	Continuous	Continuous	Continuous
Depth (m)	1.8 to 2.5	1.8 to 2.5	0.9 to 3.0	3.0 to 12.0	0.9 to 6.0
BOD removal (%)**	80 to 85	50 to 70	65 to 80	65 to 85	40 to 65
Nitrification	Intense	Partial	Partial	Limited	Absent

* Effluent recycle is usually unnecessary when treating effluents from anaerobic reactors
** Typical BOD removal ranges for TF fed with effluents from primary settling tanks. Lower efficiencies are expected for TF fed with effluents from anaerobic reactors, although overall efficiency is likely to remain similar
Source: Adapted from Metcalf and Eddy (1991) and WEF (1996)

Trickling filters can have circular or rectangular shape, the most commonly used packing material is stone, and feeding can be continuous or intermittent. Dosing siphons are usually used in the case of intermittent feeding, which is common in low rate trickling filters. The interval between loads can vary as a function of the wastewater flow, but should be short enough to avoid drying of the biofilm. Effluent recirculation may be necessary to assure humidity of the medium, especially in the hours of low influent flow, although a low rate filter does not require this practice in other hours of the day.

As a result of the small load of BOD applied to the trickling filter, per unit volume, food availability is low. This leads to a partial stabilisation of the sludge (self-consumption of the cellular organic matter) and to a larger efficiency in the removal of BOD and in nitrification. This smaller BOD load per unit volume of the tank is associated with the larger area requirements, when compared to the high rate trickling filter system. One of the main problems of low rate trickling filters is the development of flies.

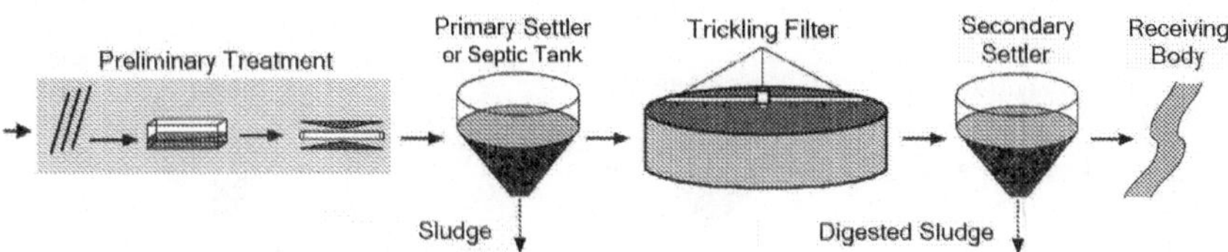

Figure 14.2. Typical flowsheet of a low rate trickling fillter

Intermediate rate trickling filter

These filters are designed with higher loading rates than those of the low rate filters. The most common type is the continuous feeding flow, although intermittent feeding can also be practised. Recirculation of the treated effluent is usually carried out, aiming at the control of the thickness of the biofilm and improvement of the efficiency of the system. The effluent produced is partially nitrified, and a reasonable development of flies is still observed.

High rate trickling filter

These filters are submitted to loading rates much higher than those applied to low rate and intermediary rate filters. As a consequence of the higher organic loading rates, high rate TFs have smaller area requirements. In parallel, there is also a slight reduction in the removal efficiency of organic matter, and the non-stabilisation of the sludge in the filter. Hydraulic loading rates can reach 60 $m^3/m^2 \cdot d$ in the peak hours, while the organic loading rates can be as high as 1.80 $kgBOD/m^3 \cdot d$, for filters with plastic medium. In filters filled with synthetic material, the depth can exceed 6.0 m.

Feeding of high rate TF is continuous and effluent recycle is regularly practised, but only when settled wastewater is applied, to have an influent BOD concentration to the filter around 100 mg/L. Effluent recycle is usually unnecessary when TFs are used for the post-treatment of effluents from UASB reactors, since the influent BOD is typically close to 100 mg/L. The high hydraulic loading rate constantly limits the thickness of the biofilm. Due to the high application rates, BOD removal in this process is lower, in the range from 70% to 80%, and the solids produced have more difficulty in settling in the clarifier. Flies do not develop and nitrification is partial with lower loading rates. Figure 14.3 presents a typical flowsheet of a high rate trickling filter system.

Super high rate trickling filter

Filters with super high rates are generally packed with synthetic granular mediums, with depths varying between 3.0 and 12.0 m. These large depths are possible due to the low density of the packing material, which results in a lower weight on the bottom slab of the filter. Flies do not develop in the filter and nitrification does not occur.

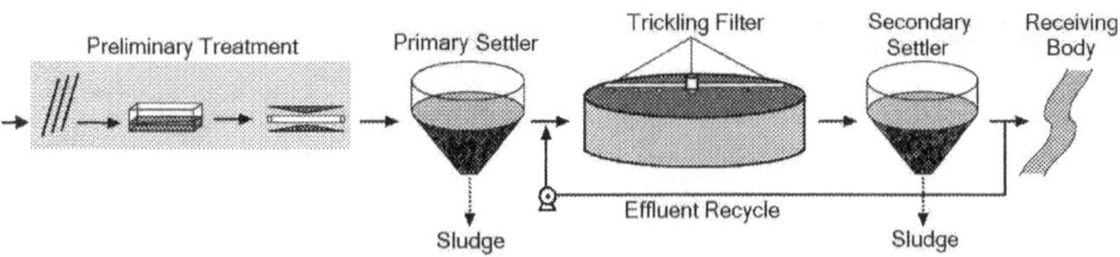

Figure 14.3. Typical flowsheet of a high rate trickling filter

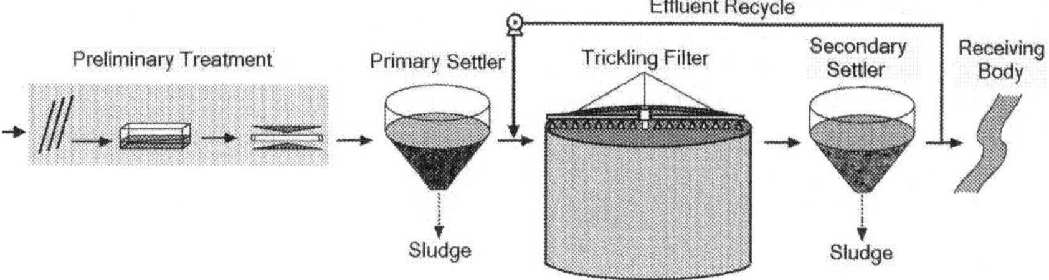

Figure 14.4.   Typical flowsheet of a super high rate trickling filter

Roughing trickling filter

Roughing trickling filters are used in the pre-treatment of wastewater, up-stream of secondary treatment. The packing material is synthetic and feeding is continuous. They are more commonly used for the treatment of highly concentrated wastewaters. Their use has been greatly reduced after the development of UASB reactors that are used in the place of the roughing filters.

A summary of the main characteristics of the different types of trickling filters is presented in Table 14.1.

14.2 DESIGN CRITERIA

The design criteria presented in this item mainly originated from the experience in the application of trickling filters for the treatment of primary effluents, that is, after the passage of the wastewater to a primary, or equivalent, settling tank (Metcalf and Eddy, 1991; WEF, 1996). The design criteria are also adapted to the application of TFs as post-treatment of effluents from UASB reactors.

(a)   Hydraulic loading rate

The hydraulic loading rate (HLR or L_h) refers to the volume of wastewater applied daily to the TF per unit surface area of the packing medium

$$L_h = \frac{Q}{A} \qquad\qquad (14.1)$$

where:
L_h: hydraulic loading rate (m³/m²·d)
Q: average influent flowrate (m³/d)
A: surface area of the packing medium (m²)

Typical values of hydraulic loading rates are presented in Table 14.1. In the case of high rate trickling filters used for the post-treatment of effluents from UASB reactors, it has been observed that TFs are capable of producing effluents with BOD and SS lower than 60 mg/L when operated with maximum hydraulic loading rates in the order of 20 to 30 m³/m²·d.

(b)   Organic loading rate

The volumetric organic load refers to the amount of organic matter applied daily to the trickling filter, per unit volume of the packing medium.

$$L_v = \frac{Q \times S_0}{V} \qquad (14.2)$$

where:

   L_v: volumetric organic loading rate (kgBOD/m^3·d)
   Q: average influent flowrate (m^3/d)
   S_0: influent BOD concentration (kgBOD/m^3)
   V: volume occupied by the packing medium (m^3)

Typical organic loading rates are presented in Table 14.1. In the case of post-treatment of anaerobic effluents, satisfactory BOD concentrations have been achieved in the effluent from TFs operating with maximum organic loading rates in the range from 0.5 to 1.0 kgBOD/m^3·d.

(c)   Influent distribution system

To optimise the treatment efficiency of the trickling filters, the growth as well as the elimination of the biofilm that grows in excess should happen in a continuous and uniform way. To achieve this, the distribution system should be designed in a way to facilitate the appropriate application of wastewater on the packing medium.

   The feeding of TF with wastewater can be accomplished through fixed or mobile (rotating) distributors. The first TFs were fitted with fixed distribution systems, composed of pipes with nozzles. This type of system is still used today, mainly in small-scale plants. However, most of the TFs have a circular shape and are equipped with a rotating distribution system.

Fixed distribution systems

Fixed distribution systems are composed of main distribution pipes and lateral pipes, both placed just above the surface of the granular medium. The nozzles are installed in the laterals, and are designed and spaced to obtain uniform feeding distribution. In general, the nozzles are made of a circular hole and a deflector.

   Most of the older fixed systems were planned considering intermittent feeding of the wastewater through a dosing tank. The flow from this device is variable, due to the variation of the water level in the dosing tank. In the beginning of the discharge period, the wastewater is discharged at a maximum distance of each nozzle, decreasing as the tank empties. The period between wastewater loads varies from 0.5 to 5 minutes.

   With the appearance of the synthetic packing mediums, the fixed distribution systems returned to be used in the deep filters and in the biotowers. In these processes, the distribution system is also equipped with mains and lateral distributors, placed immediately above the support medium, and the feeding is continually accomplished through pumping.

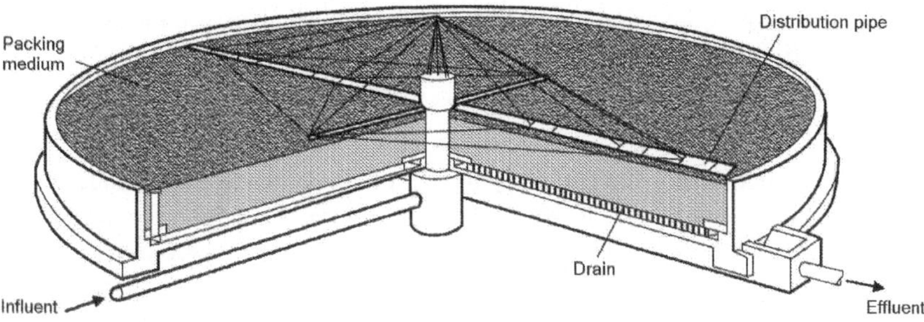

Figure 14.5. Schematics of a trickling filter with a roating distribution system. Source: Adapted from Metcalf and Eddy (1991)

The main disadvantages of this type of distribution system are the non-uniformity of the hydraulic load on the surface of the TF, the great lengths of distribution pipes the frequent blockage of the nozzles and the difficulty of maintenance of the nozzles in large TFs.

Rotating distribution systems

The rotating distribution system is composed of one or more horizontal pipes (arms), supported by a rotating central column (see Figure 14.5).

Wastewater is evenly distributed onto the packing medium by means of holes placed in one of the sides of each horizontal arm. The rotational movement of the distributor is generally assured by the energy from the jet of wastewater discharged through the group of holes. In exceptional cases, especially for control of flies and to avoid stops of the distributor arms in hours of very low influent flow rate, electric motors are also used to move the distribution system. The distributor arms usually have a circular section but can also be built with rectangular section or other quadrilateral type. A fast-opening device installed in the extremity allows the removal of coarse solids accumulated inside each arm. The area of the cross section of the arms generally decreases with the distance from the central column. The spacing among the holes is designed to guarantee a uniform distribution of the wastewater over the entire surface of the packing medium. Deflectors made of plastic or other types of non-corrosive materials are installed in front of the holes to ensure better distribution.

The arms should be designed so that the rotational velocity is between 0.1 and 2 rpm and the velocity does not exceed 1.2 m/s at the maximum flow. Filters with four-arm distributors are equipped with an overflow device in the central column, which concentrates the feeding in only two arms in periods of low flows. In periods of maximum flows, all the four arms are fed with wastewater. This procedure assures adequate discharge velocities and reaction forces for the distributor's rotation, under various flow conditions. Holes on the opposite side of the arms are also used to reduce the rotational velocity in moments of peak flow. The distributor arms also have ventilation tubes, to avoid the accumulation of air inside them. The support structure of the arms is composed of cables, which assure the stability of the support in the central column.

(d)   Packing medium

The packing medium of the trickling filter is of fundamental importance in the performance of the process. The packing material serves as support for the growth of the biomass, through which the pre-treated wastewater percolates. The air passes through the empty spaces of the medium, supplying oxygen for the aerobic reactions. The ideal packing material should have the following main characteristics:

- have the capacity to remove high BOD loads per unit volume
- have the capacity to operate at high hydraulic loading rates
- have an appropriately open structure, to avoid obstructions due to biomass growth and to guarantee an appropriate supply of oxygen, without the need for forced aeration
- have structural resistance to support its own weight and the weight of the biomass that grows attached onto its surface
- be sufficiently light, to allow significant reductions in the cost of the construction works
- be biologically inert, not being attacked by nor being toxic to the process microorganisms
- be chemically stable
- have the smallest possible cost per unit of organic matter removed

In practice, the TF is usually packed with different types of stones, such as gravel with a diameter between 5 and 8 cm, without flat and elongated stones, or blast furnace slag. These materials have a low specific surface area (55 to 80 m^2/m^3) and porosity from 55 to 60%, limiting the area for biomass growth and the circulation of air. TF with a stone bed can also present problems of blockage of the void spaces, due to the excessive growth of the biological film, especially when the filters are operated with high organic loads. In these conditions, floods and failures of the system can occur.

Sometimes, due to the need for reduction of the area required for the system and to overcome the limitations of the stone packing medium, other types of materials can be used (Figure 14.6) These materials include corrugated plastic modules and plastic rings, with very large specific surface areas (100 to 250 m^2/m^3) and with porosities from 90 to 97%, that allow a larger amount of attached biomass per unit volume of the packing material. These materials are also much lighter than stones (about 30 times), allowing the filters to be much higher, without causing structural problems. While in stone filters the heights are usually lower than 3 m, the filters packed with synthetic material can be much higher (6 m or more), decreasing, as a consequence, the area required for their installation. The use of these packing materials allows the application of much higher organic loading rates than those used for filters packed with stones, for the same treatment

(a) plastic rings

(b) 50° cross-flow block

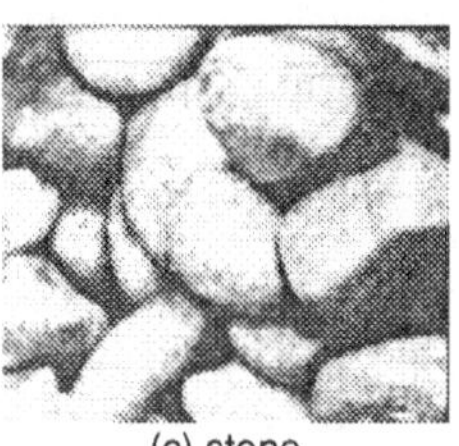
(c) stone

Figure 14.6.   Some types of packing mediums used in trickling filters

performance. However, the high costs of these materials are usually the limiting factor.

In the case of the use of TF for the post-treatment of effluents from UASB reactors, the Brazilian experience shows that high rate TFs are capable of producing effluents with satisfactory BOD concentrations when they are built with packing medium with heights between 2.0 and 3.0 m.

(e)   Underflow collection system

The underflow system of a trickling filter consists of a perforated slab, or of grids made of resistant materials, and gutters located in the lower part of the filter. The drainage system has the function of collecting the wastewater that percolates through the filter as well as the solids that are released from the packing medium, directing them to the secondary settling tank. The drainage system must be resistant enough to support the weight of the packing medium, of the attached biomass and of the wastewater that percolates through the filter. (Metcalf and Eddy, 1991).

The bottom structure should have a slope between 1 and 5%, sufficient to allow the drainage of the effluent to the centre or the periphery of the filter. The effluent collection gutters should be designed to guarantee a minimum velocity of 0.6 m/s (average feeding flow). (Metcalf and Eddy, 1991).

The bottom drainage system should be opened at both ends to facilitate inspection and occasional cleaning with water jets, should the need arise. The bottom drainage system is also responsible for the ventilation of the filter, as discussed in the following item.

(f)   Ventilation

Ventilation is important to maintain aerobic conditions, necessary for the effective treatment of the wastewater. If there are adequate openings, the difference between the air and the liquid temperatures is enough to produce the necessary aeration. A good ventilation through the filter bottom is desirable. In practice, the following measures are adopted to have adequate natural ventilation (Metcalf and Eddy, 1991):

- the drainage system and the effluent collection channels close to the bottom of the TF structure should allow free flow of air. These effluent collection

Table 14.2. Surface hydraulic loading rates for the design of secondary settling tanks after TF

	Surface hydraulic loading rate ($m^3/m^2 \cdot d$)	
Treatment level	For $Q_{average}$	For $Q_{maximum}$
BOD = 20 to 30 mg/L without nitrification	16 to 32	40 to 48
BOD $\leq$ 20 mg/L with nitrification	16 to 24	32 to 40

channels should not have more than 50% of their height occupied by the effluent
- ventilating access ports with open grating types of covers should be installed at both ends of the central effluent collection channel
- large-diameter filters should be equipped with collection channels in branches, with ventilating manholes or vent stacks along the perimeter of the filter
- the open area of the slots at the top of the underdrain blocks should not be less than 15% of the surface area of the filter
- one square metre gross area of open grating in ventilating manholes and vent stacks should be provided for each 23 m^2 of surface area of the filter

(g)   Secondary sedimentation tanks

The secondary settling tanks used downstream of the trickling filters are usually of the conventional type (Fig. 14.7), and are designed according to surface hydraulic loading rate, since the concentration of suspended solids in the effluent from the TF is relatively low. Table 14.2 lists the surface loading rates recommended for the design of secondary settling tanks after TF.

Depending on the size of the wastewater treatment plant, the secondary settling tanks can have automated or hydraulic pressure sludge removal systems.

(h)   BOD removal efficiency in TF

Several theoretical or empirical models are available for the design of trickling filters applied for the treatment of settled wastewater, and these can be found in classical wastewater treatment books. The present chapter includes one of the

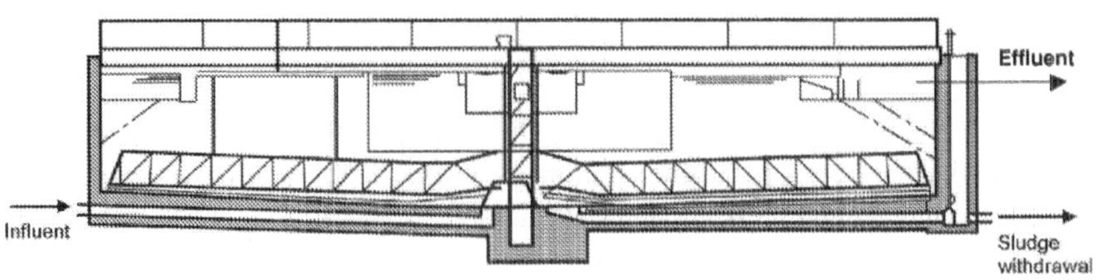

Figure 14.7.   Schematics of a secondary settling tank

traditional models, developed by the National Research Council – NRC (USA). The NRC empirical model was developed for filters with stone beds, taking into account operational data obtained in several plants operating in military facilities. The estimation of the BOD removal efficiency from a single filter or the first filter of a double-stage system fed with settled wastewater can be accomplished through Equation 14.3.

$$E = \frac{1}{1 + 0.443\sqrt{\frac{L_v}{F}}} \qquad (14.3)$$

where:

   E: BOD_5 removal efficiency (%)
   L_v: volumetric organic loading rate ($kgBOD/m^3 \cdot d$)
   F: recirculation factor

The recirculation factor F represents the average number of passes of the influent organic matter through the filter, given by Equation 14.4 (Metcalf and Eddy, 1991). Recycle ratios (R) used vary from 0 to 2.0. When R is equal to zero (which is frequently the case for TFs following UASB reactors), F is equal to 1.0.

$$F = \frac{1 + R}{(1 + R/10)^2} \qquad (14.4)$$

where:

   R: recycle ratio

In the case of the estimation of the efficiency of trickling filters applied to the post-treatment of effluents from anaerobic reactors, Equation 14.3 should be used with caution, as the information for these applications is still very limited.

(i)   Sludge production

The estimation of the sludge production in trickling filters can be made by means of the following equation

$$\boxed{P_{sludge} = Y \times BOD_{rem}} \qquad (14.5)$$

where:

   P_{sludge}: sludge production, on a dry-solids base (kgTSS/d)
      Y: yield coefficient ($kgTSS/kgBOD_{removed}$)
  BOD_{rem}: BOD load removed (kgBOD/d)

Values of Y observed in biofilm reactors, operating with high rates and without nitrification, are usually in the range from 0.8 to 1.0 $kgTSS/kgBOD_{removed}$. The VSS/TSS ratio is in the range from 0.75 to 0.85. This estimate of sludge production has been shown to be adequate for trickling filters applied for the treatment of effluents from UASB reactors. The suspended solids present in the effluent from

Table 14.3. Main design criteria for high rate trickling filters applied to the post-treatment of effluents from anaerobic reactors

Design criterion	Range of values, as a function of the flow		
	for $Q_{average}$	for $Q_{daily-maximum}$	for $Q_{hourly-maximum}$
Packing medium	Stone	Stone	Stone
Depth of the packing bed (m)	2.0 to 3.0	2.0 to 3.0	2.0 to 3.0
Hydraulic loading rate ($m^3/m^2{\cdot}d$)	15 to 18	18 to 22	25 to 30
Organic loading rate ($kgBOD/m^3{\cdot}d$)	0.5 to 1.0	0.5 to 1.0	0.5 to 1.0

TF are usually removed in conventional secondary settling tanks, as previously mentioned.

Evaluation of the volumetric sludge production is made according to

$$V_{sludge} = \frac{P_{sludge}}{\gamma \times C_{sludge}} \qquad (14.6)$$

where:

V_{sludge}: volumetric sludge production (m^3/d)

P_{sludge}: sludge production in TF (kgTSS/d)

γ : density of the sludge (usually in the order of 1,000 to 1,040 kg/m^3)

C_{sludge}: concentration of the sludge removed from the secondary settling tank (usually in the range from 1 to 2%)

(j)   Summary of the design criteria for trickling filters used as post-treatment of effluents from UASB reactors

A summary of the main criteria for the design of trickling filters applied to the post-treatment of effluents from anaerobic reactors is presented in Table 14.3.

14.3 CONSTRUCTION ASPECTS

Trickling filters are usually built with reinforced concrete, although smaller units can be made with different materials, such as steel, fibreglass, etc. Great emphasis should be placed on the longevity and integrity of the filter structure and of the packing medium, achieved with the use of appropriately selected materials, resistant to the adverse conditions imposed by the wastewater.

Particular attention should be given to the choice of the packing material and to the filling of the filter, once recurring problems of clogging of the packing medium have been reported. The recommendations contained in Section 14.2·d should be followed, particularly in the case of filters filled with stones, since improper size and shape of the stones can cause failure of the treatment system.

Another important aspect refers to the construction of the bottom drainage system that should be resistant enough to support the whole weight of the structure located in the upper part, including the packing medium, the biofilm and the wastewater. Additionally, the design recommendations regarding the slopes of the drainage system and the free areas to allow the ventilation of TF should be guaranteed.

14.4 OPERATIONAL ASPECTS

Trickling filters are characterised by their operational simplicity, as the degree of mechanisation of the system is minimum, mainly associated with the flow distribution in the tank and the sludge removal in the secondary setter. The operation of the system consists basically of routine activities, aiming at:

- the monitoring of the efficiency of the treatment system, carried out through an appropriate programme of physical-chemical analyses of the influent and effluent
- the monitoring of the sludge production in the treatment system, accomplished through measurements of suspended solids concentrations in the effluent from the TF and, principally, in the final effluent and in the sludge withdrawn from the secondary sedimentation tank
- the observation of the occurrence of flooding (ponding) on the surface of the TF, that generally occurs when the volume of the void space in the packing medium is occupied by excessive biofilm growth
- the verification of excessive proliferation of flies, usually related to the operation of the TF in an intermittent way and/or to low surface loading rates
- the verification of the bottom drainage system, eliminating any improper accumulation of solids in the lower slab and in the effluent collection gutters

Besides these basic operational items, activities of preventive maintenance should be undertaken, to guarantee the integrity of the treatment units and of all the installed equipment. Special attention should be given to the inspection of the flow distributors and of the sludge scrapers.

Example 14.1

Design a high rate trickling filter for the post-treatment of the effluent from a UASB reactor, given the following data:

Data:

- Population: $P = 20{,}000$ inhabitants
- Average influent flowrate: $Q_{av} = 3{,}000 \text{ m}^3/\text{d}$
- Maximum daily influent flowrate: $Q_{max\text{-}d} = 3{,}600 \text{ m}^3/\text{d}$
- Maximum hourly influent flowrate: $Q_{max\text{-}h} = 5{,}400 \text{ m}^3/\text{d}$
- Average influent BOD concentration to the UASB reactor: $S_{0\text{-}UASB} = 350 \text{ mg/L}$

Example 14.1 (Continued)

- BOD removal efficiency expected in the UASB reactor: 70%
- Average influent BOD concentration to the trickling filter:
  $S_{0\text{-TF}} = 105$ mg/L (0.105 kg/m^3)
- Desired BOD concentration for the effluent from the TF:
  $S_{e\text{-TF}} < 30$ mg/L
- Temperature of the wastewater: $T = 23°C$ (average of the coldest month)
- Yield coefficient (sludge production) in the TF:
  $Y = 0.75$ kgTSS/kgBOD$_{removed}$
- Expected concentration of the excess sludge wasted from the secondary settling tank: $C = 1.0\%$
- Density of the sludge: $\gamma = 1{,}020$ kgTSS/m^3.

Solution:

(a)   Adoption of the volumetric organic load (L_v)

From Table 14.3, high rate TFs following anaerobic reactors should be designed with L_v between 0.5 and 1.0 kgBOD/m^3·d. Adopted value: $L_v = 0.85$ kgBOD/m^3·d

(b)   Calculation of the volume of the packing medium (V) (Equation 14.2)

$$V = (Q_{av} \times S_{0\text{-TF}})/L_v$$

$$= (3{,}000 \text{ m}^3/\text{d} \times 0.105 \text{ kgBOD/m}^3)/(0.85 \text{ kgBOD/m}^3\text{·d}) = 371 \text{ m}^3$$

(c)   Adoption of the depth of the packing medium

From Table 14.3, high rate TFs following anaerobic reactors should be designed with packing medium heights between 2.0 and 3.0 m. Adopted value: $H = 2.0$ m

(d)   Calculation of the TF surface area (A)

$$A = V/H = (371 \text{ m}^3)/(2.0 \text{ m}) = 186 \text{ m}^2$$

(e)   Verification of the hydraulic loading rate on the TF (L_h) (Equation 14.1)

For Q average: $L_h = Q_{av}/A = (3{,}000 \text{ m}^3/\text{d})/(186 \text{ m}^2) = 16.1 \text{ m}^3/\text{m}^2\text{·d}$
For Q daily maximum: $L_h = Q_{max\text{-}d}/A = (3{,}600 \text{ m}^3/\text{d})/(186 \text{ m}^2) = 19.3 \text{ m}^3/\text{m}^2\text{·d}$
For Q hourly maximum: $L_h = Q_{max\text{-}h}/A = (5{,}400 \text{ m}^3/\text{d})/(186 \text{ m}^2) = 29.0 \text{ m}^3/\text{m}^2\text{·d}$

It is seen that the values of the hydraulic loading rates are in agreement with the recommended ranges, for the three conditions of applied flows, according to Table 14.3.

Example 14.1 (Continued)

(f)   Calculation of the TF diameter (D)

Adopt two filters, each one with an area of $186 \text{ m}^2/2 = 93 \text{ m}^2$

$$D = [(4 \times A)/PI]^{0.5} = [(4 \times 93 \text{ m}^2)/(3.1416)]^{0.5} = 10.9 \text{ m}$$

(g)   Estimation of the BOD removal efficiency of the TF (Equation 14.3)

For TF following UASB reactors, the effluent recycle ratio may be adopted as zero. Hence, the recirculation factor F is equal to 1.0 (see Equation 14.4)

$$E = 100/[1 + 0.443 \times (L_v/F)^{0.5}] = 100/[1 + 0.443 \times (0.85/1)^{0.5}] = 71\%$$

(h)   Estimation of the BOD concentration in the final effluent (Se-TF)

$$S_{e\text{-TF}} = S_{0\text{-TF}} \times (1 - E/100) = 105 \times (1 - 71/100) = 30 \text{ mg/L}$$

(i)   Estimation of the sludge production

The expected sludge production in TFs can be estimated from Equations 14.5 and 14.6.

$P_{sludge} = Y \times BOD_{rem}$
$BOD_{rem} = Q_{av} \times (S_{0\text{-TF}} - S_{e\text{-TF}}) = 3{,}000 \text{ m}^3/\text{d} \times (0.105 \text{ kgBOD/m}^3 - 0.030 \text{ kgBOD/m}^3)$
$BOD_{rem} = 225 \text{ kgBOD}_{rem}/\text{d}$

$P_{sludge} = 0.75 \text{ kgTSS/kgBOD}_{rem} \times 225 \text{ kgBOD}_{rem}/\text{d} = 169 \text{ kgTSS/d}$

Considering 75% of volatile solids:

$P_{sludge-volatile} = 0.75 \times 169 \text{ kgTSS/d} = 127 \text{ kgVSS/d}$

The volumetric sludge production is (Equation 14.6):

$$V_{sludge} = P_{sludge}/(\gamma \times C_{sludge})$$
$$= (169 \text{ kgTSS/d})/(1{,}020 \text{ kg/m}^3 \times 0.01) = 17 \text{ m}^3/\text{d}$$

(j)   Design of the secondary settling tank

From Table 14.2, the settling tanks should be designed with surface hydraulic loading rates between 16 and 32 $\text{m}^3/\text{m}^2 \cdot \text{d}$. Adopted value: $L_h = 24 \text{ m}^3/\text{m}^2 \cdot \text{d}$

$$A = Q_{av}/L_h = (3{,}000 \text{ m}^3/\text{d})/(24 \text{ m}^3/\text{m}^2 \cdot \text{d}) = 125 \text{ m}^2$$

Adopt two circular settling tanks with peripheral traction sludge scrapers, as follows:

Diameter $= 9.0$ m; useful side-wall depth $= 3.5$ m; surface area, per unit $= 63.5 \text{ m}^2$

Example 14.1 (Continued)

According to Table 14.2, the maximum hydraulic loading rate should be between 40 and 48 $m^3/m^2 \cdot d$, and the calculated value is:

$$L_h = Q_{max-h}/A = (5{,}400 \ m^3/d)/(2 \times 63.5 \ m^2) = 43 \ m^3/m^2 \cdot d$$

(k)   Sludge processing

- Sludge production in the UASB reactors

$$
\begin{aligned}
P_{sludge} &= Y \times BOD_{applied} \\
&= 0.28 \ kgTSS/kgBOD_{applied} \times 3.000 m^3/d \times 0.350 \ kgBOD/m^3 \\
&= 294 \ kgTSS/d
\end{aligned}
$$

- Total sludge production to be discharged, including the secondary aerobic sludge returned to the UASB reactors, considering 30% reduction of the aerobic sludge (VSS) in the UASB reactor:

$$P_{sludge} = 294 + (169 - 0.30 \times 127) = 425 \ kgTSS/d$$

15

Rotating biological contactors

R.F. Gonçalves

15.1 INTRODUCTION

The first commercial rotating biological contactor (RBC) was installed in Germany in 1960. The development of this process was induced by the interest in the use of plastic mediums, and it initially presented a series of advantages when compared to the classic low-rate trickling filters with stone beds.

In the 1970s, its application was expanded, due to the development of new support mediums and to the low energy requirements, when compared to the activated sludge process. Due to structural problems with shafts and support mediums, excessive growth of the attached biomass, irregular rotations and other problems of low process performance, a certain rejection of this process occurred in subsequent decades. However, progresses in technological research and new support medium systems made its application viable in certain situations, such as in small systems. In spite of the simplicity and operational stability, this process is not frequently used in developing countries. However, in the last few years, treatment plants associating UASB reactors and rotating biological contactors have become an option for the treatment of sewage from small and medium urban areas.

15.2 DESCRIPTION OF THE TECHNOLOGY

A rotating biological contactor consists of a prismatic tank, where horizontal shafts with equally-spaced coupled discs are installed. The shafts are maintained at constant rotation (1 to 2 rpm), either by mechanical action (when working with about 40% of the diameter submerged) or by air impulsion (when working with about 90% of its diameter submerged). This rotation movement first exposes the discs to the atmospheric air and then to the organic matter contained in the liquid medium. This facilitates the attachment and growth of the microorganisms onto the surface, forming a few-millimetres-thick film that covers the whole disc.

The discs are generally circular and built of low-density plastic, being installed in such a way as to be partially immersed, usually around 40%. Their main roles are:

- serve as a support medium for the development of the biofilm
- promote the contact of the biofilm with the wastewater
- maintain the excess biomass dislodged from the discs in suspension in the wastewater
- promote the aeration of the biofilm and the wastewater attached to it in the inferior part, due to the immersion of the discs

There are cases in which the discs work about 90% submerged, and in these cases introduction of air is necessary to allow enough oxygen for the aerobic process. When the biofilm reaches an excessive thickness, part of it detaches, and the organisms are maintained in suspension in the liquid medium due to the movement of the discs, increasing the efficiency of the system. However, the detached biomass and other suspended solids leave with the effluent, requiring a secondary settling tank for the removal of these solids. Well-designed biodiscs can reach secondary level treatment with respect to nitrification and denitrification.

Figure 15.1 presents a typical flowsheet of a treatment plant that uses rotating biological contactors. The primary settling tank can be substituted by a UASB reactor, substantially decreasing the organic load in the aerobic stage.

Mass transfer and substrate and oxygen diffusion, amongst several aspects, control organic matter removal in rotating biological contactors. However, due to the complexity of the transfer/diffusion phenomenon, there is no simplified model for simulating the removal of organic matter. The maximum organic matter removal rates are limited by the oxygen transfer capacity. The main source of oxygen for the system is the atmospheric air; the turbulence generated by the rotation of the discs is only an additional beneficial consequence.

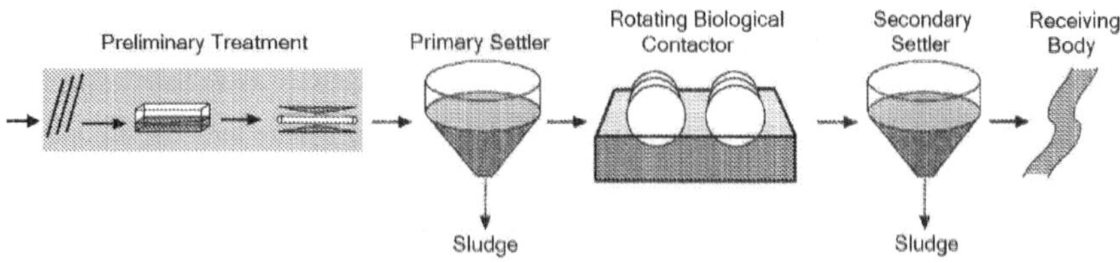

Figure 15.1.   Typical flowsheet of a treatment plant with rotating biological contactors

15.3 DESIGN CRITERIA

Rotating biological contactors are more frequently used for the treatment of settled domestic sewage, although some installations for the post-treatment of effluents from UASB reactors are already in operation. RBC units are usually designed to reach only BOD and SS removal, or to obtain a well-nitrified effluent.

(a) Hydraulic and organic loading rates

A sufficient residence or reaction time is necessary in any biological reactor. Therefore, a flow increment results in the increase in the hydraulic loading rate and a decrease in the detention time. Flow equalisation could be considered when peak daily flows are 2.5 times greater than the average flow.

To take advantage of the biological reaction rates, that are higher with larger concentrations of soluble BOD in the liquid undergoing treatment, it is usual to divide the unit into stages, operating the first stage with soluble BOD $\geq$ 50 mg/L, to approach zero-order reaction in relation to BOD, with a maximum removal rate of about 12 gBOD/m^2·day. However, the organic loading rate in the first stage is also a limiting variable for the design, due to problems with excessive loading rates leading to increases in the biofilm thickness, limitations with relation to oxygen availability, odour generation, process deterioration, structural overload, etc. In view of these observations, for settled domestic sewage, the maximum organic loading rate suggested for the first stage has been limited by some equipment manufacturers at 15 gBOD$_{soluble}$/m^2·day, or 30 gBOD/m^2·day. Metcalf and Eddy (1991) suggest maximum values from 20 to 30 gBOD$_{soluble}$/m^2·day, or 40 to 60 gBOD/m^2·day.

In general, rotating biological contactors have a minimum of two stages for secondary level treatment and three stages for BOD removal and nitrification. The organic loading rate based on soluble BOD is considered important, since the biodegradable organic matter predominantly used by the biomass attached to the disc is soluble, which is more quickly biodegraded and, therefore, the one that controls the maximum oxygen uptake rates. For settled domestic sewage there is about 50% of soluble BOD and the other 50% in suspension. For effluents from UASB reactors, the available data of the BOD$_{filtered}$/BOD$_{total}$ ratio are limited, indicating a ratio varying from 0.4 to 0.5, while the COD$_{filtered}$/COD$_{total}$ ratio is more commonly in the range from 0.4 to 0.7.

Observations on substrate concentration and hydraulic loading rate lead to the verification of the influence of these parameters in the substrate removal rate and in the efficiency of the system, converging in the concept of total organic load, as a parameter for design purposes (WEF, 1992). In an investigation of 23 treatment plants with rotating biological contactors in the USA, a curve of influent BOD$_5$ versus hydraulic load was adjusted for the first stage (Figure 15.2). Above the curve, process performance was hindered. The curve corresponds to the limit of 31 gBOD$_5$/m^2·d for the development of sulphur-oxidiser organisms. With the application of high organic loading rates, the following problems can occur: development of a heavier

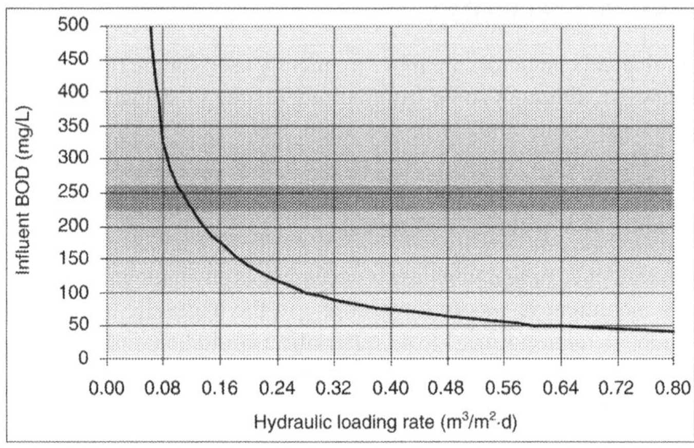

Figure 15.2.   Relationship between influent organic matter concentration and hydraulic loading rate in RBC units (adapted from WEF, 1992)

biofilm, growth of harmful organisms, DO reduction and a total deterioration of the process performance.

(b)   Characteristics of the influent wastewater

The characteristics of the influent wastewater and the impact on the biodegradability are important considerations in the design of rotating biological contactors. High concentrations of H_2S accelerate the growth of organisms that are harmful to the process. In influents with high H_2S concentration, removal systems should be included upstream, such as pre-aeration.

(c)   Temperature of the wastewater

Literature indicates that the efficiency of the process is not affected by temperatures of the wastewater above 13 °C. However, as in every biological process, there is a reduction in the process performance for lower temperatures.

(d)   Control of the biofilm

Biofilm thickness is very important for rotating biological contactors, either expressed in terms of total thickness or active thickness. Depending on the hydrodynamic conditions, the total thickness of the biofilm varies between 0.07 and 4.0 mm. However, from studies that relate biofilm thickness with removal efficiency, the part of the biofilm that contributes to the substrate removal, called the active biofilm thickness, was estimated between 20 and 600 µm. Most of these studies showed that, due to limitations of oxygen or substrate diffusion, there is a maximum active biofilm thickness, above which the removal rate does not increase.

Sufficient operational flexibility should be included for the control of the biofilm thickness. Due to the application of a larger organic loading rate in the first stages, they can have a larger biofilm growth. Devices to measure the weight of the shaft can be applied to control the growth and accumulation of the biomass. Techniques for biofilm thickness control include increases in the rotation speed (shearing forces), periodic reversal of the rotation direction, use of supplementary aeration, use of removable baffles and step-feeding for the reduction of the organic loading rate, or, as a last resort, use of chemical products for the removal of the biofilm.

(e) Dissolved oxygen levels

One of the most frequent causes of aerobic system failure is inadequate level of dissolved oxygen. Literature indicates a minimum DO level of 2 mg/L for rotating biological contactors. Low DO levels for high-rate systems lead to the production of H_2S inside the biofilm, which increases the growth of sulphur-oxidiser organisms such as *Beggiatoa* (filamentous bacteria), generating excess biomass, weight increase and a possible failure of the shafts or support medium (Metcalf and Eddy, 1991). These microorganisms compete with the heterotrophic organisms for consumption of the available oxygen and for space in the support medium, generating an increase in the biofilm thickness and a reduction in the organic matter removal efficiency.

Nitrifying organisms are more sensitive to dissolved oxygen levels than heterotrophic organisms. The DO levels necessary for nitrification vary from 0.5 to 4.0 mg/L, 2.0 mg/L being a typical value. In systems applied for nitrification, the DO level generally rises in the later stages. Combined with low BOD_5 values, this can reduce the nitrification efficiency, due to the development of protozoan predators of the nitrifying bacteria. To avoid the growth of predators, a maximum DO level of 3.5 mg/L and $BOD_{filtered}$ between 6 and 8 mg/L is suggested in the nitrification stages. The design should include ways of increasing the DO in the system, such as velocity variation control, supplementary aeration, recirculation of the effluent, step-feeding of the influent and the use of removable baffles, mainly in the initial stages.

(f) Operational flexibility

Rotating biological contactors should be provided with adequate flexibility for good operation and maintenance. The following items should be observed:

- possibility of supplementary aeration in mechanical rotation systems, aiming at counteracting possible overloads in the first stages
- means for the removal of the excess biofilm, such as air stripping, water or chemical additives, rotation control, etc
- removable baffles between all the stages
- feeding alternatives of the reactor
- recirculation of effluent from secondary clarifier
- DO monitoring in the stages
- easy access to equipment that need inspection, maintenance and replacement, such as shafts, support material, blowers, etc
- drainage of the tanks

Table 15.1. Summary of the design criteria for rotating biological contactors

Item	Treatment objective		
	BOD removal	BOD removal and nitrification	Separate nitrification
Hydraulic loading rate (m^3/m^2·day)	0.08 to 0.16	0.03 to 0.08	0.04 to 0.10
Surface organic loading rate (SOLR) (gBOD$_{soluble}$/ m^2·day)	3.7 to 9.8	2.4 to 7.3	0.5 to 1.5
Surface organic loading rate (SOLR) (gBOD/m^2·day)	9.8 to 17.2	7.3 to 14.6	1.0 to 2.9
Maximum SOLR in first stage (gBOD$_{soluble}$/m^2·day)	19 to 29 (14*)	19 to 29 (14*)	–
Maximum SOLR in first stage (gBOD/m^2·day)	39 to 59 (30*)	39 to 59 (30*)	–
Surface nitrogen loading rate (g N-NH$_4^+$/m^2·day)	–	0.7 to 1.5	1.0 to 2.0
Hydraulic detention time (hour)	0.7 to 1.5	1.5 to 4.0	1.2 to 2.9
BOD in the effluent (mg/L)	15 to 30	7 to 15	7 to 15
N-NH$_4^+$ in the effluent (mg/L)	–	<2	<2

* Typical design values
Source: Adapted from Metcalf and Eddy (1991)

(g)   Sludge production and characteristics

The production and characteristics of the sludge generated in rotating biological contactors are basically the same as those from trickling filters, around 0.75 to 1.0 kgTSS/kgBOD$_{removed}$, with a VSS/TSS ratio of 0.75 to 0.85. Equations 14.5 and 14.6 can be used in the dimensioning of the sludge treatment units.

(h)   Summary of the design criteria

The recommendations for the design of rotating biological contactors are mainly based on the BOD loading rate per unit area of support material, and also on the hydraulic loading rate per surface area available for biofilm growth. Table 15.1, adapted from Metcalf and Eddy (1991), can be used for design purposes.

15.4 CONSTRUCTION ASPECTS AND CHARACTERISTICS OF THE SUPPORT MEDIUM

The biodiscs have a shaft which supports and rotates the plastic medium that serves as support for the development of the biofilm. For high-density polythene biodiscs, the shaft length varies from 1.5 to 8.0 m and the diameter from 2.0 to 3.8 m. There are several types of corrugated surfaces (Metcalf and Eddy, 1991):

- low density (or conventional), with about 9,300 m^2 per unit, with a shaft length in the order of 8.0 m and a diameter of 3.8 m
- average or high density, with areas of about 11,000 to 16,700 m^2 per unit, with the same dimensions as previously referred

The low-density units are usually used in the first stages, while the average and high density ones are applied in the final stages of the system. The reason is that in the initial stages, with higher BOD concentrations, there is a larger biomass growth, which could lead to an excessive weight of the high-density units, harming its structure.

Some discs are composed of cylinders, with their interior made up of a beehive-type structure, with the objective of having high specific surface areas. A variant of the discs is composed of wheels with corrugated tubes that work with an immersion of about 90%, rotating and allowing the liquid to enter inside the tubes, dragging large amounts of air. The movement of the wheels is induced by the application of air that is also used to complement the oxygen requirements for the aerobic process. These wheels have a diameter varying from 1.2 to 3.3 m, with a surface area that varies from 170 m^2, for a wheel with a diameter of 1.2 m and a width of 0.9 m, to 4.000 m^2, for a wheel with a diameter of 3.3 m and a width of 2.5 m.

For discs that work with an immersion of about 40% of its diameter, it is common for the systems to be covered, to protect them against deterioration by ultraviolet radiation and also to avoid algal growth, that can lead to a substantial increase in the weight of the biomass attached to the surface of the discs.

16

Submerged aerated biofilters

R.F. Gonçalves

16.1 INTRODUCTION

Submerged aerated biofilters are nowadays a mature technology, being present at compact treatment plants that can even be buried in the sub-soil of sporting stadiums, parks and buildings in the middle of an urban area. One of the main advantages of the technology is the low environmental impact, especially when covered and deodorised, which can be done with relative simplicity (Rogalla *et al.*, 1992). Other advantages are the compactness, modular aspect, fast start-up, resistance to shock loads, absence of secondary clarification (Pujol *et al.*, 1992) and resistance to low wastewater temperatures (Gonçalves and Rogalla, 1994).

Biofilters are capable of reaching different quality objectives: oxidation of organic matter (Pujol *et al.*, 1992), secondary or tertiary nitrification (Carrand *et al.*, 1990; Tschui *et al.*, 1993), denitrification (Lacamp *et al.*, 1992), and physical-chemical phosphate removal (Gonçalves *et al.*, 1992). In warm-climate areas, biofilters can be used for the post-treatment of effluents from UASB reactors aiming at the removal of organic matter.

16.2 DESCRIPTION OF THE TECHNOLOGY

16.2.1 General aspects

In practice, a submerged aerated biofilter is constituted by a tank filled with a porous material through which wastewater and air permanently flow. In almost all of the existent processes, the porous medium is maintained under total immersion by the hydraulic flow, constituting tri-phase reactors composed of:

- *solid phase*: constituted by the support medium and by the colonies of microorganisms that develop in the form of a biofilm
- *liquid phase*: composed of the liquid in permanent movement through the porous medium
- *gas phase*: formed by the artificial aeration and in a reduced scale by the gaseous by-products of the biological activity

Submerged aerated biofilters (**SAB**) with *granular mediums* accomplish in the same reactor the removal of soluble organic compounds and suspended particles present in the wastewater. Besides serving as a support medium for the microorganisms, the granular material constitutes an effective filtering medium. In this process, periodical washings are necessary to eliminate the excess of accumulated biomass, reducing the hydraulic head losses through the medium. During washing, with or without interruption of wastewater feeding, several hydraulic discharge sequences of air and wash water are carried out.

On the other hand, the submerged biofilters with *structured beds*, also called *submerged aerated filters* (**SAF**), are classified by the same type of packing medium used for trickling filters (TF). Since they do not have granular-type packing material, as in SAB, they do not retain the suspended biomass by the filtration action, thus needing secondary settling tanks, at least for the usual hydraulic loading rates applied to trickling filters. SAF can operate with upward or downward flow and, as they need air supply for aeration, this is done through coarse bubble diffusers placed in the lower part of the filter, fed by blowers. Feeding of SAF is similar to that of SAB. When operated without sludge recirculation, they respond in a similar way to trickling filters (even though TF could be operating with final effluent recycle to dilute the influent to about 100 mg BOD/L) submitted to the same organic loading rates per unit area or unit volume of the packing medium.

The first SAB appeared at the beginning of the 1980s and were conceived for the removal of SS and the oxidation of organic matter from domestic sewage. A typical flowsheet of such a treatment plant is presented in Figure 16.1.

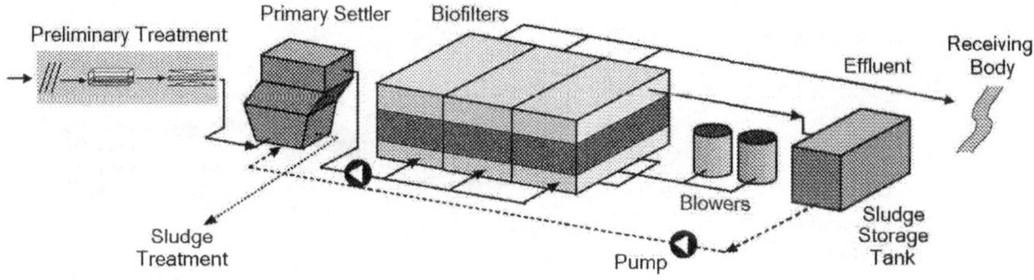

Figure 16.1.   Typical flowsheet of a treatment plant with SAB with granular medium for BOD and SS removal (Gonçalves, 1993)

Its main components are:

* *pre-treatment*: coarse screening, fine screening and grit removal
* *primary treatment*: conventional or lamellar settling tanks
* *secondary treatment*: SAB, in this case, with upward flow

The two sources of sludge generation are the primary sedimentation tanks and the washing of SAB. The wash sludge is generally retained in a storage tank, and is pumped for clarification in the primary settling tank, outside the peak flow hours. Therefore, the sludge to be treated is a mixed one, composed of primary and biological sludge.

To limit fast clogging of SAB with granular filter bed, it is imperative to have a primary sedimentation stage in the treatment of domestic sewage. The complete elimination of primary treatment is only possible in the case of very diluted wastewater, and even so with a very efficient pre-treatment (SS < 120 mg/L).

16.2.2 Treatment plants associating UASB reactors and SAB

A configuration of a treatment plant associating UASB reactors and SAB in series was developed by Gonçalves *et al.* (1994). The proposed configuration substitutes the primary sedimentation tanks by UASB reactors, which remove about 70% of the influent BOD (Figure 16.2). Post-treatment of the anaerobic effluent is accomplished in the submerged aerated biofilters, aiming at the removal of organic matter and the remaining suspended solids.

In parallel with the development of this configuration, a series of simplifications were introduced in the biofilters, compared with similar European processes. Three types of low cost, widely available commercial gravels or broken stones (grades 2, 1 and 0) are used in the composition of the packing mediums of the biofilters. The aeration system comprises Venturi tubes through which a pump sucks the aerobic effluent, captures air near the orifices, and injects water and dissolved air in the base of the biofilters. The air is captured in the vicinity of the main emission points of malodorous compounds (grit chamber, pumping station, drying bed) and is reintroduced into SAB, where biological odour removal occurs, with approximately 95% H_2S removal (Matos *et al.*, 2001). SAB units are interconnected

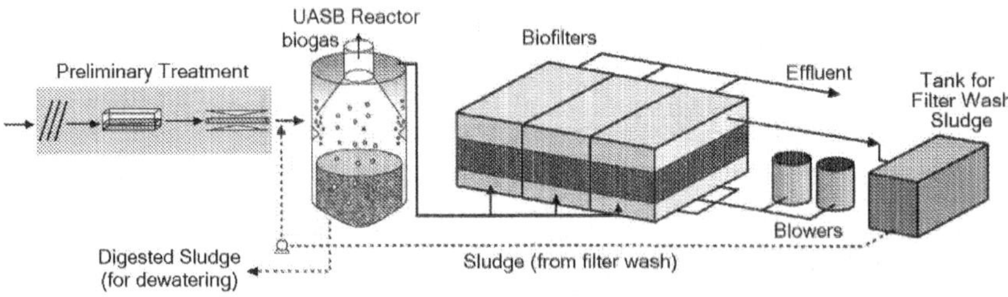

Figure 16.2.   Typical flowsheet of a treatment plant associating a UASB reactor and SAB in series

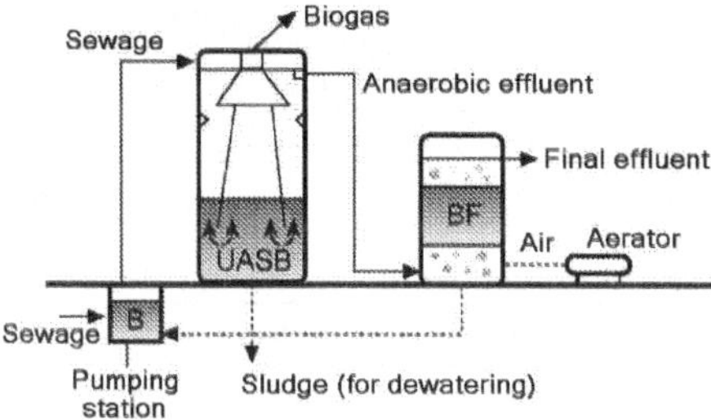

Figure 16.3.   Arrangement of a treatment plant comprising UASB + SAB

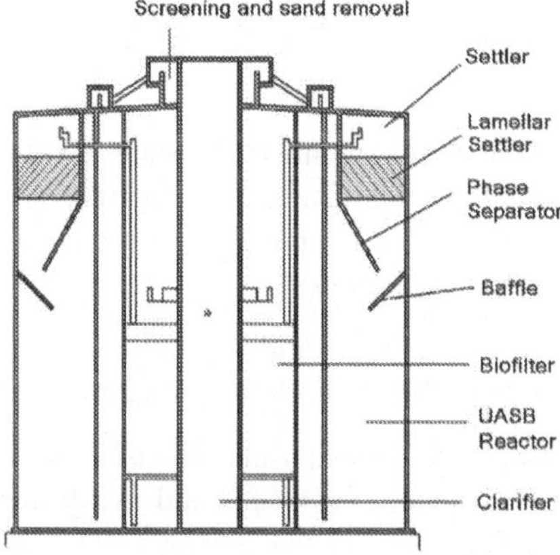

Figure 16.4.   Schematics of a compact plant with UASB + SAB in the same volume (courtesy: Sanevix Engenharia Ltda)

in their upper part, allowing the use of treated effluent in the washing operation, which is accomplished in downflow mode without air injection.

In the proposed system, the excess sludge produced in the biofilters is recirculated to the UASB reactor, where thickening and anaerobic digestion occur. The excess sludge produced in the UASB reactor is highly concentrated and stabilised, being discharged by gravity to the dewatering unit. The UASB reactors and SAB units can be built separately (Figure 16.3) or in the same volume (Figure 16.4).

16.2.3 Important aspects of the technology

(a)   Flow direction (air and water)

The flow direction (air and water) determines the main operational characteristics of a submerged aerated biofilter and directly influences the following points: SS retention, gas–liquid transfer, development of the hydraulic head loss, washing type, energy utilisation and odour production. The different options for the flow direction are presented in Figure 16.5 (Richard and Cyr, 1990). The air flow in SAB

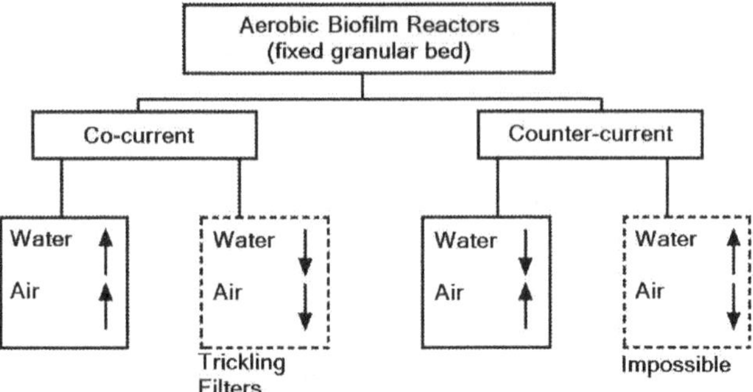

Figure 16.5.   Flow directions (air/liquid) in biofilm aerobic reactors (Richard and Cyr, 1990)

is only viable in an upflow direction, due to the state of permanent immersion of the packing medium. A downflow air current is only possible in granular mediums that are not submerged (as in the case of trickling filters), which limits the options for the flow in SAB to two possibilities:

- *co-current*: with upward liquid and air flows
- *counter-current*: with downward liquid and upward air flows

The different processes are generally classified based on the flow direction: upward or downward. The main advantages and disadvantages of each one of these options are as follows:

- *solids retention capacity*: the SS retention capacity is larger in downflow processes with heavy granular mediums (density > 1) or upflow with a floating medium (density < 1). In this case, the liquid flow proceeds in the direction of the compression of the filter bed, conferring its large filtration capacity. On the other hand, in the upflow processes with heavy packing medium, the co-current flow produces an expansion of the filter bed, which allows a better-distributed SS retention along the SAB height. SAB units with structured packing medium need secondary settling, because they possess small SS capture and storage capacity.
- *evolution of the head loss*: due to the high efficiency in SS retention, the head loss develops more quickly in downflow SAB with heavy granular beds and in upflow SAB with a floating medium. With a relatively slower head loss evolution, the upflow processes with heavy material have beds with a height that could reach up to 3.00 m. The evolution of the head loss is extremely low in structured packing mediums (honeycomb-type), with a filtration cycle that could last several days.
- *hydraulic behaviour*: the downflow processes (counter-current) can favour the formation of air bubbles trapped in the middle of the granular medium (embolism). A disadvantage of the upflow processes is the possibility of a deficient influent distribution in the granular medium, generating short circuits and reduction of efficiency.

- *aeration demand*: manufacturers state that downflow processes require less air and that the head loss is smaller, due to the relatively small height of the filter bed (Sibony, 1983). Manufacturers of upflow processes claim that, due to the larger filter bed height, the oxygen transfer efficiency is very high – from 23 to 30% (Strohmeir *et al.*, 1993). Experimental data obtained in full-scale plants indicate that this efficiency reaches, at the most, 10% in the two process types (Canler and Perret, 1993).
- *construction details*: in the downflow processes, the aeration tubing only enters in contact with the treated wastewater, and is less subject to blockages by solids present in the settled wastewater. In the case of the upflow processes, only the treated wastewater enters in contact with the atmosphere, eliminating odour problems. Upflow SAB can also be self-cleaning, when the washing is carried out in the counter-current mode.

(b) Packing material

The packing medium should accomplish two functions in SAB: to serve as a support for the attachment of the microorganisms and to physically retain SS present in the wastewater. The smaller the specific surface available for the attachment of the microorganism colonies, the smaller the SS retention capacity by filtration will be. On the other hand, materials with high specific surfaces favour fast evolution of the head loss, demanding more frequent washings of the SAB.

The choice of the characteristics of the packing medium is a compromise between quality of the treated effluent and washing frequency, within reasonable economical limits. The most commonly used packing mediums are composed of granular material with the following main characteristics:

- Grain size between 2 and 6 mm, in the case of domestic sewage treatment in downflow SAB (Gilles, 1990). Grains with an effective diameter between 1 and 2 mm are appropriate for tertiary nitrification in upflow SAB, while for carbonaceous oxidation the diameter should be larger than 2.5 mm (Richard and Cyr, 1990). In Brazil, SAB units with packing medium composed of gravel layers (grades 4, 2 and 0) have been used with success (Bof *et al.*, 2001). The introduction of sand layers significantly increases the clarification of the treated effluent. Amongst the granular materials, the more commonly used have a specific surface varying between 200 and 600 m^2/m^3.
- Density in the order of 1.5, for the case of granular materials applied to secondary treatment. Higher densities imply greater energy consumption with the expansion of the bed during washing. In the case of the packing material of the BIOSTYR process (OTV, France), polystyrene beads (density = 0.04) with specific surface greater than 1,000 m^2/m^3 are used. Denser materials, such as gravel or broken stones, can be used for the post-treatment of anaerobic effluents when the average specific surface of the filter bed does not exceed 300 m^2/m^3.
- Homogeneous grain size to avoid clogging of the void spaces by smaller particles. When using stratified beds in secondary treatment, different densities should be adopted among the layers of different materials.

- Inert, non-biodegradable and non-deforming material, so that the packing medium conserves its shape and grain size characteristics during several years of plant operation.
- Resistance to abrasion, resulting from the turbulence produced during washing of the granular medium.

The shape of the grains does not significantly influence the performance of the process. The surface characteristics that facilitate the attachment of the biomass are more important, mainly specific surface and roughness.

Granular materials of mineral origin are currently more commonly used in treatment plants. Grains of calcined clay or expanded schistus of the silicate family are the most common in Europe. Sandy materials, pozzolana and activated carbon are used less frequently. Some of these materials are patented, notably some calcined clays that receive a surface treatment with metallic salts or activated carbon. The first SAB units used in Brazil were packed with broken stone or gravel and expanded clay. However, as a result of the high cost of the expanded clay, new SAB units are packed almost exclusively with broken stone or gravel. Sandy layers of different granulometry have also been tested, aiming at the production of a highly clarified effluent.

On the other hand, the use of synthetic materials was intensified at the beginning of the 1990s and generated new processes with floating or structured medium (submerged aerated filters, – SAF). Granular mediums with a specific weight varying between 0.03 and 0.9 g/cm^3 have been used, composed of materials such as polystyrene, polypropylene, polyurethane, PVC and plastic. The price of these materials is higher than those of a mineral origin, although a part of the additional cost can be compensated by the smaller energy demand during washing. In most European SAB units, elimination of the wash water reservoirs (self-cleaning SAB) can be achieved with the use of synthetic materials. Some of the synthetic granular materials used in SAB have a porosity of 40% and specific surfaces greater than 1,000 m^2/m^3 (Tschui *et al.*, 1993).

Structured synthetic materials comprise corrugated mediums with a honeycomb or similar type, and present specific surfaces varying between 100 and 500 m^2/m^3 and porosity higher than 80%. The result of this high porosity is a smaller filtration capacity and the need for SAB units to have complementary clarification of the treated effluent. In contrast, this type of SAB presents the following advantages in relation to other SAB units: liquid velocities of up to 20 m/hour, filtration cycles of up to 1 week and the absence of hydraulic short circuits (Gros and Karl, 1993).

(c)   Aeration system – oxygen demand

Most SAB and SAF units have a direct system of artificial aeration, composed of blowers and air pipes. In older SAB units with heavy granular mediums, the aeration grid is located at the base of the filter bed, supported on the slab with diffusers. In upflow processes, the grid can be introduced inside the granular medium, allowing the creation of a non-aerated fraction at the beginning of the filter bed.

In rare cases of extremely diluted effluent, the aeration can be accomplished outside the SAB. In anoxic SAB used as denitrification reactors in wastewater treatment (Jepsen *et al.*, 1992) or in water treatment (Ravarini *et al.*, 1988), the air injection device is also non-existent. In these last two cases, the SAB operates predominantly as a two-phase reactor, in more favourable hydraulic conditions than in the case of a three-phase SAB. Tests using pure oxygen in the treatment of domestic sewage did not produce results that justified the cost increment.

(d) Washing of the filter medium

Periodic washing of the filter medium is an obligatory stage in the operation of SAB, to control the progressive clogging of the bed due to biofilm growth (microorganisms and retained SS). The duration of these cycles depends on the granulometry of the material, the applied load, the characteristics of the wastewater and the nature of the attached biomass. Most SAB units applied to secondary wastewater treatment are designed to operate for 24 to 48 hour periods between two consecutive washings.

The amount of treated water used and the energy consumption (pumps and air) are two factors to be considered in the definition of the washing procedure. The volume of wash water used in SAB with fixed granular beds is estimated, in upflow SAB, as 3 to 8% of the treated volume (Strohmeier and Schroeter, 1993) and from 5 to 10%, in downflow SAB (Upton and Stephenson, 1993). According to Pujol *et al.* (1992), the volume of water necessary to wash a SAB can be estimated as three times the filter bed volume. In the case of the association of UASB reactors with SAB, the washing can be done every 3 days, using less than 2% of the treated wastewater volume in the period between two washes.

The main manufacturers of SAB with heavy granular materials adopt washing protocols with different times, but with identical sequence of objectives, as detailed in Section 16.5. The various existing washing protocols were conceived so that the operation lasts for 20 to 40 minutes. In the case of European treatment plants, the excess sludge is pumped to the primary settling tank for combined treatment with the primary sludge.

(e) Energy consumption

The consumption of energy in the biofilters is concentrated on aeration, on the supply of air for washing and on the pumping of wash water (Table 16.1). Evaluation campaigns carried out in French treatment plants showed that the highest energy consumption is due to aeration, which consumes on average 87% of the energy related to secondary treatment (Canler and Perret, 1993). The energy balance undertaken by Kleiber *et al.* (1993) in the Perpignan treatment plant (France), covering a period of 12 months, resulted in the following consumption distribution in the secondary treatment: air-process = 83% of the total consumption, SAB washings = 17%.

Regulation devices of air supply as a function of the influent load are non-existent in most treatment plants in operation, which leads to energy consumptions

Table 16.1. Energy consumption in several treatment plants with granular SAB

Energy demand (kWh/kg removed)		Type of SAB	Treatment level	Observation	Reference
COD	BOD$_5$				
0.94		Upflow	Secondary	Overall consumption	Gilles (1990)
1.05				Consumption in the SAB	
1.30		Upflow	Secondary	BIOFOR	Partos *et al.* (1985)
1.02 to 1.25		Upflow/ downflow	Secondary	Study in 12 plants	Canler and Perret (1993)
	1.41	Downflow	Secondary with nitrification	Air for the process	Condren (1990)
	1.98			Overall consumption in SAB	
0.40		Upflow	UASB + SAB	Plants in Brazil	Bof *et al.* (2001)

that do not correspond to the real process needs. Some energy consumption values published by several authors are summarised in Table 16.1.

In the study carried out by CEMAGREF in 12 French treatment plants, installed power per unit volume of granular bed was on average 1,430 W/m^3 for upflow SAB (Pujol *et al.*, 1992). This power was split into 130 W/m^3 for process aeration, 600 W/m^3 for wash aeration and 700 W/m^3 for wash pumps. For downflow SAB, the installed power was on average 1,250 W/m^3, divided into 300 W/m^3 for process aeration, 650 W/m^3 for wash aeration and 300 W/m^3 for wash pumps.

Treatment plants associating UASB reactor + SAB in operation in Brazil present an average power of 2.0 W/inhabitant. Of this value, 50% correspond to the energy consumption in the aeration of the SAB. The other half refers to lighting and wastewater and sludge pumps. In terms of organic matter removal, the average energy consumption is 0.4 kWh/kgCOD$_{removed}$. In comparison with other data from Table 16.1, this value shows the importance of the anaerobic treatment upstream the SAB in the reduction of the energy consumption in the treatment plant.

16.3 DESIGN CRITERIA

(a) Preliminary considerations

The design of SAB and SAF is basically accomplished using empirical data, obtained through pilot- or full-scale experiments. The main design parameters are very similar to those already described in Section 14.2, related to trickling filters:

- *hydraulic loading rate*: volume of wastewater applied daily per unit area of the packing medium of the biofilter, expressed in $m^3/m^2 \cdot d$ (Equation 14.1)
- *organic loading rate*: mass of organic matter applied daily to the biofilter, per unit volume of the packing medium, expressed in $kgBOD/m^3 \cdot d$ or $kgCOD/m^3 \cdot d$ (Equation 14.2)

(b) Sludge production and characteristics

The specific sludge production in secondary treatment in upflow or downflow SAB units is of the order of **0.4 kgTSS/kgCOD$_{removed}$** or **0.8 to 1.0 kgTSS/ kgBOD$_{5\,removed}$** (Pujol *et al.*, 1992; Richard and Cyr, 1990). The excess sludge removed through washing of the bed can be estimated as 1 kg TSS/m^3 of the bed. Due to the fact that, besides biofilm growth, washing also removes SS retained by filtration, the wash sludge contains large amounts of volatile solids (>80%). Its settleability and thickening ability are relatively good.

In the case of the association of UASB reactor + SAB, the sludge production in SAB submitted to organic loading rates lower than 3.5 $kgCOD/m^3bed \cdot d$ is estimated as **0.25 kgTSS/kgCOD$_{removed}$**. In these cases, a large fraction of the rapidly biodegradable COD is removed in the anaerobic treatment stage, which allows the development of a thin biofilm with a very high sludge age inside the SAB. Volatile solids levels lower than 60% (VS/TS) are observed in the sludge discharged from SAB operating under such conditions. When the organic loading rate exceeds 4.0 $kgCOD/m^3 \cdot d$, the sludge production and characteristics resemble those described for secondary treatment.

The sludge production estimated for SAF and the design of the secondary settling tanks after SAF are identical to those described for trickling filters (Section 14.2). Considering that trickling filters lead to high head losses in the hydraulic profile of the treatment plant, SAF becomes a very attractive alternative for biological post-treatment of effluents from UASB reactors when the area available for the treatment plant is flat.

Additionally, this type of filter, when packed with material with high porosity and high specific surface area, can allow a good recirculation of sludge from the secondary settling tank, significantly increasing the biomass in the system. This configuration allows a greater organic matter removal potential per unit volume and also nitrification. However, this conception implies the use of a reactor with possible predominance of biomass in suspension, discussion of which is beyond the scope of this chapter.

(c) Aeration rates

Some values of the aeration rates practised in secondary treatment in granular SAB are presented in Table 16.2. Manufacturers of SAB with granular mediums state that oxygen transfer can reach efficiencies from 20 to 25%. However, rigorous monitoring campaigns carried out by a technical department of the French Ministry of Agriculture (CEMAGREF) showed that in full-scale treatment plants

Table 16.2. Aeration rates for secondary treatment in upflow and downflow SAB units

Aeration rate (Nm^3/kg applied)		Type of SAB	Observation	Level of treatment	Reference
COD	BOD$_5$				
20		Downflow	Whole plant	Secondary	Kleiber *et al.* (1993)
32		Downflow	Whole plant	Secondary + 50% nitrification	Rogalla *et al.* (1992)
	20	Downflow	In the SAB (efficiency = 7.7% and 1.5 mgO$_2$/L)	Secondary	Stensel *et al.* (1988)
	56	Downflow	Whole plant	Secondary + partial nitrification	Condren (1990)
	35	Upflow	UASB + SAB	Secondary	Bof *et al.* (2001)

this efficiency reaches at the most 10% (Canler and Perret, 1993). These results are equivalent to those obtained by Stensel *et al.* (1988) in a downflow SAB of the same type. For aeration rates from 10 to 40 Nm^{3}air/kgBOD$_{applied}$, oxygen transfer efficiencies varied between 9.2 and 5%. The average O$_2$ consumption calculated was 0.5 kgO$_2$/kgBOD$_{applied}$, lower than the typical values observed in conventional activated sludge (0.8 to 1.2 kgO$_2$/kgBOD$_{applied}$).

It is advisable that, for trickling filters, the influent has a BOD below about 100 mg/L, mainly due to the oxygen limitation, while for SAF such a limitation does not exist. The supply of air to reach the oxygen requirements of the aerobic process to have an effluent with BOD in the range of 20 to 30 mg/L, non nitrified, is about 35 to 40 m^{3}air/kgBOD$_{applied}$.

(d)   Summary of the design criteria for SAB units following UASB reactors

The main design criteria used for plants associating UASB reactors and submerged aerated biofilters (SAB) can be found in Table 16.3.

16.4 CONSTRUCTION ASPECTS

Among the urban wastewater treatment processes currently in operation, the submerged aerated biofilter is one of the most compact ones. SAB units can be built of concrete, fibreglass or steel with an anti-corrosion protective coating. In the case of these last two materials, and depending on the treatment capacity, the units can be pre-fabricated and transported to the plant location. Larger plants can have the pieces pre-fabricated and then transported for on-site assembly. Pre-fabrication greatly simplifies the planning and the implementation of the building site, lowering its size and duration. This aspect is in accordance with the peculiarities of the sanitation market for small localities, where, in general, there are infrastructure deficiencies for implementation of complex building sites.

Table 16.3. Main design criteria for plants associating UASB reactor + SAB

Parameter	UASB reactor	SAB	UASB reactor + SAB
Volumetric organic loading rate (kgBOD/ m^3·day)	0.85 to 1.2	3.0 to 4.0	–
Surface organic loading rate (gCOD/ m^2·day)	15 to 18	55 to 80	–
BOD removal efficiency (%)	65 to 75	60 to 75	85 to 95
SS removal efficiency (%)	65 to 75	60 to 75	85 to 95
COD removal efficiency (%)	60 to 70	55 to 65	80 to 90
Aeration rate (Nm^3/kgBOD$_{removed}$)	–	25 to 40	–
Sludge production (kgTSS/kgCOD$_{removed}$)	0.15 to 0.20	0.25 to 0.40	–
VS content in the sludge (VS/TS)	0.50 to 0.60	0.55 to 0.80	–
Aerobic sludge digestion efficiency in the UASB (% VS)	0.20 to 0.35	–	–

In the same way as for trickling filters, special attention should be given to the packing material of the filter. In the case of filter beds composed of materials of different densities and sizes, the turbulence generated by the washing operation can cause a mixture of the layers, and then loss of material or blockages. Although the aeration contributes to a significant mixing inside the biofilters, inside the filter bed the flow approaches plug flow. The positioning of the wastewater feeding points and the distribution, alignment and level of the collection gutters for the treated effluent should be thoroughly verified.

Another important aspect refers to the slab that supports the granular medium inside the biofilters. In upflow SAB treating anaerobic effluent, the slab should be built or covered with corrosion-resistant material, as it will be in permanent contact with sulphides present in the anaerobic effluent. Finally, the installation of an access window at the body of each biofilter, at the height of the aeration grid, can greatly facilitate occasional maintenance tasks.

16.5 OPERATIONAL ASPECTS

The retention of suspended solids and the growth of the attached biomass on the granular medium result in the constant increase of the head loss in biofilters with granular packing mediums. Control of this head loss is done through washings of the granular medium, accomplished in counter-current mode, just as in the rapid sand filters used in water treatment plants. The washing operation is composed of several intense hydraulic discharges of air and treated effluent. This intense turbulence temporarily expands the granular medium, promoting the removal of the excess biofilm. The washing frequency will depend on the evolution rate of the head loss, being around 1 washing/week in plants treating domestic sewage of average characteristics.

The objective of the washing operation of a SAB is to eliminate the excess biofilm accumulated in the process, during the operation between two sequential washings. Through washing, the biofilm thickness is reduced to ideal dimensions, which results in increase in the metabolic activities of the attached biomass and in the reduction of the head loss in the granular medium. The washing can or cannot consist of the total interruption of the wastewater feeding.

The washing of SAB should be conveniently dosed to preserve the integrity of the filter bed and to retain a minimum amount of biomass necessary for the immediate start after washing. The volume of wash water used in SAB is estimated as 3 to 8% of the volume of treated wastewater. The volume of water necessary to wash the SAB can also be estimated as three times the volume of the filter bed.

The washing operation may comprise the total interruption of the unit under washing, and is done during times of low flow to the treatment plant. This usually happens during dawn, when several units can be stopped without great problems. After the wastewater feed is interrupted, strong hydraulic discharge sequences of air and wash water are applied to eliminate the excess attached biomass. Generally, several hydraulic discharges are applied in the opposite direction to the wastewater flow (back-washing).

The several stages that constitute a washing operation can follow different time intervals, but always attending the following stages in sequence:

- desegregation of the material, by means of strong discharges of air
- destructuring of the excess biofilm, through strong discharges of air and water (concomitant or not)
- water discharges, to remove the excess sludge of the granular medium
- removal of the wash sludge

Washing of a biofilter requires its isolation from the others, if the plant is composed of several units. The only connection between the units will be that placed in the area above the granular medium, which guarantees a system of communicating vessels among the treated water reservoirs of each SAB (supernatant liquid layer in

Table 16.4. Stages of a SAB washing operation

Stage	Time (min)	Objective	Necessary action
1	2	Interruption of the operation	Stop wastewater and air feeding (close valves)
2	2	Intense discharge of the liquid, at rates >20 $m^3/m^2 \cdot$hour	Open the valve at the bottom of the SAB for 2 min
3	0.5	Interruption of the discharge of the liquid	Close the valve at the bottom of the SAB
4	2	Intense aeration, at rates greater than 50 $m^3/m^2 \cdot$hour	Open the valve in the aeration network of the SAB
5	0.5	Interruption of the intense aeration	Close the valve in the aeration network of the SAB
6	15	Repeat stages 2, 3, 4, and 5, in order, three more times	Follow the sequence of action described for each respective stage
7	1	Restart the operation of the SAB	Restart feeding of the SAB with wastewater and air (open the valves)

Total: 23 min
Source: SANEVIX Engenharia Ltda (1999)

Table 16.5. Main problems and possible solutions in the operation of granular
stone bed SAB

Problems	Possible causes	Possible solutions
High concentrations of suspended solids in the effluent	– Biofilm loss/washing deficiency	– Extend washings of the SAB; wash with a higher frequency; increase air and water hydraulic loads during washing
	– Biofilm loss/toxicity	– Find and eliminate the emission sources of the toxic compounds
	– High concentration of suspended solids in the influent	– Evaluate the possibility of solids removal upstream of the reactor
Excessive increase in the head loss	– Organic or hydraulic overload	– Find and eliminate the contributing sources of excessive organic material or reduce loads, by decreasing the influent flow
	– Washing deficiency	– Extend washing of the SAB; wash with a higher frequency; increase air and water hydraulic loads during washing
	– Air distribution deficiency	– Evaluate the operation of the air distribution system (possible blockage)
	– Excessive aeration	– Reduce the aeration rate
Low organic matter removal efficiency (BOD, COD and SS)	– Organic overload, high concentration of organic matter in the influent	– Find and eliminate the contributing sources of excessive organic material or reduce loads, by decreasing the influent flow
	– Hydraulic overload, peak influent flows	– Limit influent flows to the reactor or equalise flows in industries
	– Presence of toxic compounds in the wastewater	– Find or eliminate the emission sources of toxic compounds
	– Low wastewater temperatures	– Evaluate the possibility of covering the reactor

Source: SANEVIX Engenharia Ltda (1999)

the upper part of each SAB). This connection aims to guarantee the supply of treated water, introduced in downflow mode in the granular medium during washing.

As mentioned, the washing operation should be accomplished in periods in which the plant operates below its maximum treatment capacity. If automation is possible, the operation should be programmed for the period between 2:00 and 6:00 a.m., when the influent flow to the plant reaches its lowest values. The stages listed in Table 16.4 should be followed during SAB washing.

Table 16.5 summarises the main problems and possible solutions to be adopted during the operation of SAB with granular stone bed.

Example 16.1

Design submerged aerated biofilters (SAB and SAF) for the post-treatment of effluents generated in a UASB reactor, considering the same design elements of trickling filters (Example 14.1):

Data:

- Population: $P = 20,000$ inhabitants
- Average influent flowrate: $Q_{av} = 3,000 \ m^3/d$
- Maximum daily influent flowrate: $Q_{max-d} = 3,600 \ m^3/d$
- Maximum hourly influent flowrate: $Q_{max-h} = 5,400 \ m^3/d$
- Average influent BOD concentration to the UASB reactor: $S_{0-UASB} = 350 \ mg/L$
- BOD removal efficiency expected in the UASB reactor: 70%
- Average effluent BOD concentration from the UASB reactor: $S_{e-UASB} = 105 \ mg/L$
- Desired BOD concentration in the effluent from biofilter: $S_{e-SAB} < 30 \ mg/L$
- Temperature of the wastewater: $T = 23 \ °C$ (average of the coldest month)
- Yield coefficient (sludge production) in biofilter: $Y = 0.75 \ kgTSS/kgBOD_{removed}$
- Expected concentration for the sludge discharged from the secondary settling tank: $C_{sludge} = 1\%$
- Density of the sludge: $\gamma = 1,020 \ kgTSS/m^3$.

Alternatives to be considered:

- Alternative A: Use of UASB reactor followed by SAB (packing bed of stones)
- Alternative B: Use of UASB reactor followed by SAF (packing bed of stones)
- Alternative C: Use of UASB reactor followed by SAF (packing bed of plastic)

Solution:

(a)   Alternative A: Submerged aerated biofilter, SAB (packing bed of stones)

- Submerged aerated biofilters with an upward flow and a stone packing medium with a porosity of approximately 40% will be used, with the following arrangement:
  - 1st layer = 30 cm of gravel grade 3
  - 2nd layer = 30 cm of gravel grade 2
  - 3rd layer = 40 cm of gravel grade 1
  - 4th layer = 100 cm of gravel grade 0
- The final effluent is expected to have the following characteristics:

  BOD < 30 mg/L, COD < 90 mg/L and SS < 30 mg/L

Example 16.1 (Continued)

– Effluent organic load from the UASB reactor (influent to the biofilter):

$$OL_{e\text{-}UASB} = Q_{av} \times S_{e\text{-}UASB} = 3{,}000\,m^3/d \times 0.105\,kgBOD/m^3$$
$$= 315\,kgBOD/d$$

– SAB volume (V)
From Table 16.3, adopting $L_v = 4.0$ kgBOD/m³·d

$$V = OL_{e\text{-}UASB}/L_v = (315\,kgBOD/d)/(4.0\,kgBOD/m^3 \cdot d) = 79\,m^3$$

– SAB area (A)
Considering a filter bed height of 2.0 m:

$$A = V/h = (79.0\,m^3)/(2.0\,m) = 39.5\,m^2$$

Therefore, the biofilter will have a circular section with a diameter of 7.1 m, and will be divided into four equal parts.

– Upflow velocity or hydraulic loading rate (v)

$$v = Q_{av}/A = (3{,}000\,m^3/d)/(39.5\,m^2) = 75.9\,m/d = 3.2\,m/hour$$

– Air demand (without nitrification)
From Table 16.2, considering an aeration rate of 30 Nm³air/ kgBOD$_{applied}$:

$$Q_{air} = \text{aeration rate} \times OL_{e\text{-}UASB}$$
$$= (30\,Nm^3 air/kgBOD_{applied}) \times 315\,kgBOD/d$$
$$Q_{air} = 9{,}450\,Nm^3 air/d$$

The airflow per biofilter will be $9{,}450/4 = 2{,}363$ m³/day, with a pressure of 5.0 m.w.c. (metres of water column).

– Sludge production for dewatering
Sludge production in the SAB:

$$P_{sludge} = Y \times OL_{e\text{-}UASB} = 0.75\,kgTSS/kgBOD_{applied} \times 315\,kgBOD/d$$
$$= 236\,kgTSS/d$$

Considering 75% of volatile solids:

$$P_{sludge\text{-}volatile} = 236\,kgTSS/d \times 0.75 = 177\,kgVSS/d$$

Sludge production in the UASB reactor
Production due to the wastewater treatment:

$$P_{sludge} = Y \times BOD_{applied}$$
$$= 0.28\,kgTSS/kgBOD_{applied}$$
$$\times 3{,}000\,m^3/d \times 0.350\,kgBOD/m^3$$
$$= 294\,kgTSS/d$$

Example 16.1 (Continued)

Total production, including the secondary aerobic sludge returned to the UASB reactor, considering 30% reduction of the aerobic sludge (VSS) in the UASB reactor:

$$P_{sludge} = 294 + (236 - 0.30 \times 177) = 477 \text{ kgTSS/d}$$

(b) Alternative B: Submerged aerated filter, SAF (packing bed of stones)

– Submerged aerated filters with upward flow will be used. The packing medium will comprise gravel 4, with a specific surface area of 70 m^2/m^3 and 57% void spaces.

– For effluent BOD < 30 mg/L, it will be adopted a surface loading rate (L_s) of 14 $gBOD/m^2 \cdot d$ (0.014 $kgBOD/m^2 \cdot d$)

– Calculation of the volumetric organic load (L_v)

$$L_v = \text{specific surface area of the packing medium} \times L_s$$
$$= 70 \text{ m}^2/\text{m}^3 \times 0.014 \text{ kgBOD/m}^2 \cdot d$$
$$L_v = 1.0 \text{ kgBOD/m}^3 \cdot d.$$

– Calculation of the SAF volume (V)

$$V = OL_{e\text{-}UASB}/L_v = (315 \text{ kgBOD/d})/(1.0 \text{ kgBOD/m}^3 \cdot d) = 315 \text{ m}^3$$

– SAF area (A)
Considering stone bed height of 3.0 m:

$$A = V/h = (315 \text{ m}^3)/(3.0 \text{ m}) = 105 \text{ m}^2$$

Adopt two units of 52.5 m^2 each, with 7.3 m × 7.3 m, or two circular units with a diameter of 8.2 m each.

• height of the inlet compartment = 0.8 m
• height of the packing medium = 3.0 m
• water height over the packing material = 0.5 m
• total useful height = 4.3 m.

– Air demand (without nitrification)
From Table 16.2, considering an aeration rate of 30 $Nm^3 air/$ kgBODapplied:

$$Q_{air} = \text{aeration rate} \times OL_{e\text{-}UASB}$$
$$= (30 \text{ Nm}^3 \text{air/kgBOD}_{applied}) \times 315 \text{ kgBOD/d}$$
$$Q_{air} = 9{,}450 \text{ Nm}^3 \text{air/d} = 394 \text{ Nm}^3 \text{air/hour}$$

• air flow per filter = 394/2 = 197 m^3/hour or 3.3 m^3/minute (4.0 m^3/ minute will be adopted for each unit, with a pressure of 5 m.w.c.)

Example 16.1 (Continued)

- air distribution system: by coarse bubbles, through perforated tubes or coarse bubble diffusers

– Design of the secondary settling tank
From Table 14.2, the settling tanks should be designed with surface hydraulic loading rates between 16 and 32 $m^3/m^2 \cdot d$. Adopted value: $L_h = 24\ m^3/m^2 \cdot d$

$$A = Q_{av}/L_h = (3{,}000\ m^3/d)/(24\ m^3/m^2 \cdot d) = 125\ m^2$$

Adopt two circular settling tanks with peripheral traction sludge scrapers, as follows:

Diameter = 9.0 m; useful side-wall depth = 3.5 m; surface area, per unit $= 63.5\ m^2$

According to Table 14.2, the maximum hydraulic surface loading rate should be between 40 and 48 $m^3/m^2 \cdot d$ and the calculated value is:

$$L_h = Q_{max\text{-}h}/A = (5{,}400\ m^3/d)/(2 \times 63.5\ m^2) = 43\ m^3/m^2 \cdot d$$

The sludge from the secondary settling tanks will be pumped to the inlet of the UASB reactors. For sludge removed with 1% solids, the daily volume to be pumped is as follows:

$$V_{sludge} = P_{sludge}/(\gamma \times C_{sludge}) = (236\ kgSS/d)/(1{,}020\ kg/m^3 \times 0.01)$$
$$= 23.1\ m^3/d$$

– Sludge production for dewatering
Sludge production in the SAF:

$$P_{sludge} = Y \times OL_{e\text{-}UASB} = 0.75\ kgSS/kgBOD_{applied} \times 315\ kgBOD/d$$
$$= 236\ kgTSS/d$$

Considering 75% of volatile solids:

$$P_{sludge\text{-}volatile} = 236\ kgTSS/d \times 0.75 = 177\ kgVSS/d$$

Sludge production in the UASB reactor
Production due to the wastewater treatment:

$$P_{sludge} = Y \times BOD_{applied}$$
$$= 0.28\ kgSS/kgBOD_{applied} \times 3{,}000\ m^3/d \times 0.350\ kgBOD/m^3$$
$$= 294\ kgTSS/d$$

Total production, including the secondary aerobic sludge returned to the UASB reactor, considering 30% reduction of the aerobic sludge (VSS) in the UASB reactor:

$$P_{sludge} = 294 + (236 - 0.30 \times 177) = 477\ kgTSS/d$$

Example 16.1 (Continued)

(c) Alternative C: Submerged aerated filter, SAF (packing bed of plastic)

– Submerged aerated filters with upward flow will be used. The packing medium will comprise plastic, with a specific surface area of 130 m^2/m^3 and 95% void spaces.

– For effluent BOD < 30 mg/L, it will be adopted a surface loading rate (L_s) of 14 gBOD/$m^2 \cdot$d (0.014 kgBOD/$m^2 \cdot$d)

– Calculation of the volumetric organic load (L_v)

$$L_v = \text{specific surface area of the packing medium} \times L_s$$
$$= 130 \ m^2/m^3 \times 0.014 \ kgBOD/m^2 \cdot d$$
$$= 1.8 \ kgBOD/m^3 \cdot d.$$

– Calculation of the SAF volume (V)
$$V = OL_{e\text{-UASB}}/L_v = (315 \ kgBOD/d)/(1.8 \ kgBOD/m^3 \cdot d) = 175 \ m^3$$

– SAF area (A)
Considering stone bed height of 3.0 m:

$$A = V/h = (175 \ m^3)/(3.0 \ m) = 58 \ m^2$$

Adopt two units of 29 m^2 each, with 5.4 m × 5.4 m, or two circular units with a diameter of 6.1 m each.

- height of the inlet compartment = 0.8 m
- height of the packing medium = 3.0 m
- water height over the packing material = 0.5 m
- total useful height = 4.3 m.

– Air demand (without nitrification)
Identical to alternative (b)

– Secondary settling tanks
Identical to alternative (b)

– Sludge production for dewatering
Identical to alternative (b)

References

ABNT (1989) *Projeto de estações de tratamento de esgotos.* NBR-570 (in Portuguese).

Adelusi, A.A. (1989) The documentation of an expert system for the operational control of the activated sludge process. M.Sc. thesis, Imperial College, Univ. of London.

AIRE-O$_2$ (1992) Aire-O$_2$ horizontal aspirator aerator. *Technical bulletin.* Minneapolis, EUA.

Aisse, M.M., van Haandel, A.C., von Sperling, M., Campos, J.R., Coraucci, Filho B. and Alem Sobrinho, P. (1999) Tratamento final do lodo gerado em reatores anaeróbios. In *Tratamento de esgotos sanitários por processo anaeróbio e disposição controlada no solo* (coord. J.R. Campos), pp. 271–300, PROSAB (in Portuguese).

Alem Sobrinho, P. (2001) Tratamento de esgoto e geração de lodo. In *Biossólidos na agricultura*, (eds M.T. Tsutiya, J.B. Comparini, P. Alem Sobrinho, I. Hespanhol, P.C.T. Carvalho, A.J. Melfi, W.J. Melo and M.O. Marques), p. 7–40, SABESP, USP, UNESP, (in Portuguese).

Alem Sobrinho, P. and Jordão, E.P. (2001) Pós-tratamento de efluentes de reatores anaeróbios – uma análise crítica. In *Pós-tratamento de efluentes de reatores anaeróbios*, (coord. C.A.L. Chernicharo), pp. 491–513, PROSAB/ABES, Rio de Janeiro (in Portuguese).

Alem Sobrinho, P. and Kato, M.T. (1999) Análise crítica do uso do processo anaeróbio para o tratamento de esgotos sanitários. In *Tratamento de esgotos sanitários por processos anaeróbios e disposição controlada no solo*, (coord. J.R. Campos), pp. 301–20, ABES, Rio de Janeiro (in Portuguese).

Andrews, J.F. (1972) Control systems for wastewater treatment plants. *Water Research* **6**, 575–582.

Andrews, J.F. (1974) Review paper, Dynamic models and control strategies for wastewater treatment processes. Wat. Res., 8, 261–289.

Arceivala, S.J. (1981) *Wastewater Treatment and Disposal*, 892 p., Marcel Dekker, New York.

Arvin, E. and Harremöes, P. (1991) Concepts and models for biofilm reactor performance. *Water Sci. Technol.* **22**(1/2), 171–192.

ASCE (1990) *ASCE Standard. Measurement of Oxygen Transfer in Clean Water*, 2–90. 66 p. ANSI/ASCE.

Atkinson, B. (1981) Immobilised biomass – a basis for process development in wastewater treatment. In *Biological fluidised bed treatment of water and wastewater* (eds Atkinson and Cooper), Chapter 1, pp. 22–34, Ellis Horwood Limited Pub., UK.

Bailey, J.E. and Ollis, D.F. (1986) *Biochemical Engineering Fundamentals*, 2nd edn, McGraw-Hill, New York.

Barnes, D. and Bliss, P.J. (1983) *Biological Control of Nitrogen in Wastewater Treatment*, 146 p., E. & F.N. Spon, London.

Beck, M.B. (1986) Identification, estimation and control of biological wastewater treatment processes. IEE Proceedings, 133, Pt. D (5), 254–264.

Berthouex, P.M., Lai, W. and Darjatmoko (1989). Statistics – based approach to wastewater treatment plant operations. J. Environ. Eng. Div. ASCE, 115 (3), 650–671.

Bof, V.S., Sant'ana, T.D., Wanke, R., Silva, G.M., Salim, F.P.C., Nardoto, J.I.O., Netto, E.S. and Pegoretti, J.M. (2001) ETEs compactas associando reatores anaeróbios e aeróbios ampliam a cobertura do saneamento no estado do Espírito Santo *Anais do XXVII Congresso Brasileiro de Engenharia Sanitária e Ambiental (ABES, João Pessoa (PB))* (in Portuguese).

Bonhomme, M., Rogalla, F., Boisseau, G. and Sibony, J. (1990) Enhancing nitrogen removal in activated sludge with fixed biomass. *Water Sci. Technol.* **22**(1/2), 127–135.

Boon, A.G. (1980) Measurement of aerator performance. *Symposium on the Profitable Aeration of Wastewater* (London, 25 April 1980).

Boyle, W.C. (1989) Oxygen transfer in clean and process water for draft tube turbine aerators in total barrier oxidation ditches. *J. Water Pollut. Control Fed.* **61**(8), 1449–1463.

Campos, J.R. (coord.) (1999) *Tratamento de esgotos sanitários por processos anaeróbios e disposição controlada no solo*, ABES, Rio de Janeiro, (in Portuguese).

Canler, J.P.E. and Perret, J.M. (1993) Biofiltres aérés: évaluation du procédé sur 12 unités industrielles *Proceedings of the 2nd International Specialised Conference on Biofilm Reactors (Paris, France)* pp. 27–37.

Carrand, G., Capon, B., Rasconi, A.E. and Brenner, R. (1989) Elimination of carbonaceous and nitrogenous pollutants by a twin-stage fixed growth process. *Water Sci. Technol.* **22**(1/2), 261–272.

Catunda, P.F.C. and van Haandel, A.C. (1987) Activated sludge settlers: design and optimization. *Water Sci. Technol.* **19**, 613–623.

Catunda, P.F.C., van Haandel, A.C., Araújo, L.S. and Vilar, A. (1989) Determinação da sedimentabilidade de lodo ativado. In *Congresso Brasileiro de Engenharia sanitária e Ambiental*, **15** (*Belém, ABES, setembro 1989*), pp. 53–85 (in Portuguese).

CEC (1991) Council Directive of 21 May 1991 concerning urban wastewater treatment (91/271/EEC). Official Journal of the European Communities, No. L 135/40. Council of the European Communities.

Charleston, L.O., Robles, A.N. and Bohórquez, S.S. (1996) Efectos de lodos activados de purga sobre el funcionamiento de un reactor UASB piloto y las características del lecho de lodo granular. In *XXV Congresso Interamericano de Engenharia Sanitária e Ambiental (México, 3–7 November 1996)* (in Spanish).

Chernicharo, C.A.L. and von Sperling, M. (1993) Considerações sobre o dimensionamento de sistemas de lodos ativados de fluxo intermitente (batelada). In *Congresso Brasileiro de Engenharia Sanitária e Ambiental*, **17**, (*Natal, 19–23 Setembro 1993*), vol. 2, Tomo I (in Portuguese).

Chernicharo, C.A.L., von Sperling, M., Galvão J.R., A.C., Magalhaes, C.A.C., Moreno, J. and Gariglio, L.P. (2001) An innovative conversion of a full scale extended aeration activated sludge plant by using a UASB reactor as a first step treatment: case study of Botucatu city – Brazil. *World Congress on Anaerobic Digestion (International Water Association, Antuérpia, Bégica, Set 2001)*.

Chisti, M.Y., Fujimoto, K. and Moo Young, M. (1989) *Biotechnology Process – Scale up and Mixing* (eds C.S. Ho and J.Y. Oldshue), AIChE, New York.

Clifft, R.C. and Andrews, J.F. (1981a) Aeration control for reducing energy consumption in small activated sludge plants. *Water Sci. Technol.* **13**, 371–379.

Clifft, R.C. and Andrews, J.F. (1981b) Predicting the dynamics of oxygen utilization in the activated sludge process. *J. Water Pollut. Control Fed.* **53**(7), 1219–1232.

Coletti, F.J., Povinelli, J. and Daniel, L. A. (1997) Pós-tratamento por lodos ativados de efluentes provenientes de processos anaeróbios de tratamento de esgoto sanitário; Determinação de constantes cinéticas. In *Anais: Congresso Brasileiro de Engenharia Sanitária e Ambiental, (19, Foz do Iguaçu, ABES, Setembro 1997)* (in Portuguese).

Condren, A.J. (1990) *Technology Assessment of the Biological Aerated Filter.* Technical Report EPA/600/2–90/015, U.S.EPA, 103 p.

Cook, S.C. (1983) Systems identification and control of wastewater treatment processes in river basin management. Ph.D. thesis. Imperial College, Univ. of London.

Copp, J., Olsson, G., Spanjers, H. and Vanrollehem, P. (eds) (2002) Respirometry in control of the activated sludge process: Benchmarking Control Strategies. Scientific and Technical Report 11, IWA Publishing, London.

Coura, M.A. and van Haandel, A.C. (1999) Viabilidade técnica e econômica do digestor anaeróbio de fluxo ascendente (UASB) no sistema anaeróbio/aeróbio. In *Anais: Congresso Brasileiro de Engenharia Sanitária e Ambiental (20, Rio de Janeiro, 1999)*, ABES, Rio de Janeiro, pp. 973–987 (in Portuguese).

Daigger, G.T. (1995) Development of refined clarifier operating diagrams using an updated settling characteristics database. *Water Environ. Res.* **67**(21), 95–100.

Daigger, G.T. and Roper Jr, R.E. (1985) The relationship between SVI and activated sludge settling characteristics. *J. Water Pollut. Control Fed.* **57**(8), 859–866.

De Korte, K. and Smits, P. (1985) Steady state measurement of oxygenation capacity. *Water Sci. Technol.* **17**, 303–311.

Dick, R.I. (1972) Gravity thickening of sewage sludges. *Water Pollut. Control* **71**, 368–378.

Dowming, A.L. (1978) Selected subjects in waste treatment. 3rd edn. Delft, iHE.

Eckenfelder Jr, W.W. (1980) *Principles of Water Quality Management*, CBI, Boston, 717 p.

Eckenfelder Jr, W.W. (1989) *Industrial Water Pollution Control*, International edn, McGraw-Hill, New York.

Eckenfelder Jr, W.W. and Argaman, Y. (1978) Kinetics of nitrogen removal for municipal and industrial applications. In *Advances in Water and Wastewater Treatment. Biological Nutriente Removal*, (eds M.P. Wanielista and W.W. Eckenfelder Jr), pp. 23–41, Ann Arbor Science Publishers, Michigan.

Eckenfelder, W.W. and Grau, P. (1992) *Activated Sludge Process Design and Control. Theory and Practice*, 268 p., Technomic Publishing Co, Lancaster, EUA.

Ekama, G.A. and Marais, G.v.R. (1977) The activated sludge process. Part II – dynamic behaviour. *Water S.A.* **3**(1), 18–50.

Ekama, G.A. and Marais, G.v.R. (1986) Sludge settleability. Secondary settling tank design procedures. *Water Pollut. Control.* **1**, 101–114.

EPA, United States Environmental Protection Agency, Cincinatti (1987a) *The Causes and Control of Activated Sludge Bulking and Foaming.* Summary Report, United States Environmental Protection Agency, Cincinatti.

EPA, United States Environmental Protection Agency, Cincinatti (1987b) *Design Manual. Phosphorus Removal. Technology Transfer*, 116 p., United States Environmental Protection Agency, Cincinatti .

EPA, United States Environmental Protection Agency, Cincinatti (1993) *Manual. Nitrogen Control.* Technology Transfer, 311 p., United States Environmental Protection Agency, Cincinatti.

Fazolo, A. (2000) Nitrificação e desnitrificação simultânea em um único reator aeróbio/anaeróbio alimentado com efluente de reator anaeróbio horizontal de leito fixo. Projeto de pesquisa de Doutorado (EESC/USP) (in Portuguese).

Flanagan, M.J., Bracken, B.D. and Roesler, J.F. (1977) Automatic dissolved oxygen control. *J. Environ. Eng. Div., ASCE.* **103**(EE4), 707–722.

Fróes, C.M.V. (1996) *Avaliação do desempenho da estação de tratamento de esgotos do Conjunto Habitacional de Morro Alto, com ênfase na teoria do fluxo de sólidos limite.* Dissertação de mestrado, UFMG (in Portuguese).

Fróes, C.V. and von Sperling, M. (1995) Método simplificado para a determinação da velocidade de sedimentação com base no Índice Volumétrico de Lodo. In *Congresso Brasileiro de Engenharia Sanitária e Ambiental,* **18** (*Salvador, 17–22 Setembro 1995*) (in Portuguese).

Gall, R.A.B. and Patry, G.G. (1989) Knowledge-based system for the diagnosis of an activated sludge plant. In *Dynamic Modelling and Expert Systems in Wastewater Engineering* (eds G.G. Patry and D. Chapman), Lewis Publishers, EUA.

Gilles, P. (1990) Industrial scale applications of fixed biomass on the Mediterranean seaboard: design, operating results. *Water Sci. Technol.* **22**(1/2), 281–292.

Gonçalves R.F. and Rogalla, F. (1994) Biofiltros aerados para remoção de nitrogênio de águas residuárias sob diversas temperaturas. *Anais eletrônicos do XXIV Congresso Interamericano de Engenharia Sanitária e Ambiental – 30 de outubro a 4 de novembro de 1994* (*Buenos Aires, Argentina*) (in Portuguese).

Gonçalves, R.F. (1993) Elimination biologique du phosphore des eaux résiduaires urbaines par des biofiltres immergés, Thèse de Doctorat, L'Institut National des Sciences Apliquées (INSA), Toulouse, França, 169 p.

Gonçalves, R.F., Charlier, A.C. and Sammut, F. (1994). Primary fermentation of soluble and particulate organic matter for wastewater treatment. *Water Sci. Technol.* **30**(6) 53–62.

Gonçalves, R.F., Chernicharo, C.A.L., Andrade Neto, C.O, Alem Sobrinho, P., Kato, M.T., Costa, R.H.R., Aisse, M.M. and Zaiat, M. (2001). Pós-tratamento de efluentes de reatores anaeróbios por reatores com biofilme. Cap. 4. In *Pós-tratamento de efluentes de reatores anaeróbios* (coord. C.A.L. Chernicharo), 544 p., FINEP/PROSAB, Rio de Janeiro, Brasil (in Portuguese).

Gonçalves, R.F., Passamani, F.R.F., Salim, F.P., Silva, A.L.B., Martineli, G. and Bauer, D.G. (2000) Associação de um reator UASB e biofiltros aerados submersos para tratamento de esgoto sanitário. In *Pós-tratamento de efluentes de reatores anaeróbios – coletânea de trabalhos* (coord. C.A.L. Chernicharo), vol. 1, 220 p., FINEP/PROSAB (in Portuguese).

Gonçalves, R.F., Sammut, F. and Rogalla, F. (1992) High rate biofilter – simultaneous phosphorus precipitation and nitrogen removal – chemical water and wastewater treatment II (eds H.H. Hahn and R. Klute), pp. 357–372, Springer-Verlag, Berlin/Heildelberg.

Goronszy, M. (1997) Industrial application of cyclic activated sludge technology. In *IAWQ Yearbook 1997,* International Association on Water Quality, pp. 35–40

Grady, C.P.L. and Lim, H. (1980) *Biological Wastewater Treatment: Theory and Application,* Marcel Dekker, New York.

Gray, N.F. (1991) *Activated Sludge. Theory and Practice,* Oxford University Press, Oxford.

Gros, H. and Karl, V. (1993) Multilayer filtration and fixed bed systems for advanced wastewater treatment *Proceedings of the European Congress on Water Filtration* (*15–17 March, Oostend, Belgium*), pp. 3.51–3.55.

Gönenc, I.E. and Harremöes, P. (1985) Nitrification in rotating disc systems – I: criteria for transition from oxygen to ammonia rate limitation. *Water Res.* **19**, 1119–1127.

Gönenc, I.E. and Harremöes, P. (1990) Nitrification in rotating disc systems – II: criteria for simultaneous mineralization and nitrification. *Water Res.* **24**, 499–505.

Handley, J. (1974) Sedimentation: an introduction to solids flux theory. *Water Pollut. Control* **73**, 230–240.

Harremöes, P. (1982) Criteria for nitrification in fixed film reactors. *Water. Sci. Technol.* **14**, 167–187.

Hartley, K.J. (1987) Performance of surface rotors in oxidation ditches: discussion. *J. Environ. Eng. Div., ASCE.* **113**(4), 919–921.

Hatzifotiadou, O. (1989) Contribuition a l'etude de l'hydrodynamique et du transfer de matière gaz-liquide dans un réacteur a lit fluidisé thriphasique. Thèse de Doctorat, L'Institut National des Sciences Apliquées (INSA). Toulouse, 174 p.

Heath, M.S., Wirtel, S.A. and Rittmann, B.E. (1990) Simplified design of biofilm processes using normalized loading curves. *Res. J. Water Pollut. Control Fed.* **62**(2), 185–192.

Holmberg, U., Olsson, G. and Andersson, B. (1989) Simultaneous DO control and respiration estimation. *Water Sci. Technol.* **21**, 1185–1195.

Horan, N.J. (1990) *Biological Wastewater Treatment Systems. Theory and Operation*, 310 p., John Wiley & Sons, Chichester.

IAWPRC (1987) *Activated Sludge Model No. 1*. IAWPRC Scientific and Technical Reports No. 1.

IAWQ (1995) *Activated Sludge Model No. 2*. IAWQ Scientific and Technical Reports.

Jenkins, D., Richard, M.G. and Daigger, G.T. (1993) *Manual on the Causes and Control of Activated Sludge Bulking and Foaming*, 2ª edn, p. 193, Lewis Publishers, Londres.

Jepsen, S.E., Laursen, K., Jansen, J.L.C. and Harremöes, P. (1992) Denitrification in submerged filters of nitrified wastewater and chemical precipitated water. *Chemical Water and Wastewater Treatment II* (eds H.H. Hahn and R. Klute), pp. 371–387, Springer-Verlag, Berlin/Heildeberg.

Johnstone, D.W.M., Rachwall, A.J. and Hanbury, M.J. (1979) Settlement characteristics and settlement-tank performance in the Carrousel activated sludge system. *Water Pollut. Control* **78**, 337–353.

Johnstone, D.W.M., Rachwall, A.J. and Hanbury, M.J. (1983) General aspects of the oxidation ditch process. In *Oxidation Ditches in Wastewater Treatment*, (eds D. Barnes, C.F. Forster and D.W.M. Johnstone), pp. 41–74, Pitman Books Limited, London.

Kamiyama, H. and Tsutiya, M.T. (1992) Lodo ativado por batelada: um processo econômico para o tratamento de esgotos em estações de grande porte. *Revista DAE* **52**(165), 1–7 (in Portuguese).

Keinath, T.M. (1981) Solids inventory control in the activated sludge process. *Water Sci. Technol.* **13**, 413–419.

Keinath, T.M., Ryckman, M.D., Dana, C.B. and Hofer, D.A. (1977) Activated sludge – unified system design and operation. *J. Environ. Eng. Div., ASCE.* **103**(EE5), 829–849.

Kleiber, B., Roudon, G., Bigot, B. and Sibony, J. (1993) Assessment of biofiltration at industrial scale *Proceedings of the 2nd International Specialised Conference on Biofilm Reactors (Paris, France)*, pp. 271–283.

Koopman, B. and Cadee, K. (1983) Prediction of thickening capacity using diluted sludge volume index. *Water Res.* **17**(10), 1427–1431.

Kwan, A.F. (1990) Application of an expert system for the operational control of an activated sludge plant. M.Sc. thesis, Imperial College, Univ. of London.

Lacamp, B., Hansen, F., Penillard, P. and Rogalla, F. (1992) Wastewater nutrient removal with advanced biofilm reactors. *Water Sci. Technol.* **27**(5/6), 263–276.

Lazarova, V. and Manem, J. (1994) Advance in biofilm aerobic reactors ensuring effective biofilm activity control *Water Sci. Technol.* **29**(10/11), 319–327.

Lazarova, V. and Manem, J. (1995) Biofilm characterization and activity analysis in water and wastewater treatment. *Water Res.* **29**(10), 2227–2245.

Lertpocasombut, K., Capdeville, B. and Roques, H. (1988) Application of aerobic biofilm growth in a three-phase fluidized-bed reactor for biological wastewater treatment *2nd Asian Conference on Water Pollution Control in Asia (Bangkok, Thailand)*.

Lessel, T.H. (1993) Upgrading and nitrification by submerged bio-film reactors – experiences from a large-scale plant *Proceedings of the 2nd International Specialised Conference on Biofilm Reactors (Paris, France)*, pp. 231–238.

Lettinga, G., Pol, L.W.H. and Zeeman, G. (1996) *Biological Wastewater Treatment. Part I. Anaerobic Wastewater Treatment.* Department of Environmental Technology, Wageningen Agricultural University, January 1996 (Lecture notes).

Lohmann, J. and Schlegel, S. (1981) Measurement and control of the MLSS concentration in activated sludge plants. *Water Sci. Technol.* **13**, 217–224.

Lumbers, J.P. (1982) Improving the efficiency of the process. Future research and objectives for control in the water industry. In *SERC. Water and Waste Research. The Way Ahead. Proceedings of a seminar (Imperial College, London, 17–18 June 1982).*

Malina, J.F. Biological waste treatment. In *Seminário de transferência de tecnologia. Tratamento de esgotos*, pp. 153–315, (*ABES/WEF. Rio de Janeiro, 17–20 Aug 1992*)

Marais, G.v.R. and Ekama, G.A. (1976) The activated sludge process. Part I – steady state behaviour. *Water S.A.* **2**(4), 164–200.

Markantonatos, P. (1988) Modelling for the operational control of the oxidation ditch process. Ph.D. thesis, Imperial College, University of London.

Marsili-Libelli, S., Giardi, R. and Lasagni, M. (1985) Self-tuning control of the activated sludge process. *Environ. Technol. Lett.* **6**, 576–583.

Matos, T.A., Silva, H.T.M. and Gonçalves, R.F. (2001) Uso simultâneo de um biofiltro aerado submerso para tratamento secundário de esgoto sanitário e para biodesodorização de ar atmosférico contendo gás sulfídrico (H_2S). *Anais do XXVII Congresso Brasileiro de Engenharia Sanitária e Ambiental (ABES, João Pessoa (PB))* (in Portuguese).

Matsui, K. and Kimata, T. (1986) Performance evaluation of the oxidation ditch process. *Water Sci. Technol.* **18**, 297–306.

Metcalf and Eddy (1991) *Wastewater Engineering – Treatment, Disposal and Reuse*, 3rd edn, 1334 p., McGraw-Hill, Inc., New York.

Metcalf and Eddy (1991) *Wastewater Engineering: Treatment, Disposal and Reuse*, 3rd edn, 1334 pp., Metcalf and Eddy, Inc.

Okey, W. and Albertson, O.E. (1987) The role of diffusion in regulating rate and masking temperature effects in fixed film nitrification *Proceedings of the 60th Annual Conference of Water Pollution Control Federation.*

Olsson, G. and Newell, B. (1999) *Wastewater Treatment. Modelling, Diagnosis and Control*, IWA Publishing, London.

Olsson, G., Rundqwist, L., Eriksson, L. and Hall, L. (1985) Self tuning control of the dissolved oxygen concentration in activated sludge. In *Instrumentation and Control of Water and Wastewater Treatment and Transport Systems, Proceedings of the fourth IAWPRC Workshop (Houston and Denver)*, pp. 473–480.

Oodegard, H., Rusten, B. and Westrum, T. (1993) A new moving bed biofilm reactor – applications and results *Proceedings of the 2nd International Specialised Conference on Biofilm Reactors (Paris, France)*, pp. 221–229.

Orhon, D. and Artan, N. (1994) *Modelling of Activated Sludge Systems*, Technomic Publishing Co., Lancaster, EUA.

Parker, D.S., Lutz, M.P. and Pratt, A.M. (1990) New trickling filter applications in the U.S.A. *Water Sci. Technol.* **22**(1/2), 215–226.

Partos, J., Richard, Y. and Amar, D. (1985) Elimination de la pollution carbonée sur cultures fixées aérobies. Le procéde Biofor. *TSM. L'Eau.* (4), 193–198.

Passig, F.H., Vilela, L.C.H. and Ferreira, O.P. (1999) ETE – Piracicamirim – nova concepção de sistema de tratamento de esgotos sanitários – partida, operação e monitoramento de desempenho. In *Anais: 20° Congresso Brasileiro de Engenharia Sanitária e Ambiental*, (*Rio de Janeiro, Maio 1999*) (in Portuguese).

Pitman, A.R. (1984) Settling of nutrient removal activated sludges. *Water Sci. Technol.* **17**, 493–504.

Pujol, R., Canler, J.P. and Iwema, A. (1992) Biological aerated filters: an attractive and alternative biological process. *Water Sci. Technol.* **26**(3/4), 693–702.

Pöpel, H.J. (1979) *Aeration and Gas Transfer.* 2nd edn, 169 p., Delft University of Technology, Delft.

Qasim, S.R. (1985) *Wastewater Treatment Plants: Planning, Design and Operation*, Holt, Rinehart and Winston, New York.

Ramalho, R.S. (1977) *Introduction to Wastewater Treatment Processes*, 409 p., Academic Press, New York.

Randall, C.W., Barnard, J.L. and Stensel, H.D. (1992) *Design and Retrofit of Wastewater Treatment Plants for Biological Nutrient Removal*, 420 p., Technomic Publishing Co, Lancaster, USA.

Ravarini, P., Coutelle, J. and Damez, F. (1988) Le traitement d'eau potable à Dennemont – Dénitrification et nitrification biologiques à grande échelle. *TSM L'Eau.* **83**, 235–239.

Richard, Y. and Cyr, R. (1990) Les possibilités de traitement par cultures fixées aérobies. *TSM L'Eau.* **85**(7/8), 389–393.

Rittmann, B.E. (1982) The effect of shear stress on biofilm loss rate. *Biotechnol. Bioeng.* **24**, 501–506.

Rogalla, F., Roudon, G., Sibony, J. and Blondeau, F. (1992) Minimising nuisances by covering compact sewage plants. *Water Sci. Technol.* **25**(4/5), 363–374.

Saez, P.B., Chang, H.T. and Rittmann, B.E. (1988) Modeling steady-state substrate inhibited biofilms *International Conference on Physicochemical and Biological Detoxification of Hazardous Wastes.*

Sagberg, P., Dauthuille, P. and Hamon, M. (1992) Biofilm reactors: a compact solution for the upgrading of wastewater treatment plants. *Water Sci. Technol.* **26**(3/4), 733–742.

Sampaio, A.O. and Vilela, J.C. (1993) Utilização de seletores para o controle de intumescimento filamentoso em processos de lodos ativados. Programa experimental em escala real. In ABES: *Congresso Brasileiro de Engenharia Sanitária e Ambiental,* **17** (*Natal, RN, 19-23 setembro 1993*) (in Portuguese).

Sanevix Engenharia Ltda (1999) Manual de operação de ETEs do tipo UASB + BFs – Vitória (ES). 88 p. (in Portuguese).

Sawyer, C.N. and Mc Carty, P.L. (1978) *Chemistry for Environmental Engineering.* 3ª edn, 532 p., Mc Graw-Hill, Inc., New York.

Schlegel, S. (1977) Automation of the activated sludge process by oxygen and MLSS control. Prog. Water Technol., 9 (5/6), 385–392.

Sedlak, R. (ed) (1991) *Phosphorus and Nitrogen Removal from Municipal Wastewater. Principles and Practice,* Lewis Publishers, USA.

Sibony, J. (1983) Applications industrielles des cultures fixées en épuration d'eaux résiduaires *5ème Journée Scientifique: L'eau, la recherche et l'environnement* (*25–27 outubro, Lille, France*), pp. 387–397.

Silva, S.M.C.P. and Alem Sobrinho, P. (1995) Avaliação do Sistema Reator UASB e Processo de Lodos Ativados para Tratamento de Esgotos Sanitários com Elevada parcela de Contribuição Industrial. In *Anais: Congresso Brasileiro de Engenharia Sanitária e Ambiental,* **18** (*Salvador, Bahia, ABES, Setembro 1995*) (in Portuguese).

Souza, J.T. and Foresti, E. (1996) Domestic sewage treatment in an upflow anaerobic sludge blanket–sequencing batch reactor system. *Water Sci. Technol.* **33**(3), 73–84.

Souza, L.C.A. and Campos, J.R. (1999) Avaliação do consumo de energia elétrica dos sistemas de aeração por ar difuso com micro-bolhas e aeração mecânica para reator aeróbio precedido por reator anaeróbio em Estações de Tratamento de Esgotos sanitários. In *Assembléia Nacional da ASSEMAE,* **28** (Anais) (in Portuguese).

Stensel, H.D., Brenner, R.C., Lee, K.M., Meller, H. and Rakness, K. (1988) Biological aerated filter evaluation. *J. Environ. Eng.* **114**(3) 655–671.

Stephenson, R.V. *et al.* (1985) Performance of surface rotors in an oxidation ditch. *J. Environ. Eng. Div., ASCE.* **111**(1), 79–91.

Strohmeier, A. and Schroeter, I. (1993) Experiences with biological filtration in advanced waste water treatment *Proceedings of the European Congress on Water Filtration* (*15–17 March, Oostend, Belgium*), pp. 3.39–3.50.

Sveeris, S. (1969) Meting van het zuurstoftoevoervermogen. *H₂O* **2**(25), 610–642.

Takase, I. and Miura, R. (1985) Sludge flow control for activated sludge process. In *Instrumentation and Control of Water and Wastewater Treatment and Transport Systems, Proceedings of the 4th IAWPRC Workshop* (*Houston and Denver*), pp. 675–678.

Tchobanoglous, G. and Schroeder, E.D. (1985*) Water Quality: Characteristics, Modeling, Modification*, Addison-Wesley, Reading, MA.

Tschui, M., Boller, M., Gujer, W., Eugster, J., Mäder, C. and Stengel, C. (1993) Tertiary nitrification in aerated pilot biofilters *Proceedings of the 2nd International Specialised Conference on Biofilm Reactors* (*Paris, France*), pp. 109–116.

Tuntoolavest, M. and Grady, L.C.P. Effect of activated sludge operational conditions on sludge thickening characteristics. *J. Water Pollut. Control Fed.* **54**(7), 1112–1117.

Upton, J. and Stephenson, T. (1993) BAF – Upflow or downflow – the choice exists *Symposium on Wastewater Treatment by Biological Aerated Filters – Latest Developments,* (*Cranfield Institut of Technology, Cranfield, UK*).

Upton, J.B. and Green, B. (1995) A successful strategy for small treatment plants. *Water Quality Int.* (4), 12–14.

van Haandel, A.C. and Marais, G.v.R. (1999) *O comportamento do sistema de lodos ativados* (in Portuguese).

van Sluis, J.W. (undated). OC-metingen in de praktijk. H_2O.

von Sperling, M. (1990) Optimal management of the oxidation ditch process. Ph.D. thesis, 371 p., Imperial College, University of London.

von Sperling, M. (1992) Métodos clássicos e avançados para o controle operacional de estações de tratamento de esgotos por aeração prolongada. In *23° Congresso Interamericano de Engenharia Sanitária e Ambiental* (*Havana, Cuba, 22–26 Novembro 1992*) (in Portuguese).

von Sperling, M. (1993a) Determinação da capacidade de oxigenação de sistemas de lodos ativados em operação. In *Congresso Brasileiro de Engenharia Sanitária e Ambiental,* **17** (*Natal, 19–23 Setembro 1993*), vol. 2, Tomo I, pp. 152–167 (in Portuguese).

von Sperling, M. (1993b) Características operacionais específicas de sistemas de aeração prolongada. In *Congresso Brasileiro de Engenharia Sanitária e Ambiental,* **17**, vol. 2, tomo I. pp.140–151 (in Portuguese).

von Sperling, M. (1994a) Projeto e controle de decantadores secundários através de um novo método para a análise simplificada do fluxo de sólidos. In *VI SILUBESA – Simpósio Luso-Brasileiro de Engenharia Sanitária e Ambiental* (*Florianópolis-SC, 12-16 Junho 1994*), Tomo I, pp. 479–492 (in Portuguese).

von Sperling, M. (1994b) A new unified solids flux-based approach for the design of final clarifiers. Description and comparison with traditional criteria. *Water Sci. Technol.* **30**(4), 57–66.

von Sperling, M. (1994c) Variações horárias de diversas variáveis em uma estação de tratamento de esgotos por lodos ativados. In *VI SILUBESA – Simpósio Luso-Brasileiro de Engenharia Sanitária e Ambiental* (*Florianópolis-SC, 12–16 Junho 1994*), Tomo I, pp. 473–478 (in Portuguese).

von Sperling, M. (1994d) Solids management for the control of extended aeration systems. An analysis of classical and advanced strategies. *Water SA* **20**(1), 49–60.

von Sperling, M. (1996a*) Princípios do tratamento biológico de águas residuárias. Vol. 1. Introdução à qualidade das águas e ao tratamento de esgotos*, 2nd edn, 243 p., Departamento de Engenharia Sanitária e Ambiental, UFMG (in Portuguese).

von Sperling, M. (1996b*) Princípios do tratamento biológico de águas residuárias. Vol. 2. Princípios básicos do tratamento de esgotos*, 211 p., Departamento de Engenharia Sanitária e Ambiental, UFMG (in Portuguese).

von Sperling, M. (1996c*) Princípios do tratamento biológico de águas residuárias. Vol. 3. Lagoas de estabilização*, 134 p., Departamento de Engenharia Sanitária e Ambiental, UFMG (in Portuguese).

von Sperling, M. (1996d) A influência da idade do lodo no dimensionamento de sistemas de lodos ativados. *Engenharia Sanitária e Ambiental* (ABES), Ano 1, **1**(1), Jan/Mar, 18–27 (in Portuguese).

von Sperling, M. (1998) A new method for the design of sequencing batch reactors (SBR) using the concept of the hindered settling velocity of the sludge. *Environ. Technol.* **19**, 1223–1231.

von Sperling, M. (2002) *Princípios do tratamento biológico de águas residuárias. Vol. 4. Lodos ativados.* 2nd edn, 428 p., Departamento de Engenharia Sanitária e Ambiental, UFMG (in Portuguese).

von Sperling, M. and Fróes, C.M.V. (1998) Dimensionamento e controle de decantadores secundários com base em uma abordagem integrada e simplificada da teoria do fluxo de sólidos. *Engenharia Sanitária e Ambiental* (ABES). **3**(1/2), Jan/Jun, 42–54 (in Portuguese).

von Sperling, M. and Fróes, C.M.V. (1999) Determination of the required surface area for activated sludge final clarifiers based on a unified database. *Water Res.***33**(8), 1884–1894.

von Sperling, M. and Gonçalves, R.F. (2001b) Lodo de esgotos: características e produção. In *Princípios do tratamento biológico de águas residuárias. Vol. 6. Lodo de esgotos. Tratamento e disposição final*, (eds C.V. Andreoli, M. Von Sperling and F. Fernandes), pp.17–67 Departamento de Engenharia Sanitária e Ambiental – UFMG. Companhia de Saneamento do Paraná – SANEPAR. (in Portuguese).

von Sperling, M. and Lumbers, J.P. (1988) Controle operacional e otimização de custos em sistemas de aeração prolongada. In *21º Congresso Interamericano de Engenharia Sanitária e Ambiental (Rio de Janeiro, 18–23 Setembro 1988)*, pp. 171–191 (in Portuguese).

von Sperling, M. and Lumbers, J.P. (1989a) Control objectives and the modelling of MLSS in oxidation ditches. *Water Sci. Technol.* **21**, 1173–1183.

von Sperling, M. and Lumbers, J.P. (1989b) Control objectives and the modelling of MLSS in oxidation ditches (discussion). *Water Sci. Technol.* **21**(12), 1621–1622.

von Sperling, M. and Lumbers, J.P. (1991a) Optimization of the operation of the oxidation ditch process. *Water Sci. Technol.* **24**(6), 225–233.

von Sperling, M. and Lumbers, J.P. (1991b) Operational rules for the optimal management of the oxidation ditch process for wastewater treatment. In *Advances in Water Resources Technology* (ed. G. Tsakiris) (*Conference in Athens, March 1990*), pp. 387–395, Balkema, Rotterdam,

von Sperling, M., Freire, V.H. and Chernicharo, C.A.L. (2001a) Performance evaluation of a UASB – activated sludge system treating municipal wastewater. *Water Sci. Technol.* **43**(11), 323–328.

von Sperling, M., Van Haandel, A.C., Jordão, E.P., Campos, J.R., Cybis, L.F., Aisse, M.M. and Alem Sobrinho, P. (2001c) Capítulo 5: Tratamento de efluentes de reatores anaeróbios por sistema de lodos ativados. In *Pós-tratamento de efluentes de reatores anaeróbios* (coord. C.A.L. Chernicharo) (*PROSAB/ABES, Rio de Janeiro*), pp. 279–331 (in Portuguese).

Wahlberg, E.J. and Keinath, T.M. (1988) Development of settling flux curves using SVI. *J. Water Pollut. Control Fed.* **60**(12), 2095–2100.

Wahlberg, E.J. and Keinath, T.M. (1995) Development of settling flux curves using SVI: an addendum. *Water Environ. Res.* **67**, 872.

Wanner, J. (1994) *Activated Sludge Bulking and Foaming Control*, 327 p., Technomic Publishing Co., Lancaster, EUA.

WEF (1990) *Operation of Municipal Wastewater Treatment Plants*. Manual of Practice 11. Water Environment Federation, EUA.

WEF (1992) Design manual of wastewater treatment plants, vol I, Chapter 12 *Manual of Practice* n. 8 – 829 p.

WEF (1996) Operation of municipal wastewater treatment plants. *Manual of Practice* MOP 10, vol. 2, 3rd edn.

WEF and ASCE (1992) *Design of Municipal Wastewater Treatment Plants*, 1592 p., Water Environment Federation/American Society of Civil Engineers.

White, M.J.D. (1976) Design and control of secondary settlement tanks. *Water Pollut. Control* **75**, 459–467.

Winkler, M. (1981) *Biological Treatment of Waste-water*, 301 p., Ellis Horwood Publishers, Chichester.

WPCF and ASCE (1988) *Aeration. A Wastewater Treatment Process.* Manual of Practice FD-13 (WPCF), Manuals and reports on engineering practice No. 63 (ASCE), 167 p.

WRC (1984) *Theory, Design and Operation of Nutrient Removal Activated Sludge Processes.* Water Research Commission (South Africa).

WRC (1990) *Design Guidelines for Activated Sludge Process.* Chapter 2. *Bulking of Activated Sludge*, 41 p., Foundation for Water Research, UK.

Zaiat, M. (1996) Desenvolvimento de Reator Anaeróbio Horizontal de Leito Fixo para tratamento de águas residuárias. São Carlos. SP. Tese (Doutorado) Escola de Engenharia de São Carlos, Universidade de São Paulo. (in Portuguese).

IWA Publishing's authorised EU representative for General Product
Safety Regulations is Diane D'Arras, 15 rue Duret, 75116 Paris,
France, e-mail: safety@iwap.co.uk.

Printed and bound by CPI Group (UK) Ltd, Croydon, CR0 4YY
12/05/2026
02108891-0005